XIX International Scientific Conference on Metallurgy: Technologies, Innovation, Quality (Metallurgy2015)

IOP Conference Series: Materials Science and Engineering
Volume 150

Novokuznetsk, Russia
15 - 16 December 2015

ISBN: 978-1-5108-3102-5
ISSN: 1757-8981

Printed from e-media with permission by:

Curran Associates, Inc.
57 Morehouse Lane
Red Hook, NY 12571

Some format issues inherent in the e-media version may also appear in this print version.

Printed by Curran Associates, Inc. (2016)

Institute of Physics
Dirac House, Temple Back
Bristol BS1 6BE UK

Phone: 44 1 17 929 7481
Fax: 44 1 17 920 0979

techtracking@iop.org

Additional copies of this publication are available from:

Curran Associates, Inc.
57 Morehouse Lane
Red Hook, NY 12571 USA
Phone: 845-758-0400
Fax: 845-758-2633
Email: curran@proceedings.com
Web: www.proceedings.com

Table of contents

Volume 150

XIX International Scientific Conference on Metallurgy: Technologies, Innovation, Quality (Metallurgy2015)

15–16 December 2015, Novokuznetsk, Russia

Accepted papers received: 7 September 2016
Published online: 26 September 2016

Preface

Papers

Preface

XIX International Scientific Conference on Metallurgy: Technologies, Innovation, Quality (Metallurgy2015), supported by the Russian Foundation for Basic Research, was held from December 15 to December 16 2015 in Novokuznetsk, Russia (Project No. 15-08-21019).

The conference has been held since 1997 on the basis of Siberian State Industrial University. It has become a successful forum of international level, which involves leading scientists in the field of metallurgy and materials science from Russia, China, Uzbekistan, Kyrgyzstan, Kazakhstan, Ukraine, Belarus, experts from steel industry. The conference is the platform for professional discussions and introduction by the scientific and manufacturing communities of the latest research results and technological solutions in the field of production and processing of materials, energy and resources, environment and recycling of industrial wastes. It allows the dominant current trends to be determined and forecasts for the future to be justified.

The present issue of *IOP Conference Series: Materials Science and Engineering* contains 44 specially selected research papers divided into 4 topics:

- Theory, modeling, technologies and automation of metallurgical processes;
- Metallurgical heat engineering, heat-power engineering and energy saving. Environment and waste management and steel industry;
- Technological processes of metal materials treatment: foundry, metal forming, heat treatment;
- Technological processes of welding and obtainment of powder, composite materials and coatings.

The Organizing committee would like to thank all participants for the high activity and highly appreciates the contribution of all the authors and research teams into the development of technologies. We hope that the exchange of experience and thoughtful discussion of relevant scientific issues will be a powerful impetus to their effective solution, and the published proceedings of the conference will help to promote the achievements of the world industry.

Prof. Dr. Evgeniy Protopopov
Co-Chair of Metallurgy2015
Novokuznetsk, Russia

Peer review statement

All papers published in this volume of *IOP Conference Series: Materials Science and Engineering* have been peer reviewed through processes administered by the proceedings Editors. Reviews were conducted by expert referees to the professional and scientific standards expected of a proceedings journal published by IOP Publishing.

IOP Conf. Series: Materials Science and Engineering **150** (2016) 012001 doi:10.1088/1757-899X/150/1/012001

Analysis of current state and prospects of steel production development

E V Protopopov and S V Feyler

Institute of Metallurgy and Material Science, Siberian State Industrial University, 42 Kirova Street, Novokuznetsk, 654007, Russia

E-mail: feyler@rdtc.ru

Abstract. Data on world production of steel in the XXI century are provided. Analysis of current state and prospects of ferrous metallurgy development in the Russian Federation is carried out. Results of national steel production performance during 10 months of the year 2015 are given for different countries. Analysis of the main directions of metallurgical equipment import substitution aimed at technological independence in the industry is made. Russian ferrous metallurgy development predictions in its' main directions up to 2030 is provided.

1. Introduction

In accordance with modern concepts in present and foreseeable future steel production remains the main way of producing the most available construction material, which due to its physical, technical and operational properties is applicable in broad range of industries. However, if the first decade of the XXI century [1] was marked by significant growth in the world steel production (Tables 1, 2), contemporary state of iron and steel industry is marked by a decrease in production rate change.

Table 1. Steel production in the world.

Countries	Steel production, mln. t											
	2000	2004	2005	2006	2007	2008	2009	2010	2011	2012	2013	2014
China	127.2	280.9	355.8	422.7	489.5	500.3	573.6	626.7	695.5	724.7	779.0	822.7
Japan	106.0	112.7	112.5	116.2	120.2	118.7	87.5	109.6	107.6	107.2	110.6	110.7
USA	102.0	99.7	93.9	98.5	98.1	91.4	58.2	80.5	86.2	88.7	87.0	88.2
India	26.9	32.6	40.9	44.0	53.1	55.1	62.8	68.3	72.2	77.3	81.2	87.3
Russia	59.1	65.6	66.1	70.8	72.4	65.5	60.0	67.0	68.1	70.4	69.4	71.5
South Korea	41.1	47.5	47.8	48.4	51.5	53.6	48.6	58.4	68.5	69.1	66.0	71.0
Germany	46.4	46.4	44.5	47.2	48.6	45.8	32.7	43.8	44.3	42.7	42.6	42.9
Brazil	27.9	32.9	31.6	30.9	33.8	33.7	26.5	32.9	35.2	35. 9	34.7	33.9
Ukraine	31.8	38.7	38.6	40.9	42.8	37.3	29.9	32.9	35.3	33.0	32.8	27.2
Italy	26.8	28.5	29.4	31.6	31.6	30.6	19.8	25.8	28.7	27.3	24.1	23.7
TOTAL	595.2	785.5	861.1	951.2	1041.6	1032	999.6	1145.9	1241.6	1276.3	1327.4	1379.1
World production	847.7	1046.3	1138.5	1250.0	1344.3	1326.5	1229.7	1413.6	1526.9	1552.9	1607.2	1644.2

Share of 10 countries, %	70.2	75.1	75.6	76.1	77.5	77.8	81.3	81.1	81.3	79.9	82.6	83.9

Table 2. The rate of change in steel production in 2013/2014.

Region, country	2013	2014	Rate of change, %
World production, mln.t	1607.2	1644.2	2.3
including Europe (with the CIS)	313.1	311.4	–0.5
where:			
EU (28 countries)	165.6	169.3	2.2
CIS countries	108.9	105.9	–2.8
among them Russia	68.8	71.5	3.9
North America	99.5	121.1	21.7
Latin America, Africa, Middle East	108.1	87.2	–19.3
Asia	1080.9	1118.9	3.5
including China	779.0	822.7	5.6
Pacific Islands	5.5	5.5	0

For instance, in 2000 world steel production was 848 million tons, in 2004 it has exceeded 1 billion tons and reached 1.046 billion tons, and in 2014, in spite of the global crisis has risen to 1644.2 million tons. Thus, in 14 years of the XXI century world steel production has grown almost twice. This is the highest level of world steel output in the history. At the same time, in January - October 2015 metallurgical enterprises of 66 countries providing production reports to World Steel Association, have melted 1345.9mln.t, which is 31.9mln.t. (2.3%) less in comparison with the same period of the previous year.

2. Research and results

According to the World Steel Association (WSA) data [2], world steel production in October 2015 was 133.6mln.t, 3.1% less than in the same month of the previous year. Average level of capacity utilization in industry has decreased to 68.3% compared with 69.3% in September and 71.7% in October 2014. This figure, however, is higher than in August (68.0%). Average daily steel yield amounted to 4.31mln.t in October, 1.4% less than in previous month.

Reduction in steel production is observed even in China, the major world metal producer, its' share in October was 49.5% of world production, decline in comparison with January - October 2014 amounts to 16.2mln.t or 2.3%, in other countries it is 18.8mln t. or 2.7%.

According to "Metal Supply and Sales" information and publishing network [3] steel yield in China amounted to 66.1mln.t. in October 2015, which is 3.1% less than in the same period of the previous year. According to China Iron and Steel Association (CISA) consumption at the same period decreased by 5.7%. In this regard, information coming from China is unpromising for metallurgists. GDP growth in China continues, but at the same time, there is a change in its structure. Share of industry and construction is reducing as a result of reduced investment in real estate and fixed capital assets, an increase is mainly in services and other intangible spheres. Until there is no other source of growth of Chinese industry instead of export to Western countries, this trend remains unchanged.

However, in 2015 there is growth in steel production in some countries compared to the same period of the previous year (Table 3).

Table 3. Steel production in countries with positive production dynamics.

Countries	Steel production (January-October), th. t		Increase in steel production	
	2014	2015	th. t	%
India	72668.00	75075.00	2407.00	3.21

South Africa	5490.30	6334.30	844.00	13.32
Poland	7178.71	7971.27	792.56	9.94
United Arab Emirates	1939.46	2493.54	554.07	22.22
Spain	12079.65	12612.04	532.39	4.22
Hungary	918.00	1434.00	516.00	35.98
Pakistan	1939.99	2351.00	411.01	17.48
Serbia	453.90	818.80	364.90	44.57
Oceania	4614.53	4835.37	220.83	4.57
Australia	3901.75	4118.38	216.63	5.26
Finland	3182.80	3355.00	172.20	5.13
Netherlands	5769.64	5926.61	156.97	2.65
Belarus	2054.83	2208.92	154.09	6.98
Belgium	6194.78	6310.99	116.21	1.84
Germany	36094.82	36208.24	113.42	0.31
Iran	13509.94	13610.90	100.96	0.74
Slovakia	3866.23	3965.00	98.77	2.49
Moldova	302.80	375.28	72.48	19.31
Peru	893.26	948.99	55.73	5.87
Ecuador	556.21	587.30	31.10	5.29
Czech Republic	4440.45	4466.73	26.28	0.59
Morocco	433.22	455.87	22.64	4.97
BosniaandHerzegovina	649.87	671.72	21.85	3.25
Cuba	211.27	222.42	11.15	5.01
Bulgaria	512.40	522.80	10.40	1.99
Venezuela	1194.78	1204.51	9.74	0.81
Guatemala	312.81	319.05	6.25	1.96
Colombia	1034.81	1039.74	4.94	0.48
New Zealand	712.78	716.99	4.21	0.59
Uruguay	72.69	75.83	3.15	4.15
Paraguay	36.52	37.67	1.14	3.03

It is significant that India, the only country in the world, proclaiming industrial development a priority of its national policy, this year can achieve higher GDP growth rates than China. In January-October of this year, Indian steel industry has achieved the world's highest rise compared to the last year's data [4]. While national steel producers are constantly complaining on sharp rise in imports, steel production in the country for ten months gained 3.2%, or more than 2.4mln. t. compared to previous year's level.

It is interesting, that the second and the third places in the world in absolute growth over the same period took South Africa and Poland. South African steel production in January-October of this year has risen by more than 840th.t. compared to the same period of the previous year, that is due to active policy of industrial development stimulation, large construction projects, and moderate protectionism. For Poland, the main reason for nearly 800th. t production expansion compared to the first ten months of the previous year was transition to independent steel production of Huta Czestochowa plant, owned by Ukrainian group “Industrial Union of Donbass”, after the last year stop of the Ukrainian “Alchevsk Iron and Steel Works”, which provided the Polish company with slabs. AISW currently operate, but do not deliver the same amount of semi-products to Eastern Europe [5]. In addition to this, Poland this year looks the most stable EU economy in terms of construction sector development.

Generally, the number of states that have increased production of steel in comparison with the previous years equals to the number of states that cut production (Table 4), however, decrease in absolute steel production values is apparent.

First of all, the US steel production has reduced significantly. In the end of ten months it has decreased by 9.7% or 6.5mln. t., compared to the same period of the previous year, in early November, according to the American Iron & Steel Institute (AISI), the decline has reached 13%. Experts consider decline in American oil industry as a main reason for the collapse. This year investments in oil production decreased by more than 20% and, apparently, will continue to decline in future. This causes sharp drop in demand for pipes, drilling equipment, tanks and barges for oil transportation.

However, Japan, other industrial power, experiences difficulties as well, steel production in January-October has declined by 5.3% or almost 4.7mln. t. compared to the same period of the previous year. Industry in Japan and neighbouring South Korea, where steel production for the first ten months has decreased by 3.8% compared to the same period of the previous year, is suffering losses due to cut in orders for ships, drilling platforms and industrial equipment.

Table 4. Steel production in countries with negative production dynamics.

Countries	Steel production (January-October), th. t		Reduction in steel production	
	2014	2015	th. t	%
China	687632.10	675104.00	12528.10	1.86
USA	73747.98	67242.70	6505.28	9.67
Japan	92491.87	87815.23	4676.64	5.33
Ukraine	23438.50	19148.30	4290.20	22.41
South Korea	59847.42	57671.71	2175.71	3.77
Turkey	28520.04	26553.99	1966.05	7.40
Italy	20397.48	18593.19	1804.29	9.70
France	13646.98	12826.56	820.42	6.40
England	10271.77	9514.08	757.69	7.96
Taiwan	18976.97	18288.91	688.06	3.76
Egypt	5246.42	4735.02	511.41	10.80
Libya	635.70	255.89	379.81	148.43
Brazil	28605.80	28235.50	370.30	1.31
Saudi Arabia	5205.43	4866.99	338.44	6.95
Qatar	2535.00	2204.00	331.00	15.02
Argentina	4580.67	4251.53	329.15	7.74
Austria	6597.68	6386.78	210.90	3.30
Thailand	3420.34	3236.09	184.24	5.69
Mexico	16029.94	15873.14	156.80	0.99
Trinidad and Tobago	418.18	266.49	151.69	56.92
Canada	10666.29	10533.59	132.70	1.26
Kazakhstan	3102.30	2983.85	118.45	3.97
Russia	59412.80	59307.37	105.43	0.18
Uzbekistan	631.40	544.35	87.05	15.99
Sweden	3754.40	3673.74	80.66	2.20
Macedonia	187.99	110.02	77.97	70.87
Croatia	159.40	126.84	32.56	25.67
Luxembourg	1869.50	1841.08	28.41	1.54
Norway	494.39	478.72	15.67	3.27
Slovenia	531.86	519.98	11.89	2.29
Chile	905.02	900.05	4.97	0.55

Turkey has slightly reduced the last year schedule delay. 2.77 mln. t. of steel were produced in October, that is 2.0% higher than in the same period of 2014. This decrease is explained by decrease of scrap prices in the end of September and the first half of October, which reached their longstanding minimum and local mini-plants have expanded use of resource, as opposed to more expensive stock material.

Ukraine has reduced the rate of decline. According to the data for January-October it amounted to almost 4.3mln t. or 18.3% compared to the previous year. However, due to absence of active military operations in the Donbass and Alchevsk Iron and Steel Works return into operation in May, national steel production in August shows stable overrun of the last year's figures. However, recently situation in Ukrainian steel industry deteriorates again. Production declines due to collapse of domestic market, negative market conditions, energy deficit.

Russian metallurgists have completed October with a slight arrear compared to the previous year figures. In October, steel production, according to the World Steel Association [2] has decreased by 2.4% compared to the same period of the previous year, and in ten months period – by 0.18%, or nearly 100th. t. Russian mill products market continues its fall, steel producers have to conduct restrictive policy, thus by the end of the year arrear increases.

Traditionally iron and steel production is one of strategic industries in Russian economy. It makes the foundation for many other industries development, including shipbuilding, aviation, transportation and heavy engineering, defense industry, railway transport, power engineering, construction, etc. Share of iron and steel production in GDP accounts for 1.4%, it comprises 8% of national industrial production and 6% of national exports [6]. As a consumer of products and services of natural monopolies iron and steel industry uses 5.3% of total domestic consumption of electricity, 5.8% of natural gas, its share in rail freight transportation is 15% [7].

According to expert estimations [8], the main task of Russian iron steel industry is meeting demand for steel products of required assortment and quality in volumes demanded on domestic and international markets. Further development of the enterprises will be based on accelerated innovation development, improving economic efficiency, environmental safety, resource and energy saving, product competitiveness ensuring import substitution.

Over the last 10 years about 1.3 billion rubles was invested in Russian iron and steel industry [9], the industry in general corresponds to the world technology level.

There are a number of external and intra-industry challenges and constraints that handicap the industry development in Russia.

External factors constraining development of iron and steel industry are as follows:

- low demand for metal products on domestic market due to insufficient development of machine building industries;
- significant share of metal-consuming machines and equipment imports. 70-75% of metallurgical equipment, and in general, more than 50% of machines, equipment and vehicles (ad valorem) are imported;
- Chinese and other Asian countries expansion on global steel market;
- increase of energy and railway transportation costs;
- insufficient protection of national market.

Among the intra-industry factors there are the following negative trends:

- increased specific consumption of raw materials, fuel and energy per 1 ton of steel in comparison with international peer-enterprises;
- high cost of transportation component in steel production costs, due to geographic location and high production concentration.

At present, despite minor dependence on imports of metallurgical industry as a whole, compared to the other industries (Figure 1), iron and steel production shows high dependence on import of metallurgical equipment (Figure 2).

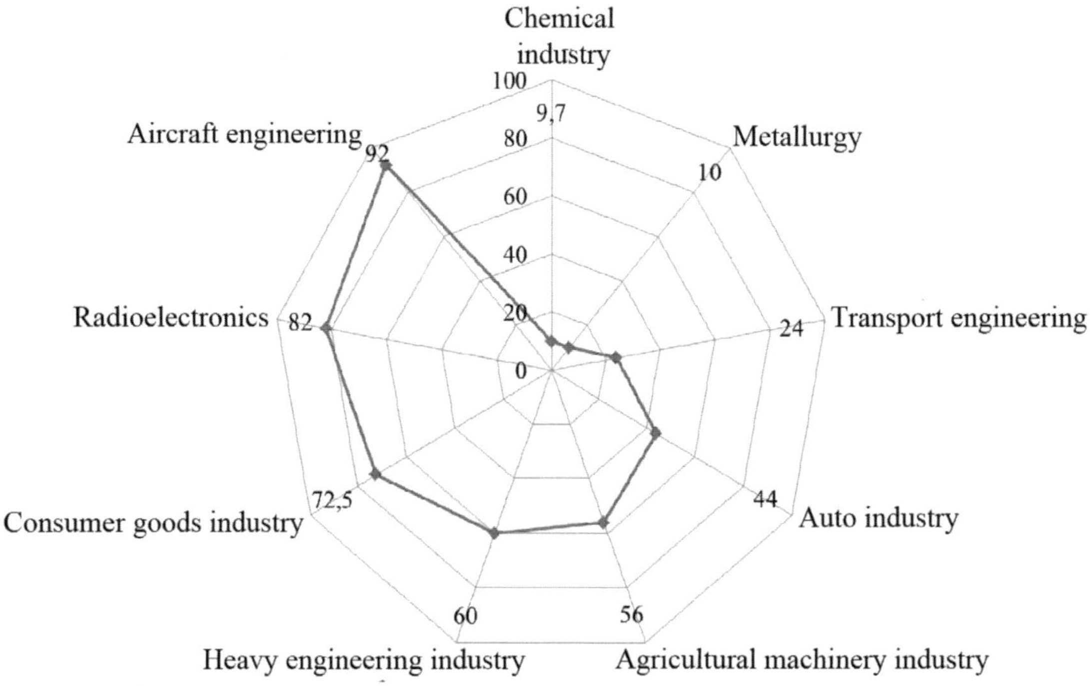

Figure 1. Import dependence of Russian industries.

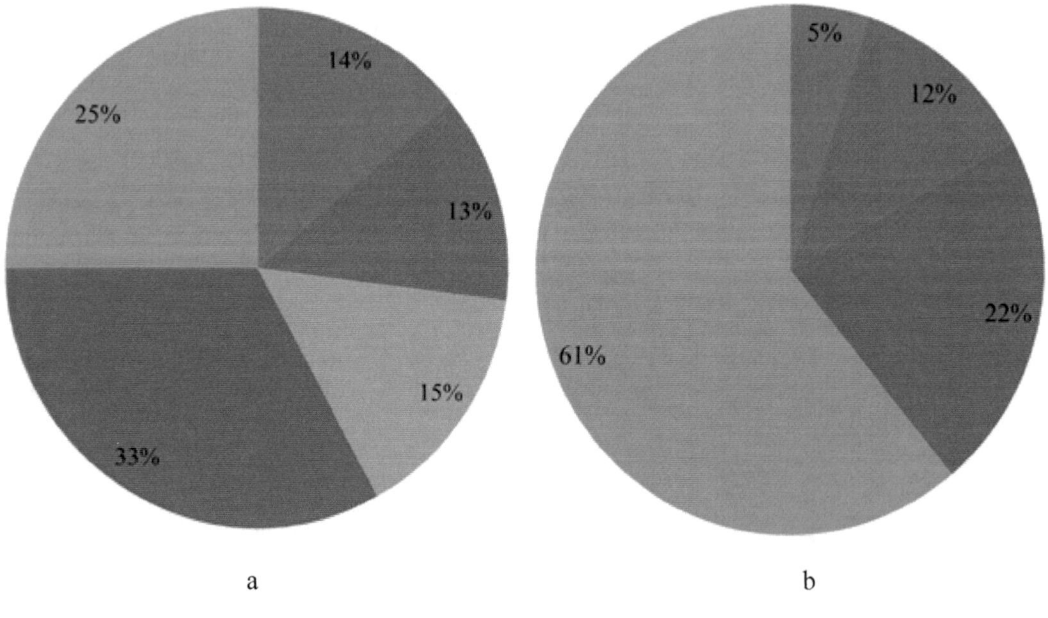

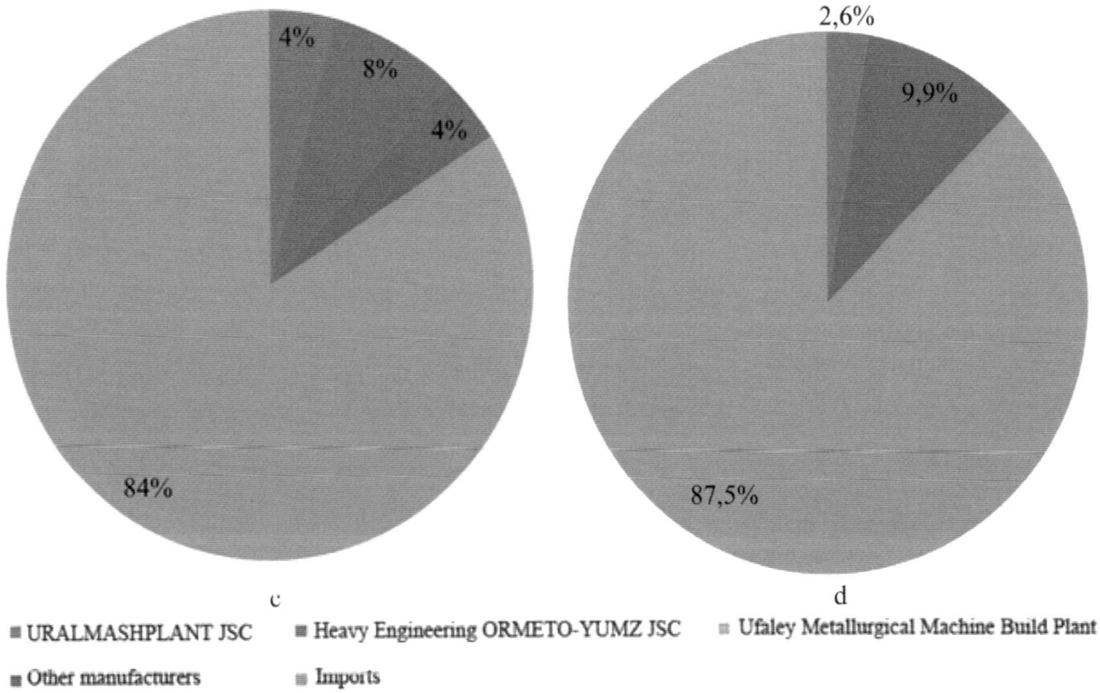

Figure 2. Metallurgical equipment producers share on domestic market (a – aglomeration and calcination equipment; b – blast furnace equipment; c – steel continuous casting equipment; d – rolling equipment).

In order to organize consistent work to ensure import substitution, particularly in metallurgical industry, an order of Ministry of Industry and Trade of Russia No. 654 dated March 31, 2015 "Approval of action plan of the Russian Federation on import substitution in heavy engineering" was issued, according to which plan of import substitution in iron and steel industry equipment was approved (Table 5).

Table 5. Plan of iron and steel industry equipment import substitution in the Russian Federation.

Technology	Project implementation period	Share of imports in consumption in 2014, %	Maximum planned share of imports in 2020, %
CCM (bloom-casters, billet casters) and accessories	2015-2017	70	55
CCMslab-casters	2015-2018	83	50
Rolling equipment	2015-2016	92	40
Pipe and tube mills	2015-2017	60	20
Isostatic equipment	2015-2017	50	20
CCMmoldtubes	2015-2017	95	20
Casting rollers	2015-2016	70	35
Oil film bearings	2015	40	0

Implementation of the plan will significantly increase the share of domestic producers of metallurgical equipment on domestic market.

3. Conclusions

In future, Russia remains a major exporter of steel products and continues to hold one of the world leading positions [10]. At the same time, absolute exports of finished steel may reduce due to a slight decrease in exports of blanks and semi-products supplied to the assets owned abroad. Exports of higher value added products will be determined by market conditions. Imports of steel products will be mainly determined by economic expediency.

Based on world metallurgical industry development prediction in the 2030 run until Russia retains leading positions on global market and will take the 5th place by steel production (after China, Japan, the USA and India), the 6th place by apparent metal consumption (after China, Japan, US, South Korea and India), and the 4th place by exports.

4. Acknowledgements

The work was performed at Siberian State Industrial University in the framework of the state task of the Ministry of Education and Science of the Russian Federation No. 2556

5. References

[1] Lindroos T J 2009 *Sectoral Approaches in the Case of the Iron and Steel Industry* (VTT Technical Research Centre of Finland)
[2] *World Steel Association* www.worldsteel.org
[3] *2015 World Steel Production Reduces Gradually* (M.: Metal Supply and Sales) www.metalinfo.ru/ru/metalmarket/analytics/531.html
[4] Agrawal A K 2014 *Indian Steel. Strategy to Ambition* (EYGM Limited) pp 1–41
[5] Auer J 2015 *German Steel Set to Benefit from Global Growth Deutsche Bank Research* May 27th pp 1–18
[6] *Ferrous Metals Industry Development Strategy for 2014-2020 and in Long Term to 2030* (Order of Ministry of Industry and Trade of Russia Dated May 5th, 2014)
[7] Revinskaya L Y 2015 *Economic Sciences* **123** 72–77
[8] Yuzov O V Sedykh A M et al 2014 *Ferrous Metallurgy. Bulletin* **7** 15
[9] Mukhatdinov N Kh et al 2014 *Proceedings of XIII Intern. Congress of Steelmakers* pp 14–18
[10] Mukhatdinov N Kh et al 2014 *Proceedings of XIII Intern. Congress of Steelmakers* pp 18–21

Analysis of the main trends in the development of rail production in Russia and abroad

A V Golovatenko[1], A A Umansky[2] and V V Dorofeev[1]

[1]JSC "EVRAZ Consolidated West Siberian Metallurgical Plant", Novokuznetsk, 654000, Russia
[2] Department of Metallurgy of Ferrous Metals, Siberian State Industrial University, 42 Kirova Street, Novokuznetsk, 654007, Russia

E-mail:umanskii@bk.ru

Abstract. The present analysis reveals the main directions in the development of rail production in the world: the transition to rolling of rail sections using continuous group of universal stands, increase of the rails length, application of differentiated hardening of rails in different media. It is shown that the technological gap between the Russian metallurgical industry and the world's leading rail manufactures of the last decades was closed in 2013-2014 when the new universal rolling mills at JSC "EVRAZ Consolidated West Siberian Metallurgical Plant" ("EVRAZ ZSMK") and OJSC "Chelyabinsk Metallurgical Plant" (company "Mechel") were launched. Currently, these steel enterprises provide complete import substitution of long rails with increased wear resistance for the construction of high-speed networks.

1. Introduction

Nowadays, Russia occupies one of the leading positions in the world as to the length of railways (86000 km). However, it is not sufficient to provide the necessary track capacity, because the share of rail transportation is about 85% of the total volume of cargo transportation. Also the low track capacity of 6000 km of the railway network should be noted due to the high degree of wear. In general, railroad infrastructure wearout reaches 85%.

In order to solve this problem the company "Russian Railways" ("RZhD") is planning in parallel with the replacement of the worn parts of railway network to construct 22000 kilometers of new railway lines by the year 2030, which conditions the annual demand in rails at the level of 1.6-1.7 million tonnes. At the same time the improvement of the railway infrastructure is possible only alongside with a substantial increase in the quality and durability of rails.

In this regard the urgent task consists in the analysis of the current state of rail production in Russia and its compliance with the level of the world's leading rails manufacturers.

2. Discussion of the problem

Currently, in the world's rail production three main trends can be distinguished:

- transition to the rolling of rail sections using universal stands;
- increase in the length of the finished rails;

- use of the differentiated hardening technologies.

Rail rolling technology in the universal passes has been actively implemented in the world steel industry since the 70s of the XX century due to the presence of a number of significant advantages over the obsolete scheme of rails rolling in the two-roll passes of stands "duo" and "trio" [1-5]:

- In contrast to the rails rolling in the two-roll pass during deformation in the universal pass the railhead is subjected to direct compression on the rolling surface (Figure 1). Because of that after rolling in the railhead a dense fine grain metal structure is detected, which leads to high mechanical properties and wear resistance of this rail element during operation.
- A decrease in the internal stresses in the rails due to the simultaneous uniform deformation takes place, which greatly reduces the possibility of defects and increases the rail straightness.
- There is a reduction in the slip of the metal relative to the rolls with respect to the vertical (idle) rolls, which reduces their wear and, as a result, the consumption of rolls for a pass goes down. Also the reduction in energy consumption and improvement of the rail quality are achieved.

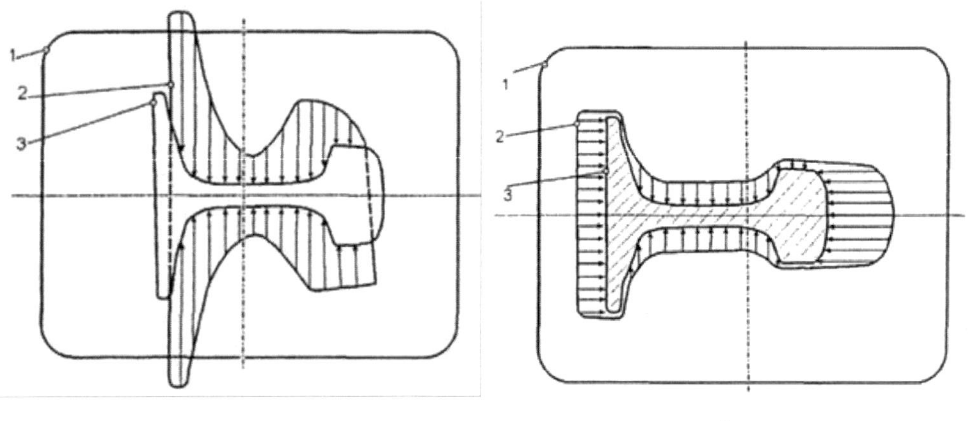

a b

Figure 1. Deformation scheme of the rail section: a – in the two-roll pass; b – the universal pass (1 – billet; 2 – an intermediate section; 3 – finished section).

For the first time the process of rail rolling in the universal mill stands passed pilot industrial testing in 1964-1968 in France [6]. Further company "Sacilor", which is the developer of this technology, sold the license for universal process of rails rolling to companies "Nippon Steel" (Japan), "Iscor" (Republic of South Africa), "Whelling-Pittsburg Steel" (USA), "Broken Hill Proprietary" (Australia), "Acominas" (Brazil) [4]. These companies in the 70-80s of the XX century built or reconstructed the operating rolling mills implementing rails rolling in the universal mill stands. Despite some technological and design features of these rolling mills the common rolling scheme consists in the deformation in one or two reversing rolling stand, then the rolled stock is transferred for rolling in the universal reversing mill stand "duo" and to the installed before and after them the auxiliary two-roll mill stands. The final pass is usually carried out in a separately installed finishing universal non-reversing rolling mill stand.

The period of the 90s in the XX century – the beginning of the XXI century is characterized by the intensive dissemination of this technology with the use of universal mill stands. From 2002 to 2009 8 universal rail rolling mills in different regions of the world were put into operation: 3 rolling mills in China, 2 – in India and one rolling mill in Austria, the USA and Saudi Arabia (Table 1).

Table 1. Commissioning of rail rolling mills.

Company (enterprise)	Country	Year of commissioning	Equipment supplier
"Steel Dynamics"	USA	2002	SMS (Germany)
"Anshan Iron andSteel"	China	2003	SMS (Germany)
"SAIL Bhilai Steel Plant"	India	2004	SMS (Germany)
"Baotou Iron and Steel"	China	2005	SMS (Germany)
"VAI Schienen"	Austria	2006	Danieli (Italy)
"Jindal Steel and Power"	India	2006	SMS (Germany)
"Wuhan Iron and Steel"	China	2008	SMS (Germany)
"Al Tuwairqi Group"	Saudi Arabia	2009	SMS (Germany)

The typical pass design of rails of the leading equipment supplier company "SMS Meer" (Germany) includes production of a T section from rectangular billet in the first three passes, then rolling in the rough closed and open two-roll passes (passes No. 4-7) with the subsequent deformation of the obtained section in the universal mill stands of the rail-finishing mill (passes from No. 8 to No. 13) – Figure 2.

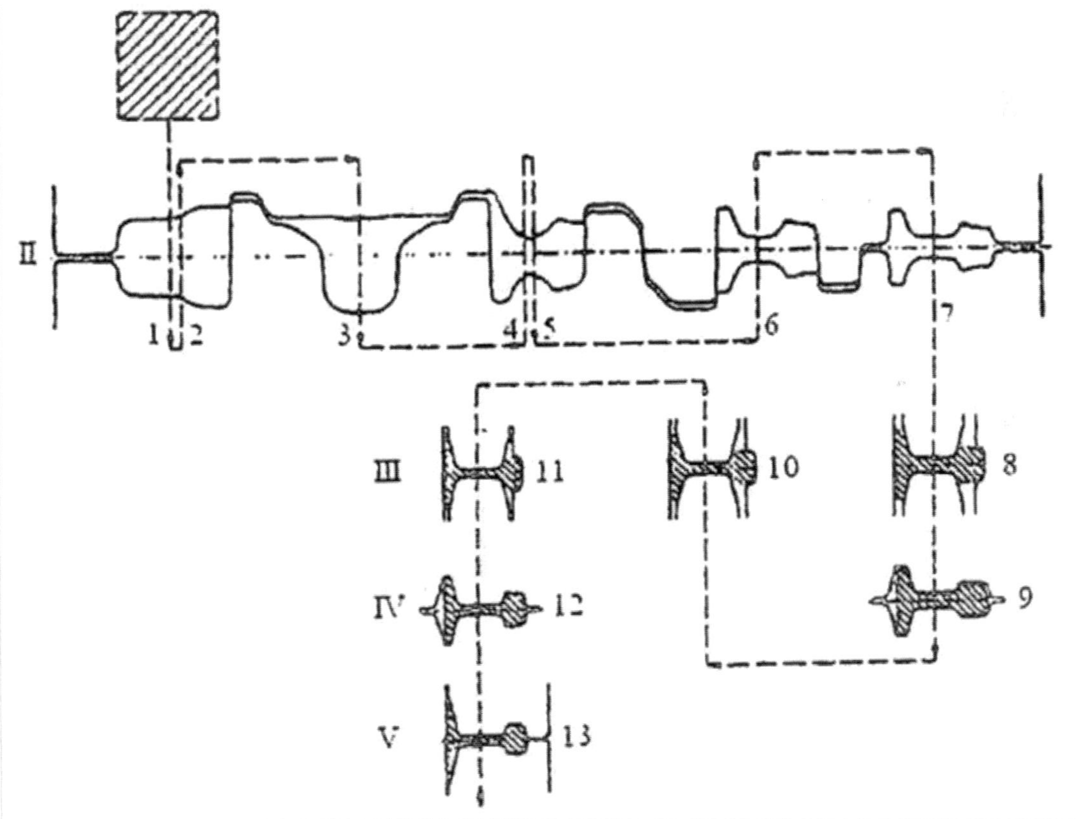

Figure 2. Typical pass design of rails for rails rolling developed by SMS.

Increase in the length of the finished rails is one of the trends in the development of rail production. When the technology of continuous rails laying is used, allowing the movement speed to be increased, the increase in the rails length would reduce the number of joint welds. Currently, everywhere on the high-speed networks rails of 100 m long are used, in Japan the production of 150-meter rails was launched and further it is planned to increase the length of the finished rails. Thus, modern rail rolling mills are intended for production of long-length rails as a main rolled product.

A characteristic feature of the technological process of rails production on the modern rolling mill is the use of differentiated hardening instead of outdated technology of volumetric hardening in oil [7-9]. At present, the technology of differentiated hardening in various media (compressed air, water, polymers) is successfully applied, as for example, during the rolling heat or separate rails heat before the heat treatment (Table 2).

Table 2. Technological features of rails produced by the world's leading manufacturers.

Country-producer of rails	Type of a unit for steel smelting	Method of rails heat treatment
France	BOF	Two-sided hardening by compressed air from a separate volumetric heat HFC
Japan	BOF	Two-sided hardening by compressed air from rolling heat
USA	BOF	Two-sided hardening by water from rolling heat
Canada (the plant is closed)	Electric furnace	Two-sided hardening by compressed air from volumetric heat HFC
Austria	BOF	One-side hardening in the polymer solution from the rolling heat

Differentiated hardening makes it possible to obtain higher wear resistance of the railhead while maintaining sufficient ductility of the remaining elements of the rail section. Comparative analysis of mechanical properties of rails obtained with the use of differential hardening and the rails with volumetric hardening shows a lower limit of endurance of the latter (Table 3).

Table 3. Comparative analysis of mechanical properties obtained by various methods of heat treatment.

Country-producer of rails	Method of heat treatment of rails	Fracture energy, tm at t = -60°C	Endurance limit, MPa
Russia	Volumetric hardening in oil	>10	400-407
France	Two-sided hardening by compressed air from a separate volumetric heat HFC	>10	477
Japan	Two-sided hardening by compressed air from rolling heat	5.5	430
Canada	Two-sided hardening by compressed air from volumetric heat HFC	>10	453
Austria	One-side hardening in the polymer solution from the rolling heat	5.5	423

The state of the rail production in the first decade of the XXI century was characterized by a significant technological gap between the Russian enterprises and the leading world's manufacturers. Production of rails by two steel mills – JSC "EVRAZ Consolidated West Siberian Metallurgical Plant" ("EVRAZ ZSMK") and "EVRAZ Nizhny Tagil Metallurgical Plant" ("EVRAZ NTMK") was carried out on the rolling mills of a line type constructed in the 30s of the XX century. 25-meter volumetric hardened rails, produced by the mills, did not comply with the world's standards on service durability, straightness and a number of other parameters. Besides, for the construction of new high-speed networks such as Moscow-St. Petersburg, long-length (100 m long) rails were needed.

This situation made the company "Russian Railways" buy rails from foreign manufacturers. If the share of rails manufactured by foreign companies ("Voestalpine Schienen Gmbh" (Austria), etc.) was 18% of the total purchases in 2011, then by 2012 it went up to 39%.

The radical change in the situation took place in 2013-2014 following the launch of the first in Russia universal rolling mill at "EVRAZ ZSMK" [10].

The composition of the main equipment for the rolling mill, supplied by the company "SMS Meer" (Germany), included (Figure 3): a heating furnace with walking beams; water descaling device for the primary and secondary scale; sequentially located two-roll reversible breakdown stands (BD1 and BD2); universal tandem mill, installed with an offset from the rolling line and consisting of two universal (UR and UF) stands and one horizontal auxiliary stand (E), separately located finishing calibrating universal stand (U0), devices for automatic measurement of the finishing section, automatic stamping machine.

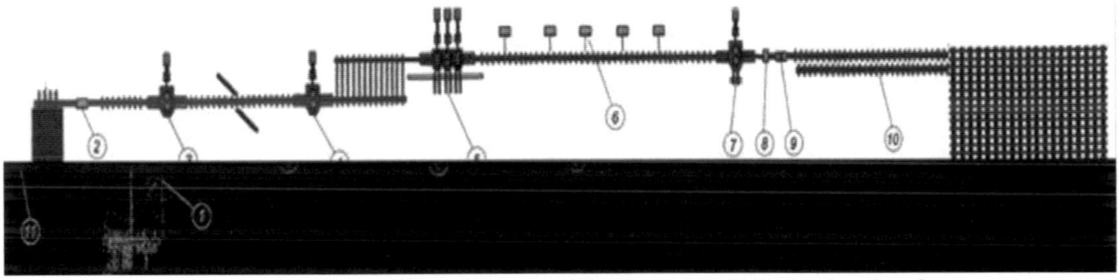

Figure 3. Arrangement of rolling mill equipment at "EVRAZ ZSMK": 1 – heating furnace; 2 – water descaling device; 3, 4 – breakdown stands BD1 и BD2; 5 – tandem-group; 6 – hot saws; 7 – finishing stand; 8 – laser measurer of section; 9 – stamper; 10 – area of differentiated hardening; 11 – refrigerator area.

The main rolled products of the universal rail and structural steel mill "EVRAZ ZSMK" are hundred-meter differentially hardened rails. One of the main features of the technological mode of rails production on this mill is the use of differential hardening by the compressed air from the rolling heat. Today, a similar technology of long-lengh rails hardening is used only in Japan (company "Nippon Steel") – Table 2.

Lack of experience in the production of rail sections using the universal rolling in the domestic steel industry and the limited amount of information in foreign sources led to significant technological difficulties during the start-up and at the beginning of rail and structural steel mill operation. In particular there was an increased rejection of rails due to the surface defects of rolling origin, low productivity of the rolling mill, increased wear of the mill rolls. The solution of these technical problems was made in parallel with the adjustment of modes of differentiated hardening [11, 12]. As a result, the first commercial batch of long rails produced by "EVRAZ ZSMK" was delivered to the "Russian Railways" in May 2014. That year the company "Russian Railways" completely abandoned the procurement of rails from foreign manufacturers.

Also in 2013 a new universal rail and structural steel mill at Chelyabinsk Metallurgical Plant (company "Mechel") was officially put into operation. However, the first industrial batch of long rails produced by this plant was purchased by the "Russian Railways" only in 2016. It should be noted that Chelyabinsk Metallurgical Plant had no previous experience in the production of rail products. Supplier of the rolling mill equipment was company "Danieli" (Italy), the construction of the rail and structural steel mill was carried out by "Minmetals Engineering" (China).

Release Organization of a hundred-meter rails production that complies with the world's standards at the Russian enterprises required significant investments. Costs for the construction of a universal rail and structural steel mill "Mechel" amounted to more than 900 million USD, "EVRAZ" spent on the reconstruction of its own rail production approximately 600 million USD.

3. Conclusions

The analysis of directions in improvement of the production technology of rails sections identified three major trends in the world's steel industry: transition from the rolling of rail sections in the two-roll passes of stands "Duo" and "Trio" to the rolling with the use of continuous groups of universal mill stands; increase of the rails length up to 100 m and more; introduction of the technologies of differentiated rails hardening in different media instead of volumetric hardening. It is shown that the technological gap between the Russian metallurgical industry and the world leading rail manufactures of the last decades was closed in 2013-2014 when the new universal rolling mills at JSC "EVRAZ Consolidated West Siberian Metallurgical Plant" ("EVRAZ ZSMK") and OJSC "Chelyabinsk Metallurgical Plant (company "Mechel") were launched. Currently, these steel plants ensured a complete import substitution of long-length rails with increased wear resistance used in the construction of high-speed highways.

4. References

[1] Shilov V A, Shvarts D L and Litvinov R A 2008 *Steel in Translation* **38** (3) 214–216
[2] Sveykovsky U and Nerzak T 2006 *Ferrous Metals* **2** 50–56
[3] Nikitina L A 2000 *Rolling Production* **11** 2–10
[4] Smirnov V K, Bondin A R and Mikhaylenko A M 2002 *Rolling Production* 12 24–30
[5] Desvallees J, Faessey A, Gouth G and Mennel G 1987 *Iron and Steel Engineer* March 25–31
[6] Samoilovich Yu A 2012 *Metallurgist* **55** (11) 903–911
[7] Sinel'nikov V A and Filippov G A 2001 *Metallurgist* 45 9 403–407
[8] Stalinskii D V and Rudyuk A S 2011 *Steel in Translation* **41** (5) 73–77
[9] Golovatenko A V, Volkov K V, Aleksandrov I V, Kuznetsov E P, Dorofeev V V and Sapelkin O I 2014 *Ferrous Metallurgy* **6** 32–38
[10] Polevoy E V, Volkov K V, Golovatenko A V, Antonova O P and Yunusov A M 2013 *Problems of Ferrous Metallurgy and Material Science* **4** 26–28
[11] Golovatenko A V, Volkov K V, Dorofeev V V, Stepanov C S and Dobryansky A V 2014 *Rolling Production* **2** 25–39

A study on the production processes of granulated iron

O I Nokhrina, I D Rozhikhina and I E Khodosov

Department of Metallurgy of Ferrous Metals, Siberian State Industrial University, 42 Kirova Street, Novokuznetsk, 654007, Russia

E-mail: khodosov@mail.ru

Abstract. The results of theoretical and experimental studies on the production process of high-quality granulated iron using hematite-magnetite iron ore and coal of different brands as a raw material are presented. According to the research the optimal coal consumption, temperature and time parameters of the metallization process, necessary for obtaining metallized materials with the specified composition and metallization degree, are defined. The conditions for the formation of metal granules with high content (over 98%) of primary iron are found. The process includes two stages: initially the solid-phase reduction of iron from oxides with production of metallized sponge material is carried out, further as the temperature increases the separation of slag and metal phases takes place.

1. Introduction

Most modern steel grades have rigid restrictions on the content of non-ferrous metal impurities. In addition, in the initial raw material – metal scrap the impurities content is growing steadily (Figure 1), and to date there are no effective methods for their removal from the melt. In this regard, production and application of metallized materials, which do not contain non-ferrous metals and can replace metallic scrap, has become a perspective direction. Production volumes of metallized raw materials are steadily growing. [1] Currently, most metallized materials are produced with the use of converted gas as a reducing agent, but a faster rate has the developing technology of metallized material production using a solid carbonaceous component as a reductant [2]. The most progressive way of advanced processing of iron ore materials with coal is the production of granulated iron [3].

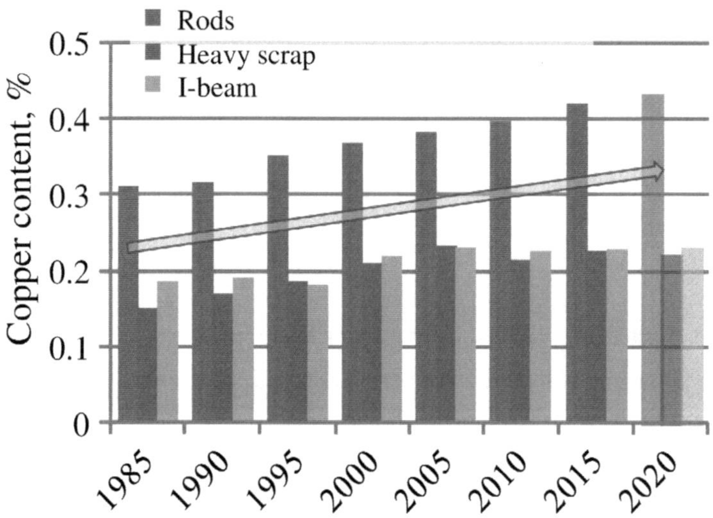

Figure 1. Dynamics of copper accumulation in the scrap.

The granulated iron (Figure 2) is a new kind of metallurgical raw material significantly improving the quality of material metal products and reducing production costs. Production technologies of granulated iron are called "third-generation of iron production" [4]. The objective of the conducted research was to study the process of obtaining the granulated iron using iron ores and coals from Kemerovo region.

Figure 2. Granulated iron.

At present there are several technologies for the production of granulated iron. According to the patent of the company "Kobe Steel – ITmkS" (Japan), metallic iron is produced by direct reduction of material containing iron oxides, using a solid carbon-containing reductant. The charge is fed into a rotary furnace, where at the beginning in the direction of hearth movement, the solid-phase restoration occurs, and in the end of hearth travel carbonization and melting take place. Between the solid reduction zone and the zone of carburization and melting there is a zone of reduction with the adjustable within the range 1473-1773 K. Another way to obtain granulated metallic iron is the feeding of ore smelting mixture into the rotary furnace where initially the solid phase reduction of iron from oxides takes place with subsequent melting and separation of slag from metallic granules. The quantity of CaO, MgO and SiO_2, contained in the charge mixture is adjusted in such a way that a slag, having a basicity from 1.3 to 2.3, is formed (the basicity is calculated according to the ratio – (CaO + MgO)/SiO_2). The content of MgO in the slag should be in the range from 5 to 13%, which provides a reduced sulfur content in the granulated iron [5].

The appearance of a liquid phase promotes the formation of large granules. Formed as a result of reduction particles of metallic iron are quickly carbonized due to the excess of carbon reductant in the material mixture. As a result of carburization the temperature of their melting is rapidly reduced, which ensures the concentration of reduced iron fine particles in the form of coarse granules.

The authors [6] developed schemes for metal production in the liquid state. These schemes involve a preliminary reduction of iron from ores or concentrates by gaseous or solid reducing agents in the rotary furnaces, furnaces with a suspended layer or other units, and a subsequent separating smelting in the rotary or electric furnaces. Selectivity of reduction is realized at the first stage of the process at moderate temperatures; stage of separating smelting is carried out in the manner so that the reduction processes would not further develop. Schemes, involving the iron extraction from complex ore in the solid state, presuppose production of metallized concentrates on the basis of solid-phase reduction of metal from oxides. The subsequent processing is carried out at temperatures of 1473-1623 K, the conditions are created for the development of coagulation of metal grains. One of the alternative schemes of metallized materials production is the method of pellets production in which metal is coagulated inside the slag shell.

As appears from the described methods the granulated iron is the product of the liquid phase separation of metallization products. Initially a solid-phase reduction of iron from oxides is carried out; as a result of temperature increase the separation of solid-phase reduction products takes place. Thus, the granulated iron is formed by macroscopic, visible separation of metal and slag. To ensure the conditions for separation of metal and slag phases it is necessary for one of the phases (metal or slag) to be liquid, and the rapid formation of a fluid phase occurs as temperatures go up.

2. Theoretical research

Thermodynamic modelling of solid-phase reduction of iron from oxides with use of different coal brands as reducing agents showed that in the result of interaction of FeO, SiO_2, Al_2O_3, CaO during solid-phase reduction of iron from iron ore oxides in the ranges of the studied temperatures, the formation of compounds shown in Table 1 is possible.

Table 1. Possible oxide compounds with iron, silicon, calcium.

Mineral	Formula	Name	Melting temperature*, K
Iron silicates	$(CaO)_{0.5}\cdot(FeO)_{1.5}\cdot SiO_2$	–	1403
$(CaO)_x\cdot(FeO)_{2-x}\cdot SiO_2$	$CaO\cdot FeO\cdot SiO_2$	–	1483
	$FeO\cdot SiO_2$	Fayalite	1478
Calcium ferrites	$CaO\cdot 2Fe_2O_3$	Semicalcium	1503
	$2CaO\cdot Fe_2O_3$	Dicalcium	1709
	$CaO\cdot Fe_2O_3$	Monocalcium	1489
Calcium silicates	$CaO\cdot SiO_2$	Wollastonite	1813
	$3CaO\cdot 2SiO_2$	Rankinite	1748

*the indicated minerals can form eutectics that melt at lower temperatures

It is known [1] that the formation of primary fluid phases is preceded by interaction between SiO_2 and FeO. It should be borne in mind that the mass fraction of FeO changes in the course of the reduction processes, and the system is heterogeneous. The most low-melting compound of these oxides are minerals from olivine group. Probably, the main role in the formation of the initial slag phase belongs to iron silicates, namely fayalite. According to the known data about the temperatures of the beginning of chemical interaction between the solid phases, the temperature of the beginning of fayalite formation during heating in the presence of just reduced iron is 1263 K.

With the help of software package FactSage 6.4 the analysis of interaction between SiO_2 and FeO with change in molar concentrations and temperatures was performed. In the diagram FeO-Fe_2O_3-SiO_2 (Figure 3) the zone of iron silicates formation at temperatures 1373 K and 1573 K is established. The analysis of the received data allowed us to determine the boundary zones of concentrations FeO and SiO_2 favorable for the formation of low-melting phases at different temperatures [7].

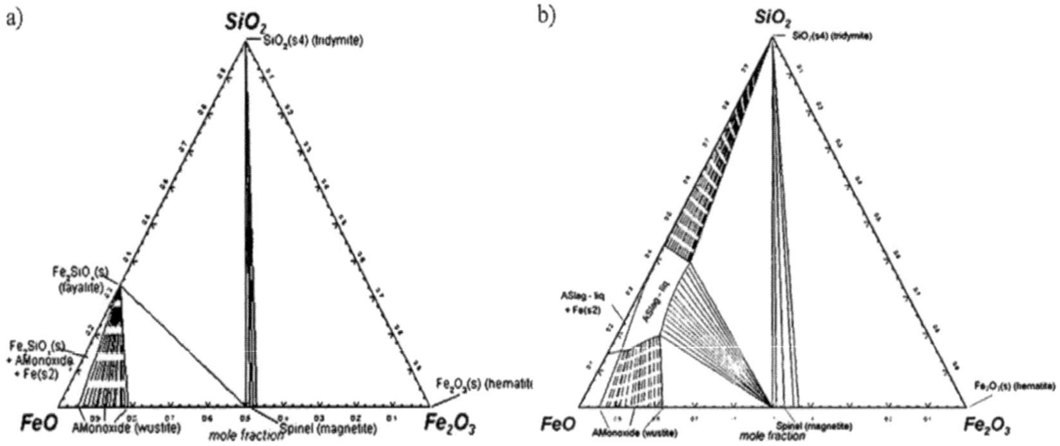

Figure 3. Diagram FeO-Fe_2O_3-SiO_2 at temperatures 1373 K (a) and 1573 K (b).

The results of calculations, performed using the software complex "TERRA" for materials in our experimental studies (Table 2, 3), allowed us to establish the optimum ratios of components in the ore-coal mixtures needed for the production of metallized products with the given composition [8, 9].

3. Experimental research

The task of experimental studies consisted in production of granulated iron using coal as a reducing agent, determination of conditions for formation of liquid phases and coagulation of metal droplets on the basis of theoretical studies.

For the experiments ore-coal briquettes of various compositions were produced. As an iron component ores from Kemerovo region were used. As solid carbonaceous reductants coals of different technological marks were applied. Composition of iron ores and reducing agents characteristics are shown in Tables 2 and 3.

Table 2. Composition of iron ores.

Chemical composition of iron ore ZhR-1, % mass							
$Fe_{tot.}$	SiO_2	Al_2O_3	P_2O_5	CaO	MgO	S	moisture
44.19	13.74	3.99	0.117	9.63	2.29	2.65	6.3
Chemical composition of iron ore ZhR -2, % mass							
52.3	19.15	3.31	0.072	0.94	0.76	0.42	5.78
Chemical composition of iron ore ZhR -3, % mass							
61.2	6.75	2.2	0.02	1.96	2.31	0.41	7.2

Table 3. Characteristics of reducing agents.

Characteristic	Coal grade			
	Lignite B2	Long-flame	Low-caking	Lean coal
C_{fix}(content of the fixed carbon), %	49.1	55.6	70.2	76.3
V^r (volatile yield on the working mass),%	42.1	36.0	20.6	9.34
W^r(moisture on the working mass), %	24.7	10.4	6.7	5.1
A^d(ash on the dry basis),%	7.83	5.64	6.41	8.34
Chemical composition of ash, %:				
Fe_2O_3	10.30	8.17	7.4	16.4
Al_2O_3	5.40	25.6	16.3	25.0
CaO	45.30	9.70	2.5	5.38
MgO	5.90	3.70	1.58	1.66
P_2O_5	0.20	1.80	0.37	0.75
SiO_2	26.90	50.0	46.6	47.9
S	0.68	0.61	0.02	
*Apparent density, γ_a, g/cm^3	1.09	1.15	1.20	1.28
*Real density, γ_r, g/cm^3	1.2	1.31	1.41	1.50
*Porosity, P, %	39.4	15.4	14.7	11.3
*Reactivity on CO_2 at 1000 °C, K, cm^3/(g·s)	5.47	4.48	0.62	1.56
Elemental composition, %:				
C^{daf}(carbon content on dry ash-free basis), %	70.41	72.02	85.30	91.6
H^{daf}(hydrogen content on a dry ash-free basis), %	3.88	6.12	4.78	3.20
N^{daf} (nitrogen content on dry ash-free basis), %	0.98	1.74	2.16	1.80
O^{daf} (oxygen content, on a dry ash-free basis), %	22.58	20.12	6.52	2.3
S^{daf} (sulfur content in the dry ash-free basis), % the	1.15	0.34	0.14	0.32

*averaged reference data

The components ratio in the mixture was determined according to the thermodynamic calculations. Experiments were performed on the basis of the developed techniques of solid-phase reduction of iron from oxides with the use of solid carbonaceous reducing agents [10]. The metallization was performed by reductive roasting of ore-coal briquettes in the laboratory furnace at temperature 1273 K for 40 min. The metallization mode was selected based on previous studies. Spongy metalized materials were produced.

The study on the liquid phase separation of products of solid phase metallization was carried out by stepwise increase of the temperature within the range 1273-1673 K and analysis of the obtained melt products. Isothermal exposure at each temperature was 10 minutes. These materials were studied by chemical, spectral, X-ray diffraction analysis.

4. Results and discussion

The results of experiments determined that by changing the composition of the initial ore-coal mixture, temperature and time of reduction roasting metallized materials can be obtained with a specified amount F_{met}, FeO and SiO_2. The results of melt separation (Table 4) for metallized materials with a high degree of metallization (over 95%) showed that the formation process of fluid phases and coagulation of metal droplets occurs less intensively at higher temperatures than for metallized materials with a lower (less than 95%) degree of metallization. It is found that when using coals of various technological grades as reducing agents the coagulation of metal granules takes place differently, probably due to the influence of different compositions of reducing ash.

Table 4. Results of the experiment.

Reducing agent	Granule formation temperature (> 1 mm), K	Mass of the granulated iron*, g	Spectral analysis of the granules
Lignite	1573	412	Fe_{met}> 99 %;
Long-flame coal	1573	402	C< 1 %;
Low-caking coal	1623	420	Si < 0.5 %;
			Al < 0.5 %;
Lean coal	1673	415	P < 0.1 %;
			S < 0.1 %

*amount of iron obtained by using 1 kg of iron ore (ZhR-2).

The spectral analysis of the obtained granules showed a high content of primary ore and low content of residual elements in the granules. Metal composition is determined by the process parameters and depends on the composition of iron ore and the used reducing agent. The lowest temperature of the beginning formation of visible, larger than 1 mm, metal granules was 1473 K for the mixture with brown coal as a reductant.

5. Conclusions

The results of the research theoretically proved and experimentally confirmed the possibility of obtaining high-quality granulated iron using iron ores and coals from Kemerovo region as raw materials. It was found that the process includes two stages. Initially the solid-phase reduction of iron from oxides is performed obtaining metallized sponge material, then as temperature increases the separation of slag and metal phases occurs.

6. References

[1] Yusfin Yu S and Pashkov N F 2007 *Metallurgy of Iron* (M.: IKTs "Akademkniga") p 464
[2] Naito M 2006 *Shinnittetsu Giho* **384** 2
[3] Zhang Y Y et al 2013 *Advanced Materials Research* **746** 533–538
[4] Nagano K I 2004 *Tetsu-to-Hagane* **90** (2) 51–60
[5] Seki K and Tanaka H 2008 *Changes in Paradigm Development of Iron & Steel Industry by Applying Coal Based Processes: FASTMELT® & itmk3®* (Tokyo Kobe Steel Ltd)
[6] Tsuge Osamu and Ito Suzo *Method of Production of Granulated Iron* Patent No. 2301834
[7] Bale C W et al 2013 *Thermfact and GTT-Technologies*
[8] Nokhrina O I, Rozhihina I D, Proshunin I E, Hodosov I E and Osipova V G 2015 *Applied Mechanics and Materials* **770** 8–13
[9] Nokhrina O I, Rozhihina I D, Proshunin I E and Hodosov I E 2015 *Steel in Translation* **45** (5) 295–300
[10] Nokhrina O I, Rozhihina I D and Hodosov I E 2015 IOP *Conference Series: Materials Science and Engineering* (IOP Publishing) **9** (1) 012045

Metallization of oxide-ore-containing wastes with the use of brown coal semicoke from Berezovsky deposit of the Kansk-Achinsk Basin

A E Anikin[1], G VGalevsky[1], E V Nozdrin[2], V V Rudneva[1] and S G Galevsky[3]

[1] Institute of Metallurgy and Material Science, Siberian State Industrial University, 42 Kirova Street, Novokuznetsk, 654007, Russia
[2] JSC "Mittal Steel", 34 Zavodskaya Street, Temirtau, 101400, Republic of Kazakhstan
[3] National Mineral Resources University, 2 21st line V.O, Saint Petersburg, 199106, Russia

E-mail: kafcmet@sibsiu.ru

Abstract. The research of the metallization process of the roll scale and sludge after gas treatment in the BOF production with the use of brown coal semicoke mined in Berezovsky field of the Kansk-Achinsk Basin was carried out. A flow diagram of "cold" briquetting using a water-soluble binder was offered. The reduction of iron from its oxide Fe2O3 with brown coal semicoke in the laboratory electric-tube furnace in the argon atmosphere was studied. The mathematical models of dependence of the metallization degree on variable factors were developed. The optimal values of technological factors and essential characteristics of the obtained metallized products were revealed.

1. Introduction

Currently, metallization of oxide-iron-containing wastes is an important trend in the industry. This is especially true for the mini-plants with the absence in their structure of sinter production allowing the full cycle metallurgical enterprises to process mill scale, sludges from the BF and steelmaking production. The urgency of the problem is due to a significant number of mini-plants in the world – approximately one thousand [1], and, accordingly, a significant amount of generation of oxide-iron-containing wastes. High dispersion of these wastes determines the need for their agglomeration before use. It conditions the technological feasibility of engaging into the processing highly dispersed carbonaceous materials of technogenic origin and specifically manufactured. Brown coal semicokes (BCSs), in particular BCSs of Berezovsky mine in the Kansk-Achinsk Basin (KAB), in the long run can be used as reducing agents [2-6].

The aim of this research is to develop the scientific and technological bases for the use of BCSs in the process of metallization of oxide-containing technigenic raw material.

2. Current state of technologies used for metallization of oxide-iron-containing raw materials

All metallization processes (direct iron reduction) are based on the following principle [7]. The initial oxide-iron-containing raw material is reduced to the state of solid sponge iron, also known as iron of direct reduction (DRI – Direct Reduced Iron), or to hot briquetted iron (HBI – Hot Briquetted Iron) having a degree of metallization 85-95%. At a less degree of metallization the resulting product is called pre-reduced (partially metallized).

Reduction processes depending on the used type of reducing agent are classified into gas and solid phase reduction. At an industrial scale the gas phase processes are carried out in the shaft furnaces, retorts and fluidized (boiling) layer. Solid reduction occurs in rotary furnaces, furnaces with a rotary hearth or multi-hearth furnaces.

The metallized products need to meet the requirements in granulometric composition, degree of metallization, carbon content and the compressive strength [8]. Table 1 shows requirements for metallized products depending on their function.

Table 1. Requirements to metallized products depending on their use.

Properties of metalized products	Use of metallized products	
	Blast furnace	Arc steel furnace
Size of pellets and briquettes, mm	16-22	8-14
The degree of metallization, %	85-92	> 90
Carbon content, %	4-7	1.5-3.0
Compressive strength, H	> 250	> 200

According to the authors [9], the most important characteristics of metallized products include the following: the degree of metallization – 90-94%; content of rock refuse (SiO_2) – not more than 5%; sulfur content – not more than 0.010%; phosphorus – not more than 0.015%; carbon content – 1-2%; bulk weight – not less than 1.8 tonne/m3; nitrogen content – 0.02-0.14%; the resulting product must be passivated, i.e. the secondary oxidation should be suppressed.

3. Experimental research: methodology, results, discussion

While conducting the research the non-roasted briquetted compositions produced by "cold" briquetting (Figure 1) were tested, consisting of oxide-iron-containing component, carbonaceous reducing agent and a binder. The weight ratio between C and Fe_2O_3 briquettes was 4.44: 1.0 (i.e., 81.6% Fe_2O_3 and 18.4% C).

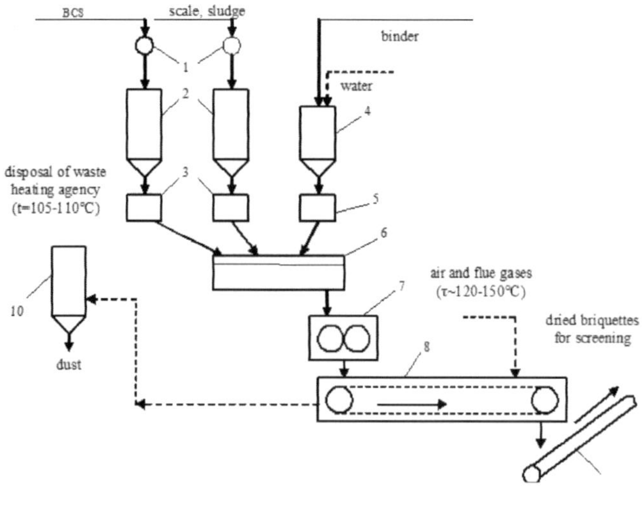

Figure 1. Technological scheme for production of non-roasted briquettes (1 – milling unit; 2 – proportioning bunker; 3 – dispensers; 4 – storage container for aqueous binder solution; 5 – binder dispenser; 6 – double spiral mixer; 7 – roll press; 8 – belt drier; 9 – conveyor belt; 10 – cyclone).

Consumption of binder (molasses) in all cases was 10% by weight of the mixture of oxide-iron-containing. As oxide-iron-containing components in experiments we used iron oxide (III) (chemically pure), as well as mill scale and sludge from gas treatment of BOF production at JSC "EVRAZ ZSMK". As carbonaceous reducing agents we used BCSs from Berezovsky field of KAB, small coke of JSC "Koks" (SC) and the dust of dry coke quenching from JSC "EVRAZ ZSMK" (CD).

The investigation of the metallization process was carried out in two stages. At the first stage in the laboratory electric-tube furnace in the protective argon atmosphere (Figure 2) we studied the reduction of iron from its oxide Fe_2O_3 with BCS. The study was conducted using the method of the planned experiment [10]. The optimization parameters, factors, influence of which was taken into account, and the range of their variation are given in Table 2.

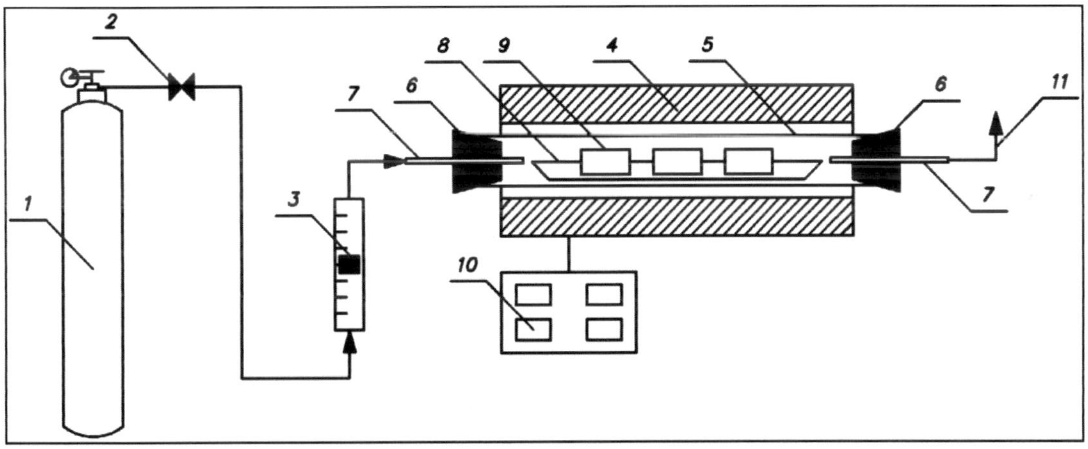

Figure 2. Installation for the study of metallization of oxide-iron-containing technogenic raw material (1 – tank with; 2 –valve for fine regulation of argon flow; 3 – rotameter RS-3A; 4 – electric-tube furnace SUON 0.3.2 / 12.5-I1; 5 – corund tube with diameter 0.032 m, 0.36 m long; 6 – rubber stopper; 7 – metal tubes for argon supply (left) and for argon, gaseous products of metallization outflow (right); 8 – metal boat; 9 – briquettes; 10 – furnace control panel; 11 – release of argon and gaseous products of metallization into the atmosphere).

Table 2. Variable factors selected for the study of metallization process.

Factors	Levels of factors			Variation interval
	–1 (low)	0	+1 (upper)	
First series				
x_1 – violates yield from BCS, V^{daf}, %	0.6	5.05	9.5	4.45
x_2 – temperature, t, °C	600	750	900	150
x_3 – duration of metallization, τ, min	25	37.5	50	12.5
x_4 – pressing pressure during briquetting, P, MPa	10	25	40	15
Second series				
x_1 – reactivity of the reducing agent with CO_2, K, $cm^3/(g{\cdot}s)$	0.33	2.025	3.72	1.695
x_2 – temperature, t, °C	700	800	900	100
x_3 – duration of metallization, τ, min	15	27.5	40	12.5
x_4 – pressing pressure during briquetting, P, MPa	10	25	40	15

The mathematical models in the form of the following equations are obtained:

$$\eta_1 = -183.40 + 0.62\ V^{daf} + 0.37\ T + 1.86\ \tau - 0.008\ V^{daf} \cdot \tau - 0.002\ T \cdot \tau; \qquad (1)$$

$$\eta_2 = -430.28 + 10.52\ K + 0.55\ T + 8.80\ \tau - 0.08\ K \cdot \tau - 0.01\ T \cdot \tau, \qquad (2)$$

whereV^{daf}– violates yield from BCS (0.6 – 9.5 %);
T – temperature (600 – 900 °C);
τ – duration of metallization (15 – 50 min);
K – reactivity of the reducing agent with CO_2 (0.33–3.72 $cm^3/(g \cdot s)$).

Figures 3-6 shows graphical dependences of the degree of metallization on volatiles yield from the reducer and the temperature at metallization duration 40 minutes (Figure 3), on the reducer reactivity and temperature at metallization duration 50 minutes (Figure 4), on the violates yield from reducer and duration of metallization at 900 °C (Figure 5), as well as on the reducer reactivity and the duration of the metallization at a temperature 900 °C (Figure 6).

The first most important factor is a temperature. Thus, for the same duration of metallization if the temperature is increased from 700 to 900 °C, then η increases from 1.1 to 96.7%. The second most important factor is the reactivity of the reducing agent. For example, during the reduction of iron oxide (III) (ch. p.) using a carbonaceous reducing agent with reactivity 3.72 $cm^3/(g \cdot sec)$ at a temperature 900 °C and duration of the metallization 40 minutes η = 96.9%, whereas the use of a carbonaceous reductant with the reactivity 0.33 cm3/(g·s) under the same conditions – 47.8%. The third factor is the value of the metallization duration (when it is changed from 15 to 40 min).

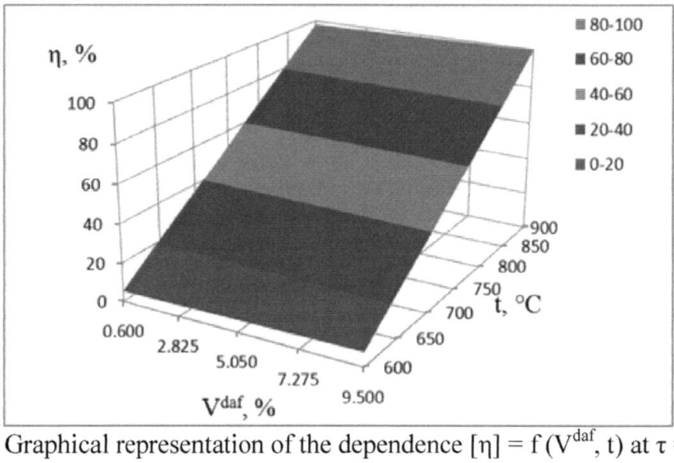

Figure 3. Graphical representation of the dependence [η] = f (V^{daf}, t) at τ = 40 min.

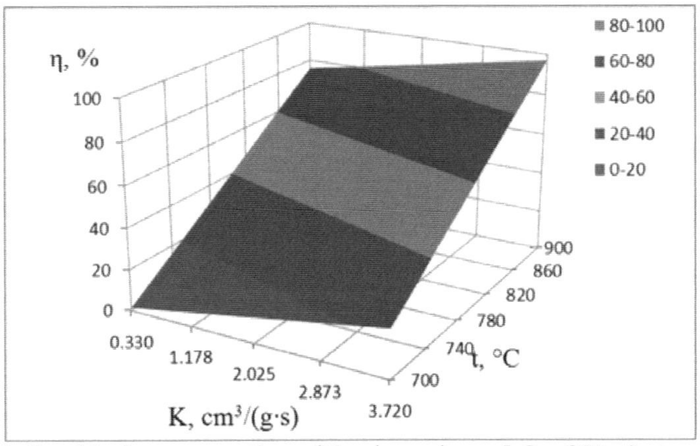

Figure 4. Graphical representation of the dependence [η] = f (K, t) at τ = 5 min.

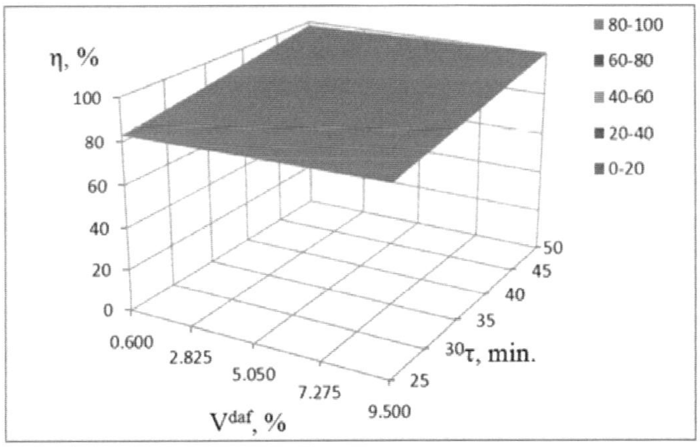

Figure 5. Graphical representation of the dependence [η] = f (V^{daf}, τ) at t = 900 °C.

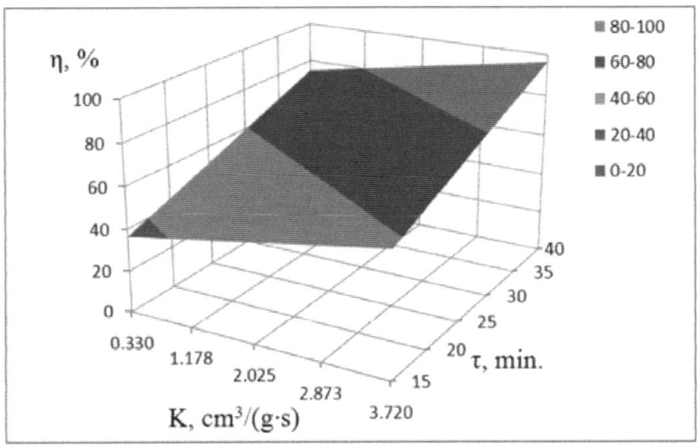

Figure 6. Graphical representation of the dependence [η] = f (K, τ) at t = 900 °C.

Thus, to obtain the metallized product with the desired quality the metallization process should be performed at a temperature 900 °C (x2 factor), the duration of the metallization 40 min (x3 factor) and the reactivity of the carbonaceous reductant CO_2 3.72 cm3/(g·s) (x1 factor).

The optimum values of technological factors and essential characteristics of the obtained metallized products are given in Table 3.

At the second stage of the research under these conditions in the laboratory electric furnace the metallization of briquetted batches of 6 compositions was done: from scale and BCS, small coke (SC) and dust (CD), and from slurries with these carbonaceous reductants. To approximate conditions of metallization to industrial ones argon was not fed into the reaction area in the furnace. A comparative analysis of its indicators was performed and the physical and chemical certification of products was carried out (Table 4).

Table 3. The optimum values of technological factors of metallization process and the main characteristics of the obtained metallized products.

Process parameters of metallization and metallized product characteristics	Value
Metallization temperature, t, °C	900
Duration of metallization, τ, min	40
Reactivity of the reducing agent CO_2, K, $cm^3/(g \cdot s)$	3.72
Pressure pressing during briquetting, P, MPa	40
Degree of product metallization, η, %	96.7
Size of metallized briquettes, mm	20
Content in the metallized product, %:	
$Fe_{tot.}$; $Fe_{met.}$; FeO	86.1; 83.4; 3.5
S, P, C	0.04; 0.001; 3.0
CaO, MgO	0.9; 0.1

Table 4. Quality of metallized products.

Content of briquetted compositions	η, %	Content in metallized products, %							
		Fe_{tot}	Fe_{met}	FeO	S	P	C	CaO	MgO
Scale + BCS	97.5	92.5	90.2	3.0	0.07	0.017	1.8	1.1	0.3
Scale + SC	70.7	87.5	61.9	33.0	0.14	0.025	3.8	0.3	0.3
Scale + DC	71.1	88.0	62.6	32.8	0.11	0.026	3.6	0.4	0.3
Slurry + BCS	97.5	73.1	71.3	2.3	0.21	0.121	1.8	17.4	0.4
Slurry + SC	68.9	70.2	48.4	28.1	0.28	0.130	3.8	16.6	0.3
Slurry + DC	69.2	69.7	48.2	27.7	0.25	0.131	3.6	16.7	0.3

The metallization degree was from 69 to 97.5% with the best results achieved by using BCS as a reductant. Metallization products from the scale burden and BCS in their degree of metallization, content of sulfur, phosphorus, carbon and waste rock meet the requirements concerning metallized products for the production of electric steel.

4. Conclusions

The technological scheme for obtaining non-roasted strong briquettes from charges with the compositions mill scale – semicoke, slurry – semicoke with 10% of water-soluble binder – molasses was developed and implemented. Drop strength of oxide-iron-containing briquettes is 98-99% and conserved during exposure in sand at 900 °C for 30 min. We experimentally studied the metallization processes of batches with compositions oxide-containing components (scale, sludge) – (BCS, SC, DC). Carried out metallization at a temperature of 1173 K and a duration of 40 min and made a comparative analysis of its indicators for briquetted batches of six types: scale – BCS, SC and DC, sludge – BCS, SC and DC. Determined the degree of metallization and the content of metallic iron, which with the use of BCS are 97.5 and 90.2% for scale, 97.5 and 71.3% for sludge, DC 70.7 and 61.9% for the scale 68, 9 and 48.4% for sludge, DC 72.1 and 62.6% for scale, 69.2 and 48.2% for sludge. Developed mathematical models describing the dependence of the degree of metallization on temperature, its duration, reducer reactivity and content of volatiles in it.

Found the optimum temperature and time conditions for metallization of briquetted batch scale – brown coal semicoke with reactivity 3.72 cm3/(g·s): temperature 1173 K with the duration of 40 minutes. Performed physical and chemical attestation of metallization products, which includes the study of phase, chemical, particle size distribution and morphology. Metallization products of briquetted batch scale – BCS on the degree of metallization, content of sulfur, phosphorus, carbon, waste rock meet all necessary requirements.

5. References

[1] Yaroshenko Yu G, Gordon I Yu and Khodorovskaya I Yu *2012 Energy-efficient and Resource-saving Technologies in Iron and Steel Production* (Ekaterinburg: JSC "UIPTs" p 670

[2] Ulanovskaya M A et al 2000 *Coke and Chemistry* **4** 14–20

[3] Shkoler M B 2007 *Coke and Chemistry* **12** 18–24

[4] Strakhov V M 2008 *Coke and Chemistry* **9** 44–49

[5] Anikin A E and Galevsky G V 2014 *Proc. of SibSIU* **3** (9) 52–59

[6] Anikin A E and Galevsky G V 2014 *Proc. of XVIII All-Russian Conf. on Metallurgy: Technology, Control, Innovation, Quality* (Novokuznetsk: SibSIU) pp 115–120

[7] Lungen Kh B, Peters M and Shmels P 2010 *Ferrous Metals* **9** 52–66

[8] Degel R et al 2000 *Ferrous Metals* **7** 31–39

[9] Yusfin Yu S, Gimmelfarb A A and Pashkov N F 1994 *New Processes of Metal Production* (M.: Metallurgiya) p 320

[10] Anikin A E and Galevsky G V 2015 *Proc. XI Int. Conf. on Foundry. Metallurgy* (Zaporozhye: ZTPP) pp 271–273

IOP Conf. Series: Materials Science and Engineering **150** (2016) 012005 doi:10.1088/1757-899X/150/1/012005

Development of scientific and technological bases for application of brown coal semi coke in the technology of non-milled silicon carbide

A E Anikin[1], G V Galevsky[1], V V Rudneva[1], E V Nozdrin[2] and S G Galevsky[3]

[1]Institute of Metallurgy and Material Science, Siberian State Industrial University, 42 Kirova Street, Novokuznetsk, 654007, Russia
[2] JSC "Mittal Steel", 34 Zavodskaya Street, Temirtau, 101400, Republic of Kazakhstan
[3] National Mineral Resources University, 2 21st line V.O, Saint Petersburg, 199106, Russia

E-mail: kafcmet@sibsiu.ru

Abstract. Thermodynamics is investigated, and the optimum temperature and time modes of carbonization of a briquetted silica fume batch– brown coal semi coke are defined. The complete carbonization of the batch in the conditions of heat treatment is achieved at a temperature of 1923 – 1973 K within 15 – 20 minutes. The conditions and indicators of the chemical enrichment of carbonization products are established. After enrichment, the carbide content is more than 90%. Silicon carbide micro-powder is obtained with a specific surface area 8000 – 9000 m^2/kg.

1. Introduction

Production of silicon carbide is one of the most important in the structure of modern metallurgy; its annual output is about 800 thousand tonnes. The major fields of use are metallurgy industry (about 45 % of the world demand), production of abrasives (up to 30 %) and refractory materials (25 %). Specific nature of thermo-mechanical, electro-physical, physical-chemical properties of silicon carbide determines the following areas of its traditional and advanced application: bonded materials; structural and functional ceramics; composite materials and coatings; surface and volume modification of materials. Scientific and technological substantiation and development of production of domestic non-milled silicon carbide is a task of great practical importance. In this regard, the aim of the present work was to study the processes of carbonization of technogenious silica fume by semi coke from brown coal mined in Berezovsky deposit of the Kansk-Achinsk Basin, technological properties of which are described in detail in [1-6].

2. Thermodynamic process modelling of carbide formation in Si–O–C, Si–O–C–H systems

Thermodynamic modeling was carried out in order to predict the optimal parameters of silicon carbide production (ratio of the components, temperature and pressure in the system), to define the equilibrium parameters of the process (conversion degree of raw materials to carbide, composition of gaseous and condensed products), to evaluate the contribution of individual reactions providing an effective processing of raw materials in the investigated conditions to the processes of carbide formation.

Taking into account use of brown coal semi coke with content of volatiles up to 10% in the process as a reducing agent, the objects for research were systems Si–O–C, Si–O–C–H. Calculation of equilibrium compositions of Si–O–C and Si–O–C–H systems was made by the “constant” method.

The calculations were carried out using the software for simulation of high-temperature complex chemical equilibria “PLASMA” with a built-in database for reaction products of oxide-, boride-, carbide- and nitride-forming systems. In the calculations we considered the temperature range from 800 to 2000 K at different pressures in the system: from 0.1 to 0.0001 MPa. The calculations were carried out in increments of 100°. The equilibrium constants of reactions of compounds formation from the elements served as initial data. During the calculations the possibility of formation was taken into account in the gas phase in the system Si–O–C Si, Si+, Si, Si_2, Si_3, O, O_2, C, C+, C_2, C_3, C_4, C_5, SiO, SiC, SiC_2, Si_3C, CO, CO_2, in the system Si–O–C–H Si, Si+, Si□, Si_2, Si_3, O, O_2, C, C+, C_2, C_3, C_4, C_5, H, H+, H_2, CH_2, CH_3, CH_4, C_2H, C_3H, C_4H, C_2H_2, C_2H_4, SiO, SiC, SiC_2, Si_3C, CO, CO_2, in the condensed phase in two systems of Si, SiO_2, C, SiC. The choice of the ratio of initial elements in the investigated systems was determined by the stoichiometry of reaction of silicon carbide formation and composition of the batches.

The calculation results of the equilibrium compositions of systemsSi–O–C and Si–O–C–H are given in Figures 1, 2 and in Tables 1, 2.

It can be seen that in both systems the process of carbide formation is dominant:

$$SiO_{2(s)} + 3C_{(s)} = SiC_{(s)} + 2CO_{(g)} \quad (1)$$

In the batch of stoichiometric composition the maximum content in the silicon carbide products is achieved at 1700 K, and with10% of carbon deficiency – at 1900 K.

Introduction of hydrogen into the system practically does not affect the process of carbide formation due to the extremely low (less than 0.001 mol) content of hydrocarbons and hydrocarbon radicals in the temperature range of carbide formation.

In the system Si–O–C the equilibrium degree of silicon conversion to carbide is not more than 0.97%, which corresponds to the content of silicon monoxide in the gas phase 0.02 mol. Thus, it is impossible to obtain a single phase silicon carbide with no free carbon (Table 2) from the batch of stoichiometric composition (SiO_2+3C). Use of a batch with some (~ 10 %) carbon deficiency – reducing agent can help to avoid it (Table 2).

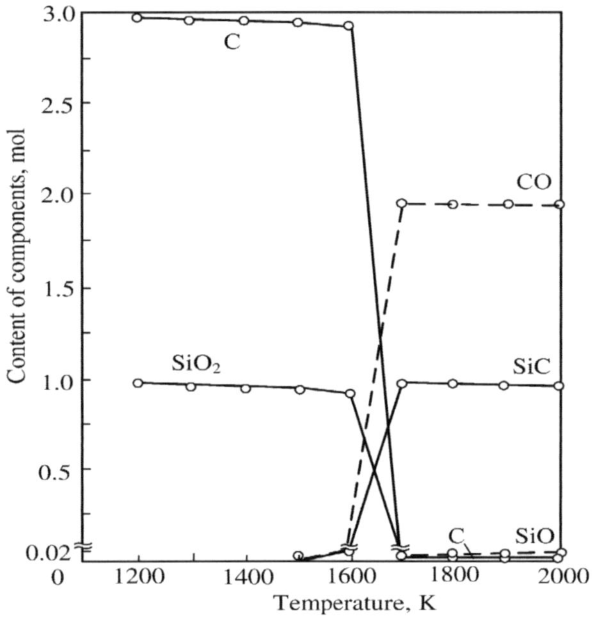

Figure 1. Dependence of the equilibrium composition of Si–O–C system on the temperature with the ratio of Si:O:C = 1:2:3 and pressure P = 0.1 MPa (—condensed, --- gaseous products).

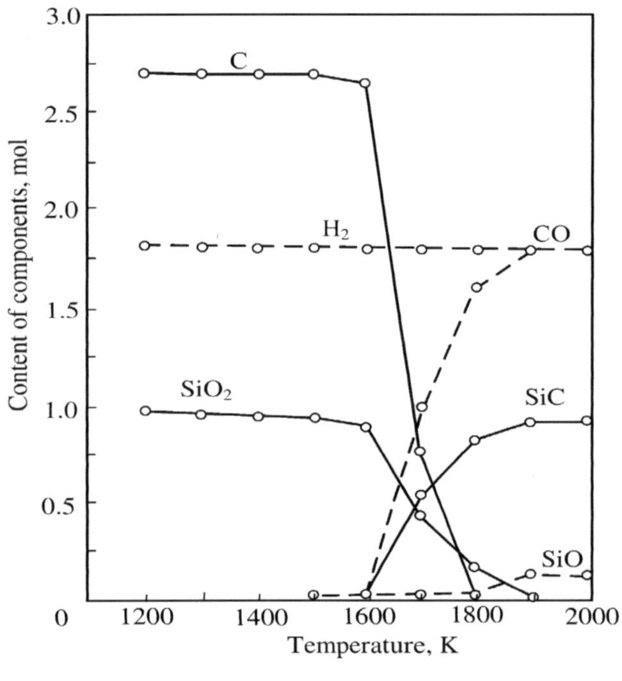

Figure 2. Dependence of equilibrium compositions of Si–O–C–H system on temperature with the ratio of Si:O:C:H = 1:2:2.7:3.6 and the pressure P = 0.1 MPa (—condensed, --- gaseous products).

In this case, the area of coexistence of silicon carbide and excessive silica fume appears in the system. And the wider the temperature range of these phases coexistence, the higher the pressure in the system. Above these temperatures the silicon carbide itself is a reducing agent for silica fume, i.e. the following interaction becomes possible:

$$2SiO_{2(s)} + SiC_{(s)} = 2SiO_{(g)} + CO_{(g)} (2)$$

Table 1. Equilibrium concentration (in%) of the condensed reaction products in the system Si–O–C with the ratio of components Si:O:C = 1:2:3.

T, K	P = 0.1 MPa			P = 0.01 MPa			P = 0.0001 MPa		
	SiO_2	C	SiC	SiO_2	C	SiC	SiO_2	C	SiC
800	62.5	37.5	0	62.5	37.5	0	62.5	37.5	0
900	62.5	37.5	0	62.5	37.5	0	62.5	37.5	0.02
1000	62.5	37.5	0	62.5	37.5	0.01	62.2	37.3	0.52
1100	62.5	37.5	0	62.4	37.4	0.2	0	0.3	99.71
1200	62.5	37.5	0	61.0	36.7	2.3	0	0.4	99.60
1300	62.5	37.5	0	0	0.2	99.8	0	0.6	99.40
1400	61.8	37.1	1.1	0	0.2	99.8	0	1.0	99.0
1500	61.0	36.8	2.2	0	0.3	99.7	0	1.3	98.7
1600	53.7	32.5	13.8	0	0.4	99.6	0	2.0	98.0
1700	0	0.6	99.4	0	0.6	99.4	0	2.8	97.2
1800	0	0.8	99.2	0	0.9	99.1	0	3.7	96.3

Table 2. Equilibrium concentration (in %) of the condensed reaction products in the system Si–O–C–H with the ratio of components Si:O:C:H = 1:2:2.7: 3.6.

T, K	P = 0.1 MPa			P = 0.01 MPa			P = 0.0001 MPa		
	SiO_2	C	SiC		SiO_2	C	SiC		SiO_2
900	64.94	35.06	0	64.94	35.06	0	64.94	35.06	0
1000	64.94	35.06	0	64.94	35.06	0	64.94	35.06	0.001
1100	64.94	35.06	0	64.94	35.06	0.002	64.84	35.00	0.16
1200	64.94	35.06	0	64.74	36.04	0.04	62.54	33.86	3.60
1300	64.94	35.06	0	64.54	35.94	0.30	12.47	0	87.53
1400	64.92	35.06	0.03	62.84	34.53	2.63	8.95	0	91.05
1500	63.97	34.51	1.52	51.27	26.61	22.12	0	0	100.00
1600	60.75	32.15	7.10	11.35	0	88.65	0	0	100.00
1700	25.57	8.89	65.54	0	0	100.00	0	0	100.00
1800	9.33	0	90.67	0	0	100.00	0	0	100.00
1900	0	0	100.00	0	0	100.00	0	0	100.00

As a result of this reaction silicon carbide is cleaned from the excessive silica fume. It might be particularly effective in the vacuum.

3 Research methodology of silica fume carbonization

The study of silica fume carbonization was conducted by the discrete method that includes heat treatment of the briquetted batch at given temperatures and duration and quenching of carbonization products with their subsequent chemical analysis.

Heat treatment of the batches was carried out in the electric resistance furnace with a cylindrical graphite heater with the inner diameter of 0.075 m (Tamman furnace). Temperature control in the furnace was performed by the tungsten-rhenium thermocouple VR 5/20. Cylindrical briquettes weighing about 2.5 kg·10-3 and with an outer diameter 0.015 m were placed into the graphite crucibles. Three crucibles were simultaneously installed in the workspace of the heated electric furnace, were heated in it for 3 – 4 min up to the target temperature and kept for the specified time period – from 5 to 90 minutes. After the specified time the graphite crucibles were taken out of the electric furnace, installed into the corundum crucibles that were tightly stoppered and placed for 10 minutes into the cold water to cool the carbonization product. This prevented oxidation of the carbonization product with oxygen and atmospheric moisture.

After cooling the carbonization products were quantitatively, i.e. lossless extracted from the crucibles, weighed, placed into the platinum containers filled with a solution of nitric acid (concentration 7.5 – 12.5M) and hydrofluoric acid (concentration 2.0 – 3M) taken in a volume ratio of 1:(3 – 5), and boiled for 2.5 – 3.0 hours. Treatment of silicon carbide with a mixture of nitric and hydrofluoric acids removes impurities of aluminum, magnesium, calcium, iron, manganese, nickel, chromium, copper, sodium, potassium, silicon and their oxides and is recommended for the silicon carbide powders with grain size from 5 to 200 μm [7, 8]. In the presence of nitric acid the metallic impurities and their oxides form soluble nitrate complexes, whereas treatment with hydrofluoric acid solely may cause formation of insoluble fluorides. After cooling the insoluble residue – silicon carbide with free carbon impurity was filtered through a paper filter, washed with distilled water and acetone, dried to a constant weight, and weighed. Further, the undissolved residue was subjected to oxidizing burning for refining from free carbon at a temperature of 1073 – 1123 K during 1.5 – 2.0 hours. In the carbonization products the content of free carbon and silicon carbide was calculated according to the mass loss. Based on the average content of silicon carbide in three crucibles its yield was calculated as the ratio of the obtained carbide mass to the theoretically possible mass.

4. Results and discussion

During the research silica fume was used, generated in the production of silicon (SF-S) and high-silicon ferrosilicon (SF-FS), containing silica dioxide 93.92 and 93.00 % by weight with a specific surface area of 25000 m2/kg [9-11], as well as semi coke from brown coal (BCSC) of Berezovsky deposit in the Kansk-Achinsk Basin containing, %: carbon – 81.9; volatiles – 9.5; ash – 8.6 with a specific surface area 264000 m2/kg. In all cases the batches of stoichiometric composition were used to produce carbide.

Investigation of time-temperature conditions for carbonization of the briquetted silica fume mono-batches SF-S, SF-FS – BCSC was conducted at temperatures 1883, 1923 1973 K with the duration of heat treatment 5, 10, 15, 20, 25, 30 min. The research results are given in Figures 3 and 4.

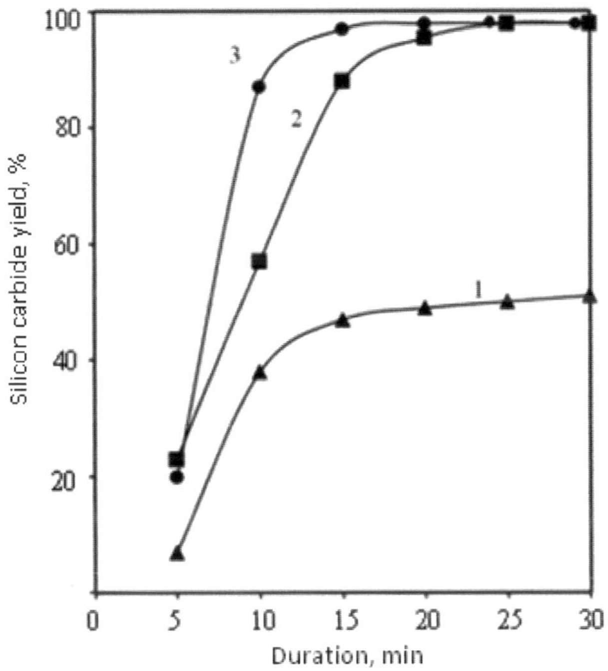

Figure 3. Dependence of silicon carbide yield on the temperature and duration of heat treatment for the silica fume batch SF-FS – BCSC. 1 – 1873 K: (Δ) = (7.21 – 53.13) ± (0.20 – 1.13); 2 – 1923 K: (□) = (17.06 – 96.82) ± (0.55 – 1.35); 3 – 1973 K: (○) = (20.20 – 97.01) ± (0.61 – 1.45).

In both cases the process of carbide formation begins from the first minute, and by the 5th minute the carbide yield at carbonization temperatures 1873, 1923, 1973 K is 5.23 – 7.21, 17.06 – 22.30, 20.20 – 23.51%. The carbonization process is completed at a temperature of 1973 K by the 15th minute, at 1923 K – by the 20th minute. Carbide content in the products of heat treatment of batches is 92.50 for SF-FS and 94.90% by weight for SF-S. Such time-temperature conditions of carbide formation are confirmed by the study results of the phase composition of these products.

Thus, it was found that in both cases by the 10th minute silicon carbide of cubic structure (β-SiC) becomes the prevailing phase, partially turning into a hexagonal one (α-SiC_{II}) at a temperature 1923 – 1973 K and duration of heat treatment of 50-90 min. The carbonization process of the batch SF-FS – BCSC is accompanied by alpha iron. In all test samples the glassy phase is also present formed, presumably, by silicates of calcium, magnesium and iron.

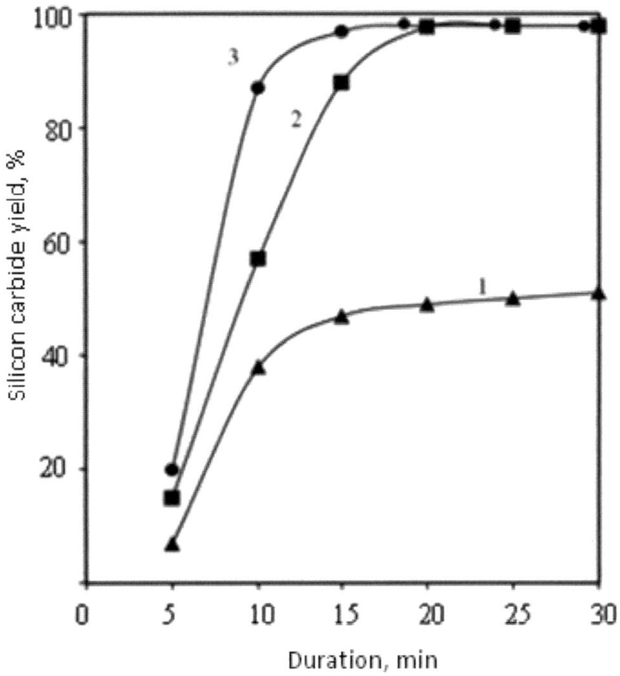

Figure 4. Dependence of silicon carbide yield on the temperature and duration of heat treatment for the silica fume batch SF-S – BCSC. 1 – 1873 K: (Δ) = (5.23 – 46.26) ± (0.19 – 1.11); 2 – 1923 K: (□) = (22.30 – 96.78) ± (0.59 – 1.33); 3 – 1973 K: (○) = (23.51 – 97.62) ± (0.67 – 1.38).

As a result of chemical processing the content of SiC in carbide exceeds 90%, i.e. the level specified for abrasive micro-powders with grain size 1-2 μm. The efficiency of enrichment from iron and oxides impurities is sufficiently high 87 – 95 %. Silicon carbide is characterized by high silica content – more than 7%, which allows us to consider it as a promising material for production of silicon carbide refractories with a silica bond, usually containing 84 – 89% of SiC, 6 – 12% of SiO_2 [7]. Optimal conditions for production and enrichment of silicon carbide and variation range of its basic characteristics for the series of five parallel experiments are provided in Table 3. It can be seen that the examined technological variants ensure production of silicon carbide with the reproducible phase, chemical and granulometric compositions and can be regarded as a technological basis for the design of industrial production of non-milled silicon carbide from briquetted high-dispersed batch of silica fume – brown coal semi coke.

Table 3. Optimal conditions for obtaining and enrichment of silicon carbide and its characteristics.

The production conditions and enrichment of silicon carbide and its main characteristics	SiC (SF-FS + BCSC)	SiC (SF-S + BCSC)
Batch composition, %:		
– silica fume (SF)	58.80	58.04
– brown coal semi coke (BCSC)	41.20	41.96
Specific surface area of the batch components, m^2/kg:		
– silica fume (SF)	20000 – 22000	
– brown coal semi coke (BCSC)	255000 – 265000	
Temperature of the batch carbonization, K	1923 – 1973	

Duration of the batch carbonization , min	25 – 20	
Chemical enrichment of the carbonization products:		
– concentration of the hydrochloric acid solution, %	35	
– ratio Solid : Liquid	1:2	
– temperature, K	383	
– duration, hour	3	
Phase composition	β-SiC	
Chemical composition, %:		
– SiC	90.42–90.86	90.94–91.18
– Al_2O_3	0.10 – 0.25	0.10 – 0.15
– CaO	0.32 – 0.53	0.37 – 0.49
– MgO	0.12 – 0.18	0.03 – 0.10
– Fe	0.08 – 0.12	0.05 – 0.11
– SiO_2	7.32 – 7.94	6.97 – 7.13
– C free	0.41 – 0.54	0.63 – 0.81
– Si free	0.13 – 0.17	0.23 – 0.27
Silicon carbid cyield, %	96.84–97.12	97.31–97.63
Fineness and morphology of silicon carbide particles:		
– specific surface area, m^2/kg	8000 – 9000	
– size range, μm	0.2 – 1.0	
– shape	irregular	

5. Conclusions

Thermodynamic and experimental investigations of reductive processing of technogenious silica fume into silicon carbide using brown coal semi coke are carried out.

It is found that in both systems Si–O–C and Si–O–C–H chosen for the thermodynamic analysis the process of carbide formation is the dominant one. In the batch of stoichiometric composition the maximum content of silicon carbide in the reduction products is achieved at 1700 K and with 10% of carbon deficiency at 1900 K. The second composition of the batch is more technologically preferable since it allows carbide free of impurities to be obtained.

Carbonization of the briquetted mono-batches silica fume (SF-FS and SF-S) – brown coal semi coke is carried out at temperatures 1883, 1923, 1973 K with the duration of 5 – 30 minutes. The optimum time-temperature conditions of carbonization are defined: temperature 1923 – 1973 K with the duration of 15 – 20 minutes.

The expediency is established, and the optimal conditions and the parameters of chemical enrichment of carbonization products are found: the effect of hydrochloric acid with concentration not less than 35% at a temperature of 353 K; ratio S:L= 1:2; duration 3 hours, content of carbide in the enrichment products more than 90 %; removal of oxides and iron impurities by 87 – 95 %. The phase, chemical, granulometric compositions are investigated, and morphology of silicon carbide particles for both technological variants are received. It is found that during carbonization silicon carbide of cubic structure (β-SiC) is formed, partially turning into a hexagonal one (α-SiC_{II}) at a temperature 1923 – 1973 K and duration of heat treatment of 50-90 min. The content of carbide after enrichment when using SF-FS is 90.42 – 90.86 % and when using SF-S is 90.94 – 91.18 %. In both cases, silicon carbide as a micropowder with a specific surface area 8000 – 9000 m2/kg, particles of irregular shape within the size range 0.2 – 1.0 μm is obtained.

6. Acknowledgements

The work is performed in Siberian State Industrial University according to the state task of the Ministry of Education and Science of the Russian Federation No.11.1531/2014/K.

References

[1] Romanov S M 2009 *Coal* **1** 15–17
[2] Strakhov V M Surovtseva I V and Dolinsky V A 2007 *Coke and Chemistry* **8** 20–26
[3] Shkoller M B 2007 *Coke and Chemistry* **12** 18–24
[4] Strakhov V M 2008 *Coke and Chemistry* **9** 44–49
[5] Golovin K S and Krapchin S S 2006 *Coal* **6** 64–67
[6] Anikin A E and Galevsky G V 2014 *Bulletin of Siberian State Industrial University* **3** 52–59
[7] Garshin A P, Shumyacher V M and Pushkarev O I 2008 *Abrasive and Construction Materials on the Basis of Silicon Carbide* (Volgograd: VSUACE) p 189
[8] Ordanyan S S and Chupov V D 2004 *Refractories and Technical Ceramics* **8** 28–30
[9] Polyakh O A, Rudneva V V, Yakushevich N F, Galevskii G V and Anikin A E 2014 *Steel in Translation* **44** 565–572
[10] Anikin A E, Polyakh O A, Galevskii G V and Rudneva V V 2014 *RAEN Bulletin of the Mining and Metallurgical Section* **33** 104–113
[11] Anikin A E, Galevskii G V and Rudneva V V 2015 *Innovations in Metallurgy and Materials science: Proc. IV Int. Conf (Ekaterinburg, UFU)* pp 393–396

IOP Conf. Series: Materials Science and Engineering **150** (2016) 012006 doi:10.1088/1757-899X/150/1/012006

A study on reduction processes of elements in the system V_2O_5-Si

M A Golodova[1], I D Rozhihina[2], O I Nokhrina[2] and I A Rybenko[3]

[1] Institute of Fundamental Education, Siberian State Industrial University, 42 Kirova Street, Novokuznetsk, 654007, Russia
[2] Institute of Metallurgy and Material Science, Siberian State Industrial University, 42 Kirova Street, Novokuznetsk, 654007, Russia
[3] Institute of IT and Automation Systems, Siberian State Industrial University, 42 Kirova Street, Novokuznetsk, 654007, Russia

E-mail: golodova_ma@mail.ru

Abstract. The calculation of parameters of elements reduction in the system V_2O_5–C –Si was made at temperatures of metallurgical processes by the method of thermodynamic simulation. Dependences of vanadium reduction from its oxides with carbon and silicon oxides separately with each reducing agent, as well as in combination, on the temperature and reducing agents consumption were considered. The obtained results led to the conclusion that the vanadium reduction from vanadium pentoxide in combination with carbon and silicon takes place with the predominant formation of vanadium compounds: vanadium carbide and vanadium silicides, the main reductant is carbon.

1. Introduction

Currently, the actual problem is a complex usage of primary and secondary raw materials in the iron and steel industry. Extraction of components – valuable alloying additives from such materials can improve the production efficiency and at the same time the products quality. Among the most needed alloying elements, the introduction of which can significantly increase the metal performance properties, is vanadium [1-3]. The basic alloying material in the manufacture of vanadium-containing steels is high-priced ferrovanadium. For the further expansion of production and competitiveness assurance of vanadium-containing steels the alloying method, based on the metal processing by oxide vanadium-containing materials including a converter vanadium slag, has a special importance, while ensuring the conditions for vanadium reduction from these materials.

Currently, aluminum, silicon, calcium of high value are used as reducing agents. In this regard, the development of steel processing technology by converter vanadium slag with application of silicon and carbon as reducing agents is an urgent task. Low value of carbon application would significantly reduce the cost of steel products. However, to determine the optimal parameters for carbon-silicon thermal reduction of vanadium from converter vanadium slag the comprehensive examination of processes, including reduction of vanadium pentoxide with carbon and silicon, should be performed.

2. Research and results

The aim of this study was to consider the processes taking place during vanadium reduction with carbon and silicon, as well as vanadium pentoxide reduction with each reducing agent separately and in combination.

Currently, thermodynamic simulation is widely used to study elements reduction from the oxide systems [4-6]. The problem solution targeting determination of conditions for vanadium reduction in the system V_2O_5–nC –mSi was performed with the use of methods of thermodynamic modeling based on the equilibrium conditions calculation in the model thermodynamic systems. While implementing the thermodynamic simulation, a ready software complex "Terra", developed in Moscow State Technical University, was used. On the basis of maximum entropy principle it finds the equilibrium composition of a multicomponent heterogeneous thermodynamic system for high temperature conditions [7].

The process research was carried out by solving model problems for determination of the conditions for vanadium reduction in the systems, the input flow of which consists of a combination of elements V-O-C-Si, represented by a group of substances V_2O_5–nC –mSi, where n and m – mass of carbon and silicon per 1 kg of vanadium pentoxide. The initial composition of the system was formed by specific values of the parameters n, m.

At the first stage of calculations the possible compositions and the boundaries of concentration areas of reduction processes flow were identified, as well as the temperature influence on these parameters. The list of substances, that can be formed from the given elemental composition of the mixture for temperature range from 1573 to 2273 K, were determined by numerical simulation using the software package "Terra". Thermodynamically the possible quantity of substances, generated from the indicated elements for a selected temperature range, was approximately 20. All concentration values of the substances in the final state were divided into significant and insignificant with a difference threshold 10-4 mol/kg of the mixture. The significant substances were the condensed phase consisting of C, V, VC, VO, V_2O_4, V_2O_3, V_2O_5, Si, SiO_2, SiC and vanadium silicides and a gas phase consisting of CO, CO_2, SiO.

To investigate the temperature influence on vanadium reduction process in the system V_2O_5–C –Si the calculations of equilibrium states at temperatures from 1673 K to 2273 K were made. The minimum and maximum temperatures were conditioned by the temperature values characteristic to the out-of-furnace steel treatment, and vanadium melting point 2173 K. The calculation results are presented in Figure 1.

Analysis of the results showed that the vanadium reduction process from its oxides does not dependent on the temperature in the range from 1673 to 2073K. The total mass of vanadium oxides in the reaction products at a temperature below 2073 K goes to zero during the increase in the carbon consumption from 0 to 0.5 kg/kg V_2O_5, during increase of silicon consumption from 0 to 0.7 kg/kg V_2O_5. The vanadium reduction process from vanadium pentoxide in the system V_2O_5–C ends at a specific carbon consumption 0.5 kg/kg V_2O_5, and the system V_2O_5–Si – at a specific silicon consumption 0.7 kg/kg V_2O_5.

Condensed and gas phases were selected for a better study of the vanadium reduction in the process products. The analysis of these phases was performed.

The composition of the condensed phase comprises vanadium oxides, carbide and vanadium silicides, silicon carbide, silicon dioxides, metallic vanadium, carbon and silicon. The gas phase, formed during the vanadium oxides reduction, consists of reaction products of vanadium oxides reduction with carbon and silicon (CO, CO_2, SiO) and dissociation product of vanadium oxides – O_2.

Results of thermodynamic calculations for the system V_2O_5–C –Si are shown in Figures 2 and 3.

From these dependences it follows that the mass of the condensed phase (Figure 2a) increases with the increase of silicon consumption and decreases with the increase of carbon consumption. For the gas phase (Figure 2b) the relationship is reverse.

As can be seen from the curves in Figure 3 the basic component of the gas phase is carbon monoxide. Its mass increases with the increase of carbon consumption and decreases with the increase of silicon consumption. SiO mass increases with the increase of carbon consumption, while at the increase of silicon consumption at first goes up (at a specific silicon consumption not more than 0.3 kg/kg V_2O_5) and then decreases. CO_2 mass does not exceed 0.00003 kg, O_2 mass does not exceed 0.03 kg with the reducing agents mass is zero, further in the gas phase it is present in the form of tracks.

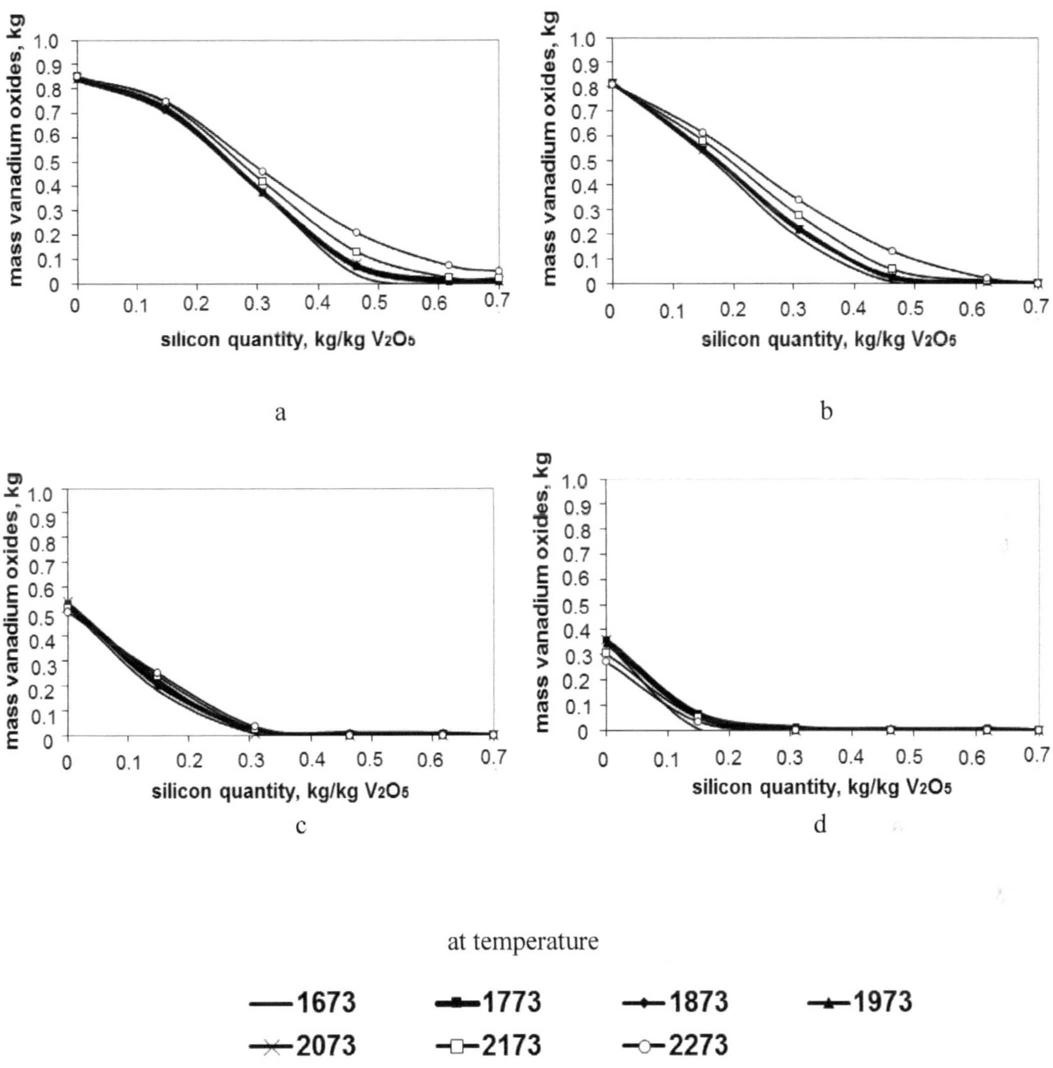

Figure 1. Dependences of vanadium oxides mass on the temperature and reducing agents consumption (a – carbon consumption 0.05 kg/kg V_2O_5; b – carbon consumption 0.15 kg/kg V_2O_5; c – carbon consumption 0.25 kg/kg V_2O_5; g – carbon consumption 0.35 kg/kg V_2O_5).

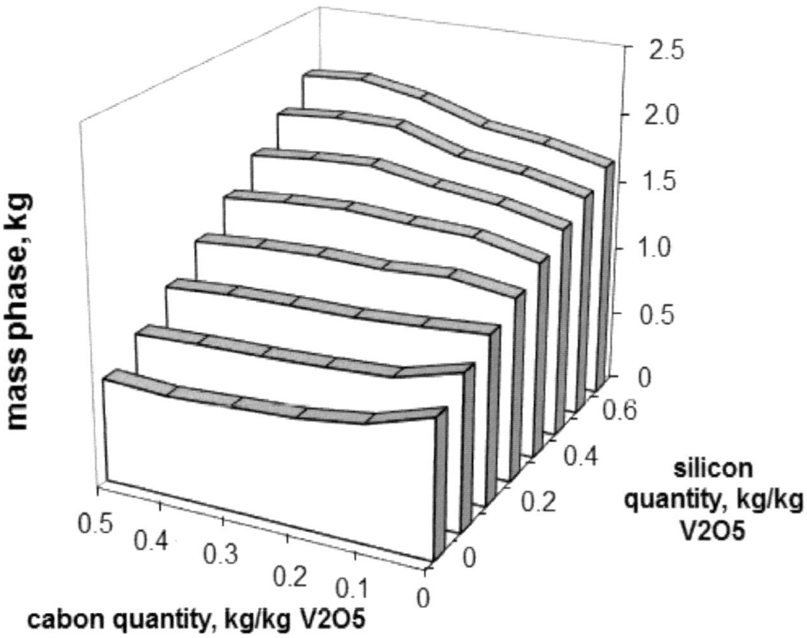

a

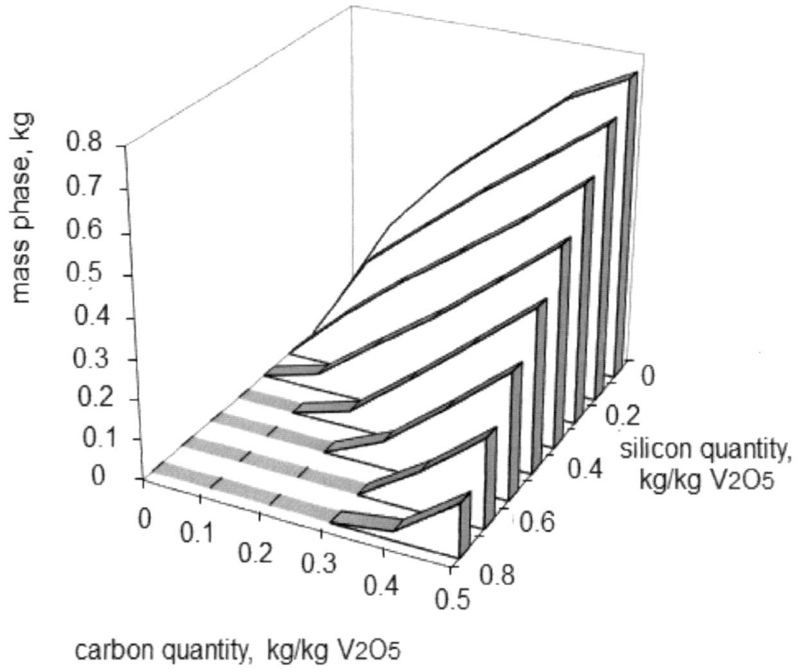

b

Figure 2. Mass dependences of the condensed and gas phases on the reducing agents consumption in the system V_2O_5–C –Si system at 1873K (a – condensed phase; b – gas phase).

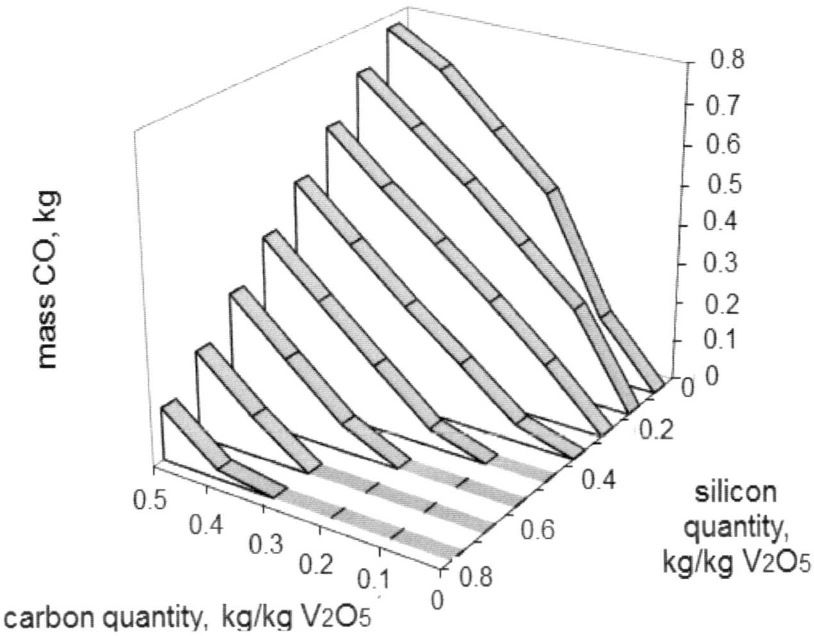

a

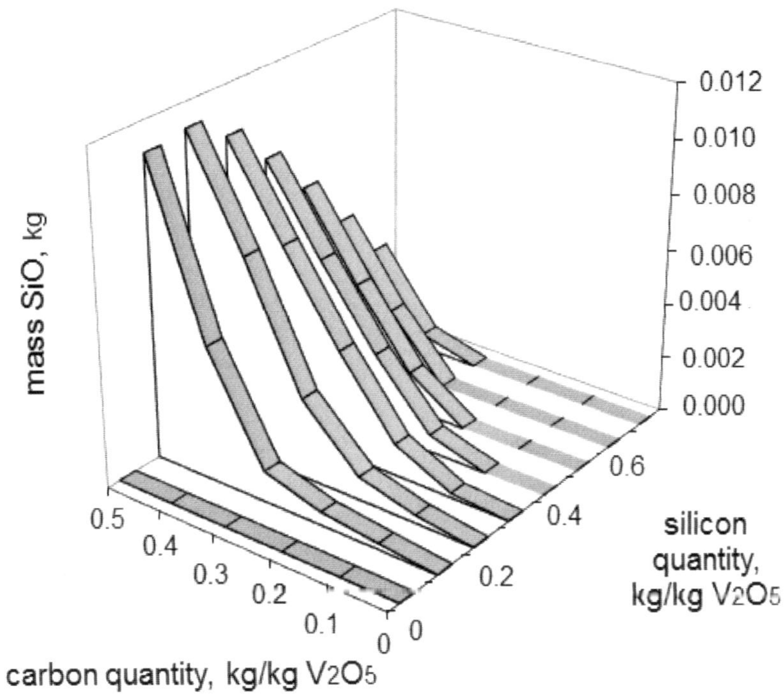

b

Figure 3. Mass dependences of the gas phase components on the consumption of reducing agents in the system V_2O_5–C –Si at 1873K (a – mass of CO, kg; b – mass of SiO, kg).

From the data obtained it can be concluded that in the system V_2O_5–C –Si system the direct reduction of vanadium pentoxide with carbon takes place. In addition, carbon reduces the formed silica.

The calculation results of reduction products masses for the system V_2O_5–C –Si given in Figures 4, 5 and 6, showed that vanadium, reduced from the vanadium pentoxide during carbon-silicon thermal reduction, is mainly found in the form of vanadium carbide and vanadium silicides.

The results of thermodynamic calculations of the vanadium coreduction with carbon and silicon for the system V_2O_5–C –Si shows that vanadium reduction from vanadium pentoxide is initially occurs with formation of vanadium carbide (Figure 4), and further as the silicon consumption increases more than 0.3 kg / kg V_2O_5, vanadium silicides VSi_2, V_5Si_3 and V_3Si (Figure 5) are also formed.

The mass of the metallic vanadium according to the calculated data is insignificant and does not exceed 8% of the reduced vanadium mass. With the increase in the specific consumption of silicon this value decreases due to the formation of vanadium silicides (Figure 6).

Thermodynamic modeling showed that carbon is a predominant reducing agent during carbon-silicon thermal reduction of vanadium from pentoxide vanadium.

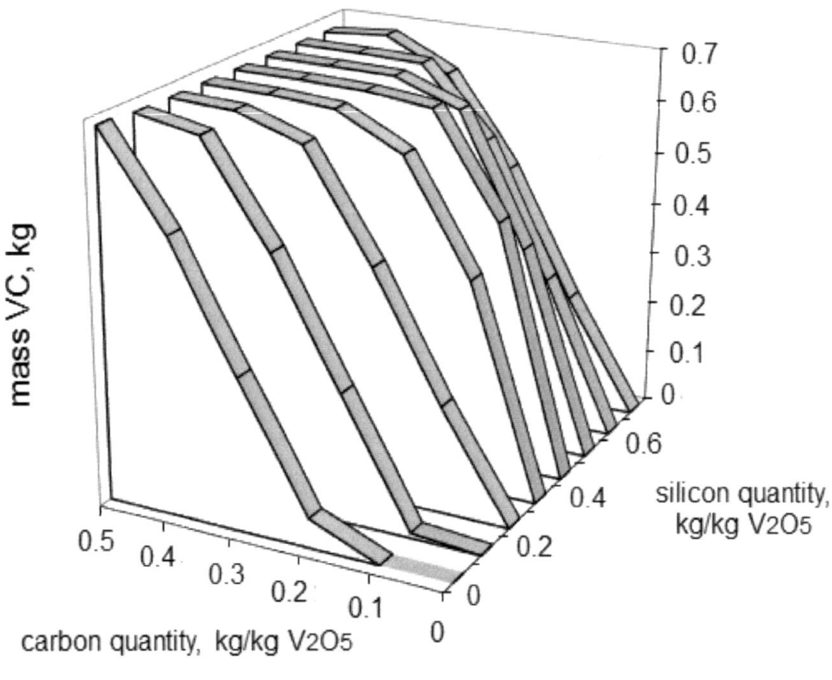

Figure 4. Mass dependences of vanadium carbide on reducing agents consumption in the system V_2O_5–C –Si at 1873 K.

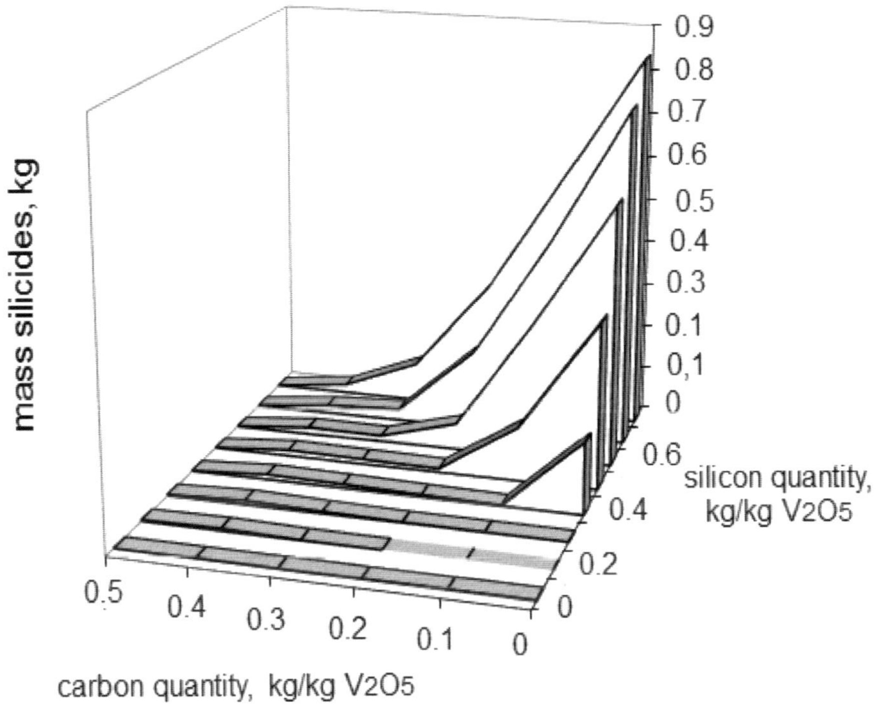

Figure 5. Mass dependences of vanadium silicides on the consumption of reducing agents in the system V_2O_5–C –Si at 1873 K.

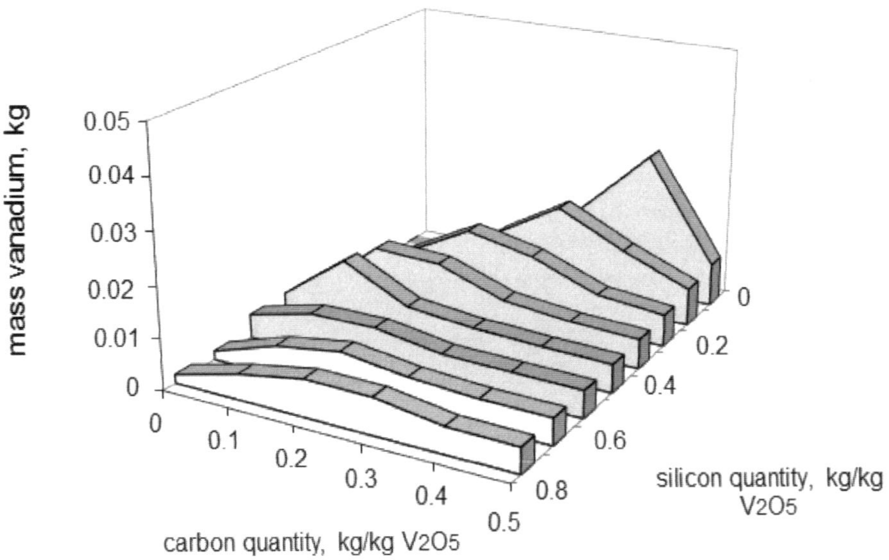

Figure 6. Mass dependence of metallic vanadium on the consumption of reducing agents in the system V_2O_5–C –Si at 1873 K.

The performed thermodynamic calculations showed that in the examined system the reduction products contain silicon carbide formed from SiO_2 in the reaction: $SiO_2 + 3C = SiC + 2CO$.

According to the dependency given in Figure 7 the amount of silicon carbide increases linearly with the increase in the amounts of the introduced reducing agents.

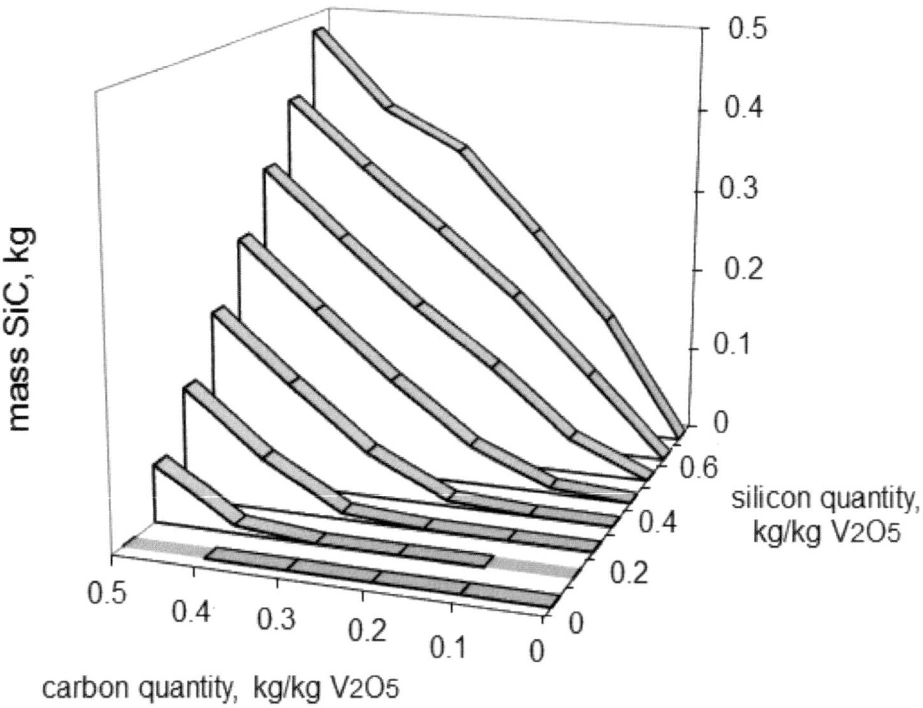

Figure 7. Mass dependences of silicon carbide on the consumption of reducing agents in the system V_2O_5–C –Si system at 1873 K.

3. Conclusions

The performed thermodynamic calculations of vanadium reduction from its oxides in the system V_2O_5 – C – Si demonstrated:

- at a temperature from 1673 to 2073 K this process depends only on reducing agents consumption. The temperature variation does not have any influence.
- vanadium reduction from vanadium pentoxide in combination with carbon and silicon takes place with predominant formation of vanadium compounds: vanadium carbide and vanadium silicides. Mass of metallic vanadium is not significant.

These results confirm the previous findings [8-10], and suggest that in the carbon-silicon thermal reduction vanadium carbon plays the basic role.

4. References

[1] Mitchel P V 2007 *Proc. Int. Scient. Tech. Seminar on the Problems of Production and Use of Vanadium Steel* (Ekaterinburg) pp 53–80
[2] Langeborg et al 2001 *The Role of Vanadium for Micro-alloyed Steels* (Ekaterinburg: SSC of the Russian Federation "Ural Institute of Metals")
[3] Smirnov L A, Panfilova L M and Belenkiy B Z 2005 *Steel* **6** 108–115
[4] Okhotsky V B 2004 *Proc. of Higher Educ. Inst. Ferrous Metallurgy* **4** pp 13–17
[5] Chernova L A and Mikhaylov G G 2004 *Proc. XI Ruusian Conf. on Structure and Properties of Metallic and Slag Melts* (Chelyabinsk) **3** pp 130–131.
[6] Marshuk L A and Zhuchkov V I 2004 *Proc. XI Ruusian Conf. on Structure and Properties of*

Metallic and Slag Melts (Chelyabinsk) **3** pp 114–118

[7] Trusov B G 2005 *III Int. Symp. on Combustion and Plasma-chemistry* (Alma Ata, Kazakhstan) pp 52–57

[8] Golodova M A et al 2015 *Proc. XLVIII-XLIX Int. Scient. Pract. Conf. on Technical Sciences – From Theory to practice* (Novosibirsk) **44** pp 70–77

[9] Golodova M A et al 2011 *Natural and Intellectual Resources of Siberia. Sibresurs* (Tomsk) pp 57–60

[10] Golodova M A et al 2010 *Steel in Translation* **40** 310–313

Theoretical foundations of ferrosilicium disintegration

N F Yakushevich[1], A A Maksimov[2], A Yu Pronakin[2] and O A Polyakh[1]

[1] Department of Non-ferrous Metallurgy and Chemical Technology, Siberian State Industrial University, 42 Kirova Street, Novokuznetsk, 654007, Russia
[2] JSC "Kuznetsk Ferroalloys", Novokuznetsk, 654000, Russia

E-mail: kafcmet@sibsiu.ru

Abstract. The crystallization processes and solid-phase transformations in alloys FS-75, FS-65, FS-45 and the influence of impurities (aluminum) on the possibility of disintegration of ferroalloys were considered on the basis of state diagrams analysis of Fe–Si and Fe–Si–Al systems. It was shown that the mechanisms of crystallization of melts of different compositions differ quite significantly, both in the chemistry of the process, and in the number and composition of the crystallization products. Process of ferroalloys disintegration depends on many factors: the alloy composition, the degree of superheat, the cooling rate, completeness of solid phase reactions, etc.

1. Introduction

The phenomenon of ferrosilicon disintegration reduces the output of standard quality products of ferrosilicon grades; is accompanied by the release of toxic and flammable in certain conditions and explosive gases [1]. The main causes of this phenomenon are considered to be the occurrence of thermal stresses during polymorphic transformations of leboite, accompanied by a significant increase in volume ($\Delta V \approx 17\%$), the influence of impurities (Al, Ca, As, P, S) on formation in the crystallization process of multicomponent low-melting eutectics, which are located on the grain boundaries of silicon and iron silicides and forming with the target components of the alloy (Fe, Si, C) the complex chemical compounds (carbides, phosphides, arsenides, sulfides), interacting with the atmospheric moisture and forming hydrates ($Ca(OH)_2$, $Al(OH)_3$) and gaseous compounds (PH_3, AsH_3, H_2S, C_2H_2) [2, 3].

2. Research and results

A more detailed analysis of crystallization processes and solid phase transformations can be done on the basis of the modern state diagrams (Fe-Si, Fe-Si-Al, and others.) [2, 4-6]. A fragment of the state diagram Fe-Si in the range of concentrations 33–100 wt%. Si is given in Figure 1.

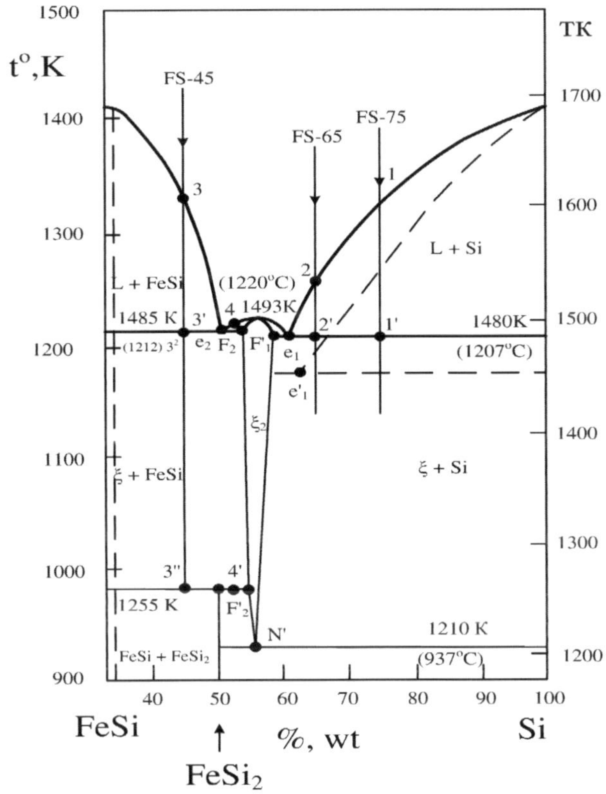

Figure 1. Fragment of the state diagram Fe-Si.

Within the range of silicon concentrations 33.3-50.1% by weight monosilicide of iron FeSi is crystallized during cooling, within the range of silicon concentrations 61-100% by weight during the primary crystallization the silicon crystals are formed, in the intermediate range of Si concentrations 50.1-61% by weight leboite is formed from the liquid phase during cooling – ξ phase $FeSi_{2,33}$, congruently melting at 1493 K. Leboite density is 5.06 g/cm^3. In solid form, it forms solutions with silicon and monosilicides, the zone of which converge to zero values at 1210 K. With silicon and monosilicides leboite forms eutectics – e_1 (1480K, 61% wt. Si) and e_2 (1485K, 50% wt. Si). Leboite composition corresponding to the congruent melting point (1493 K, 55 % wt. Si), is identified as $FeSi_{2,33}$, however, taken into account solutions non-stoichiometry, it is presented as phase $FeSi_2$, critical in the iron content $Fe_xSi_{2,33}$ (x=0.77-0.87). Leboite participates in two eutectic reactions:

$$L_{e_1} \rightleftarrows Si + \xi \text{ (1480 K)}, \quad (1)$$

$$\mathrm{L}_{e_1} \rightleftarrows \mathrm{FeSi} + \xi\text{(1485 K)}. \quad (2)$$

When xFe = 0.77-0.82 from the supersaturated with silicon solution of leboite during cooling from 1480 to 1210 K the silicon crystals are formed:

$$\xi_{xFe=0,77} \rightarrow \mathrm{Si} + \xi_{xFe=0,82} \quad (3)$$

At 1210 K leboite breaks according to peritectoid reaction forming stoichiometric disilicide $FeSi_2$ and silicon:

$$\xi_{x_{Fe}=0,82} \rightarrow FeSi_2 + Si. \quad (4)$$

This transformation proceeds with the increase in volume (to 17%) and is a major cause for alloys disintegration with silicon content 65% wt. and more. Since in these alloys leboite is formed during crystallization of eutectic e_1 in the form of a fine-disperse eutectic mixture (ξ + Si), and the grain growth practically does not occur during cooling, furthermore, a partial disintegration of leboite takes place with formation of a new phase (silicon). Even at this stage the local stresses and microcracks appear in the eutectic structure of a solid mixture. During crystallization of alloy FS-75 the amount of liquid eutectic melt $e_1\approx50\%$. The amount of silicon crystals in the eutectic mixture is less than 10%, thus, the amount of leboite during equilibrium crystallization may reach ≈40% of the initial melt mass.

During crystallization of alloy FS-65 the amount of silicon crystals, formed within the temperature range 1570-1480K (point 2-2'), is significantly larger (≈25%), and eutectic mixture (ξ+Si), formed during the melt crystallization of composition e_1 (61% Si), – is significantly higher (75%), resulting in a large, compared with FS-75, propensity to disintegrate (leboite amount in the alloy during equilibrium crystallization may reach 65%). It should be noted that during a substantial alloy overheating above the liquidus surface and subsequent rapid cooling (which often takes place during casting of ferrosilicon) the temperature of the beginning of crystallization reduces, and subsequently, due to a more intense crystallization heat release the temperature of the end of the crystallization also decreases; and the eutectics parameters change (dotted lines in Figure 1).

During crystallization of FS-45 alloy as a primary crystallizing phase in the temperature range 1600-1485K (point. 3-3') the monosilicide silicon crystals are formed, their quantity at the eutectic temperature e_2 (1485 K) is ≈33% (≈67% respectively eutectic e_2) at the equilibrium (without supercooling) crystallization.

At 1485 K eutectic reaction (2) proceeds, in the result of which from a liquid melt with composition e_2 evolve iron monosilicide FeSi and solid solution of leboite composition of point F_2 (50.1% Si). The crystallized eutectic mixture (FeSi+leboite F_2) contain ≈19% of silicon crystals and, ≈81% of leboite, i.e. ≈54% of leboite phase.

During cooling to 1255K leboite of F_2' composition reacts with monosilicide:

$$FeSi + \xi_{F'} \rightarrow xFeSi_2 + (1-x)FeSi^0. \quad (5)$$

Leboite completely disappears, and monosilicide remains in abundance ($FeSi^0$). After the reaction the alloy must contain ≈30% of residual monosilicide and 70% of disilicide $FeSi_2$.

During cooling of alloys containing 50-54% of Si (e_2 -F_2), for example, the alloy of composition point 4, at 1255K interaction of monosilicide FeSi with leboite leads to monosilicide disappearance, and leboite of F_2' composition remains in abundance (ξ^0):

$$FeSi + \xi_{F_2'} \rightarrow FeSi_2 + \xi^0{}_{F_2'}. \quad (6)$$

With further cooling, the amount and composition of the mixture components disilicide-leboite changes slightly, and at 1210K the residual leboite of composition point N' reacts with disilicide forming silicon, and leboite disappears:

$$FeSi_2 + \xi_N \rightarrow Si + FeSi_2. \quad (7)$$

Reaction (7) proceeds with the increase in volume, but since the quantity of leboite in point 4′ is low, it does not significantly affect the alloy disintegration.

The alloy crystallization is of particular interest, corresponding in its composition to the point of leboite congruent melting point. In this case, in the absence of impurities during slow cooling the pseudo-single crystalline crystallization can be observed with formation only large single crystals of leboite, which during cooling to 1210 K dissociate into silicon and disilicide and $FeSi_2$ (eutectoid reaction):

$$\xi_{N/} \rightarrow Si + FeSi_2. \quad (8)$$

This reaction, which was previously identified as polymorphic transformation of leboite, proceeds with the increase in volume, but large single crystals of leboite can withstand stresses for a long time without fracture, and a small zone of intergrain boundaries minimize the impurities influence on the process of disintegration. In this case the system is thermodynamically unstable and in the presence of precipitating factors (long-term storage, the presence of impurities, high humidity) such alloys may also disintegrate.

The impurities Al, Ca, P, As, S, C, which income into the ferrosilicon with raw materials (quartzites, sands, cokes, coals, iron turnings, scale, etc.), are of particular importance for the issue of ferrosilicon disintegration. The influence of impurities on crystallization process with leboite is considered on the example of a well-studied system Fe-Si-Al, the state diagram of which is given in Figure 2 [4].

Aluminium is present in ferrosilicon as an impurity in the largest compared with other impurities concentrations (up to 2% or more) [7-9] and its effect on the crystallization process is most significant.

In accordance with the state diagram of the system Fe-Si-Al, in addition to silicon and silicides in the final phases of crystallization ternary chemical compounds might be formed, melting incongruently: $Al_3Fe_3Si_2$, $Al_{12}Fe_6Si_5$, $Al_9Fe_5Si_5$, Al_3FeSi_2, $Al_{15}Fe_6Si_5$, Al_4FeSi. Directly to the crystallization fields of silicon, leboite and monosilicide the crystallization fields of compounds $Al_3Fe_3Si_2$, $Al_9Fe_5Si_5$ adjoin, which are present in the final crystallization phases.

During crystallization of alloy FS-75, containing ≈1,5% of Al, the alloy composition point (p. 1) lies in the crystallization filed of silicon in 1–1′. From the liquid melt the crystals of silicon are formed. Further, along the boundary curve of monovariant equilibrium e_1–u_6 from the melt simultaneously two crystalline phases are crystallized – silicon and leboite. Crystallization finishes in the point u_6 at a temperature 880°C according to peritectic reaction:

$$L + FeSi_2 \rightarrow Al_9Fe_5Si_3 + Si. \quad (9)$$

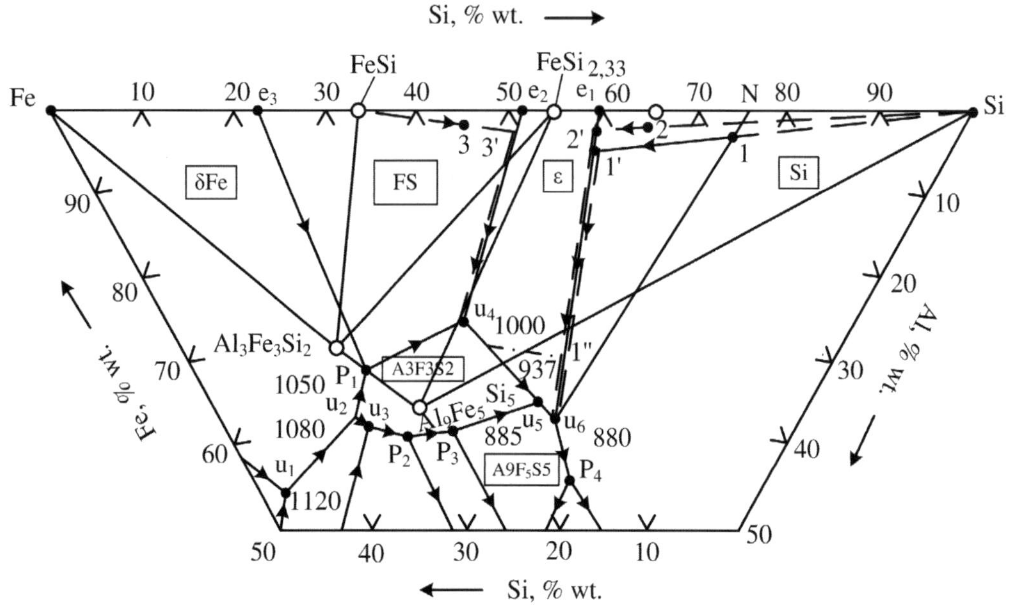

Figure 2. Fragment of the state diagram of the system Fe-Si-Al.

Leboite transformation into disilicide $FeSi_2$ takes place along the isotherm 937°C, and specifically for the alloys FS-75 and FS-65 in the point 1" on the boundary curve e_1–u_6. Therefore, in the section 1'–1" the eutectic reaction proceeds:

$$L \rightarrow \xi + Si, \tag{10}$$

and in the section 1"–u_6:

$$L \rightarrow FeSi_2 + Si. \tag{11}$$

Crystallization ends in the invariant point of the four-phase equilibrium u_6 at 880°C, and the aluminum content in the liquid mother melt of composition p. u_6 increases to 38%, and silicon content, respectively, is reduced to 36%. The amount of liquid, crystallizing in p. u_6, is characterized by the segment 1–N_1 on the line N_1 – u_6 and is ≈4.5%, the proportion is Si:$FeSi_2$≈1:1.

Crystallisation of alloy FS-65 (Al≈1%) – p. 2, proceeds similarly, differing only in the fact that the quantity of silicon crystals, formed on the section of primary crystallization (2-2') will be considerably lower (≈20%), and crystals $FeSi_2$, formed by the reaction (11), – higher (≈78%), which largely determines the increased ferrosilicon disintegration.

Crystallization of alloy FS-45 (Al≈1%) occurs otherwise. The point of alloy composition 3 is in the elementary phase triangle FeSi–ξ– $Al_3Fe_3Si_2$, thus, the melt completes crystallization in the invariant point of four-phase equilibrium u_4 (Al 26%, Si 32%) at a temperature ≈1000°C. Since the composition point of alloy FC-45 lies in the crystallization field of monosilicide FeSi, then in the result of primary crystallization in the section 3-3' the crystals of monosilicide are formed in the quantity of ≈33% of the melt mass. During further cooling the crystallization follows monovariant boundary curve e_2–u_4 with the appearance in the section 3'– u_{11} of two solid phases – monosilicide FeSi and leboite Fe_xSi_2. Aluminum concentration in the remaining mother melt rises to 26%, and silicon concentration is reduced to 32%. In the final invariant point of crystallization u_4 (35% Si, 26% Al) peritectic reaction proceeds:

$$L_{u_4} + FeSi \rightarrow FeSi_2 + Al_3Fe_3Si_2. \quad (12)$$

The amount of the final liquid melt of composition t. u_4, crystallizing at 1000°C with the initial aluminum content of 1%, is ≈7%.

During crystallization of alloys containing 55-56% Si, the composition point of which lies in the crystallization field of Leboite, for any aluminum content in the alloy (3-2% or more) the monocrystalline leboite crystallization takes place, ending in the point u_5 according to peritectic reaction:

$$L_{u_5} + \xi \rightarrow Al_3Fe_2Si_2 + Al_9Fe_5Si_5. \quad (13)$$

Quantity of the finite crystallized liquid u_5 (35% Si, 37% Al) and, respectively, the secondary crystallized phases depends on the aluminum content in the initial melt. If the aluminum content in the alloy is ≈2%, the amount of liquid will be ≈3%.

It should be noted that the eutectoid and peritectoid reaction (5)-(7) in the presence of impurities (Al, Ca, P) proceed in the presence of a liquid phase, and the part of the stresses arising due to the increase of volume are accepted by the liquid, which leads to a significant increase of pressure, and after crystallization – to a stress concentration on the grain boundaries of matrix crystals (FeSi, $FeSi_2$, Si) and abundant desorbing phase (complex compounds with impurities).

The presence of calcium and phosphorus in the structural phases of ferrosilicon complicates the analysis, since the calcium content is by 4-5 times and phosphorus content is by 40-50 times lower than the aluminum content, the compounds proportion with calcium and phosphorus is very low and their quantitative participation in crystallization process is insignificant.

3. Conclusions

Based on the analysis of state diagrams of the systems Fe-Si and Fe-Si-Al, it is shown that the mechanism of the process of ferrosilicon disintegration is complex and depends on the alloy composition, the overheating degree, cooling rate and degree of completion of solid phase eutectoid and peritectoid reactions, relative to the amount of eutectic structures and, respectively, the area of intergrain surfaces in the crystallized mixtures, the presence of impurities, providing phase and chemical transformations in the presence of a liquid phase up to a final crystallization temperature, which is 300-400 K below the crystallization temperature of double eutectics e1 and e2, and leading to a significant increase of the amount of intergrain desorbing phases, their composition and structure.

4. References

[1] Zubov V L and Gasik M I 2002 *Electrometallurgy of Ferrosilicon, Physico-chemistry and Technology of Ferrosilicon Production* (Dnepropetrovsk: Sistemnye Tekhnologii) p 703

[2] Geld P V and Gasik M I 1971 *Silicides and Transition Metals of the Fourth Period* (M.: Metalluriya) p 584

[3] Radugin V A 1997 *Proc. on Improvement of Ferrosilicon Production* (Novokuznetsk) pp 28–32

[4] Bannykh O A and Drits M E 1986 *Handbook on State Diagrams of double and Multicomponent Systems Based on Iron* (M.: Metallurgiya) p 439

[5] Lyakisheva N P 1997 *Handbook on the State Diagrams of Double Metallic Systems* (M.:Mashinostroyeniye) p 1024

[6] Radugin B A et al 1969 *Proc. on Improvement of Ferrosilicon Production* (Kemerovo) vol 2 pp 318–327

[7] Yakushevich N F et al 2015 *Proc. of Higher Educ. Inst. Ferrous Metallurgy* **58** (5) 316–321

[8] Yakushevich N F et al 1998 *Proc. of All-Russian Meeting on Computer Methods in Control of Electro-technological Modes of Ore-thermal furnaces* (St. Petersburg) pp 59–65

[9] Yakushevich N F and Kovrova O A 1997 *Proc. of Higher Educ. Inst. Ferrous Metallurgy* **8** 3–8

Modern trends in improvement of steel heating technology in continuous furnaces

V I Timoshpolskiy[1], M V Temlyantsev[2] and I A Trusova[3]

[1] CJSC "Neftegazstroyizolyatsiya", 14 Marshal Grechko, Kiev, 04136, Ukraine
[2] Institute of Metallurgy and Material Science, Siberian State Industrial University, 42 Kirova Street, Novokuznetsk, 654007, Russia
[3] Mechanical and Technological Faculty, Belarusian National Technical University, 65 Nezavisimosty Ave., Minsk, 20013, Belarus

E-mail: uchebn_otdel@sibsiu.ru

Abstract. The principles and approaches in the development and improvement of steel heating technology in the furnaces of rolling manufacture of various structural design, based on the systematic study of thermal physical and technological processes, including mathematical modeling, industrial experiments, development of rational temperature-thermal modes.

1. Introduction

This paper contains the results of the research we have been carrying out for many years in the field of technology of metal heating in the continuous furnaces. This study is based on the experience and knowledge obtained within the Ukrainian (Dnepropetrovsk) and Ural-Siberian metallurgical schools founded by the leading theoreticians and practitioners in the field of metallurgical heat: Semikin I.D., Taits N.Yy., Nevsky A.S., Timofeev V.N., Nazarov I.S., Kavaderov A.V., Kitaev B.I., Samoylovych Yu.A. et al. [1].

It is noted that in fact the trendsetters in the field of investigations of thermal performance of rolling mill furnaces today are the Ural, Siberian and Belarusian schools.

The purpose of this paper is to summarize the research for the last 10-15 years, performed by the scientists of Siberian Industrial Institute and the Belarusian National Technical University.

The study is focused primarily on experimental methods for studying the thermal operation of the furnaces, which functions in the lines of modern high productive rolling and tube mills (walking beam furnace of mill 850 and a furnace with walking hearth and walking beams of mill 320/150 of the Belarusian Metallurgical Plant, furnaces of pusher-type of mill 810 at Novosibirsk Metallurgical Plant, annular furnace axis-rolling mill 250 and continuous furnaces of mill 900/750 at Dneprovsky Metallurgical Combine n.a. F.E. Dzerzhinsky, pusher furnaces of mills 600 and 2250 at Alchevsk Metallurgical Combine [2-5], pusher furnace of mill 2800 at Orsk-Khalilov Metallurgical Combine, furnace with walking beams of mill 2000 at JSC "Severstal" [6], etc.).

The important role is played by theoretical studies of heat- and mass-transfer processes in fuel furnaces, including mathematical modeling and numerical methods for calculations of heating furnaces of modern design, first of all, with walking beams and crown heating, as well as of pusher-type for medium- and heavy-section mills [2-4, 6-10].

It should be noted that the Belarusian scientists and experts paid great attention to the study of nonlinear thermal and thermo-mechanical processes using different mathematical apparatus (different versions of numerical methods, approximate methods (method of equivalent sources)). Of course, these researchers made a significant contribution to the formation in effect of a new fundamental metallurgical science – nonlinear thermal mechanics and thermal physics of metallurgical production (Postolnik Yu.S., Timoshpolskiy V.I. et al.) [8].

The scientists of the Siberian scientific school paid much attention to the creation of resource saving thermal technological processes during the heating of large slabs and blooms in the continuous furnaces of various designs [6, 9, 10]. In detail they studied the mass-transfer processes with regard to the processes of decarburization and scale formation, as well analyzed numerical methods for solving nonlinear tasks and generalized the obtained results to ensure high performance of furnaces.

As an example, we provide results of the research aimed at improvement and development of energy and resource saving thermal technological processes in the heating furnaces of different structural design.

2. Research and results

In accordance with the system approach, in which the obligatory component was the experimental study of temperature fields inside the heated blanks during their displacement in heating devices, we conducted numerous experiments in conditions of real production in order to analyze the thermal operation of the furnaces, to evaluate the possibility of productivity increase and heat-transfer enhancement.

Figures 1 and 2 show the results of experimental studies of thermal treatment of special steel grades (for example, steel ShH15SG) during organization of blanks hot charge in the furnace of tube-rolling mill 850 (with walking beams) at the Belarusian Metallurgical Plant (BMP) and during heating in the furnaces of pusher-type mill 810 at Novosibirsk Metallurgical Plant [3].

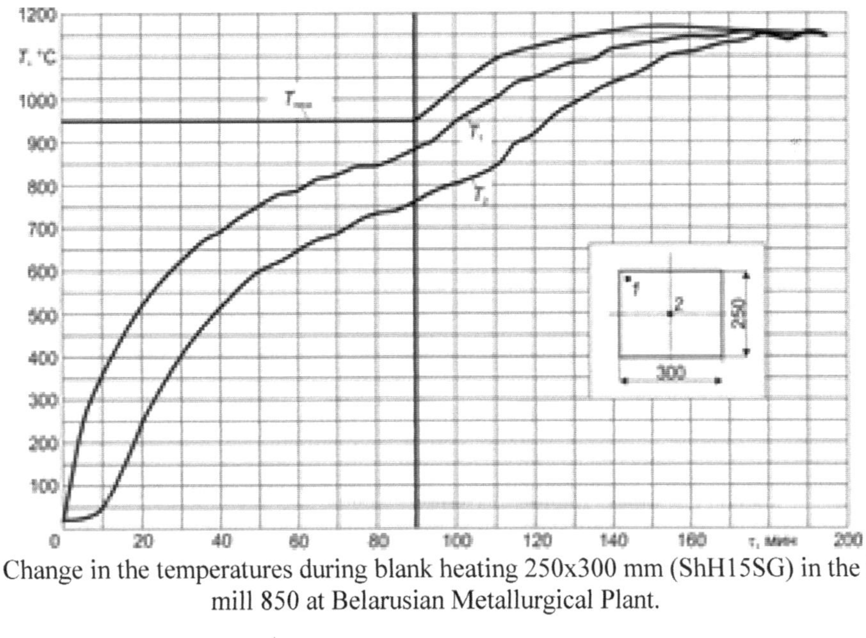

Figure 1. Change in the temperatures during blank heating 250x300 mm (ShH15SG) in the furnace of mill 850 at Belarusian Metallurgical Plant.

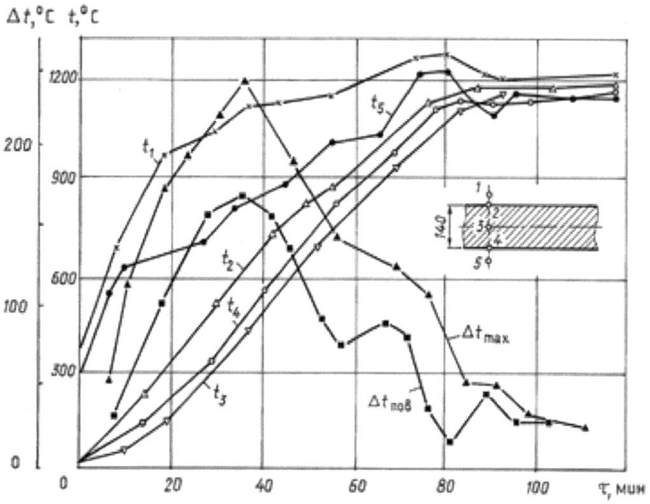

Figure 2. Temperatures change in the characteristic points of cross-section during slab heating (12H18N10T) according to the developed mode in the furnace of mill 810 at Novosibirsk Metallurgical Plant.

Figures 3 and 4 show the results of experimental studies of slab blanks heating in the heating furnace of a pusher-type and the furnaces with walking beams [6].

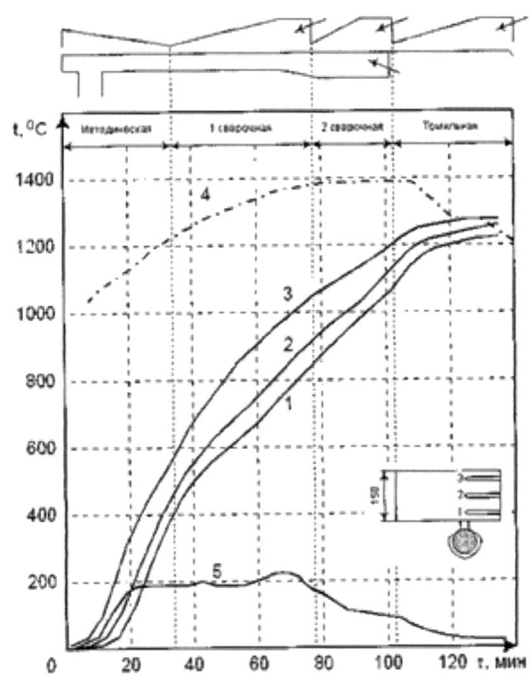

Figure 3. The results of experimental studies of slabs heating in the continuous furnace of plate mill 2800 at Orsk-Khalilov Metallurgical Combine.

In accordance with the research methodology further we carried out mathematical modelling of heat- and mass-transfer processes in the heating furnaces, and depending on the research purpose developed and used a wide spectrum of mathematical models: the traditional metal heating model with boundary conditions of III type [2-6]; model of external heat-transfer [6]; coupled heat transfer (internal and external) [2, 3]; methods of optimal control of heating with a minimum scale formation [2, 4] and under the terms of the heat resistance of the metal [3]; methods for solving the thermal mechanics problems [3, 8]; modelling of interrelated processes of oxidation and decarbonization of steel [6, 9, 10].

As an example, some simulation results of heat- and mass-transfer processes in heating furnaces are given.

In the case of conjugated heat-transfer in the furnaces of pusher-type the system of equations for calculation of zones of II type (heat flows are defined) has the form:

$$\sum_{k=1}^{N^{u(l)}} \left(\alpha \Sigma_{ki}^{u(l)} T_k^{u(l)^4} + g_{ki}^{u(l)} T_k^{u(l)} \right) + g_i^{0^{u(l)}} + Q_i^{v^{u(l)}} = 0, i \in N_2^{u(l)},$$

where N – total number of zones;

N_2 – set of zones numbers of type II;

to calculate the heat flows on the metal surfaces (zones of the I type)

$$Q_i^{v^{u(l)}} = \sum_{k=1}^{N^{u(l)}} \left(\alpha_{ki}^{u(l)} T_k^{u(l)} + g_{ki}^{u(l)} T_k^{u(l)} \right) + g_i^{0^{u(l)}}, \; i \in N_1,$$

where N_1 – set of zones numbers of type I (zones on the upper and lower metal surfaces, where the temperature is given);

a_{ki}, g_{ki}, g_i^0, Q_i^v – coefficients of radiation (effective taking into account the radiation selectivity) and convective heat-transfer, as well as the quantities of volumetric heat generation; “u” and “l” – the upper and lower furnace structure.

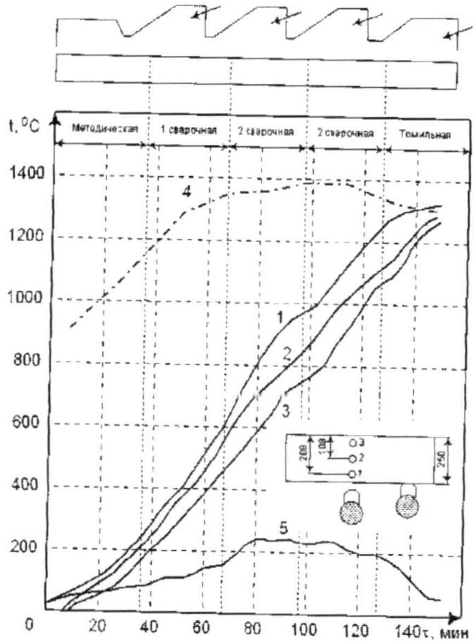

Figure 4. The results of experimental studies of slabs heating in the furnace with walking beams of broad-strip mill 2000 at JSC “Severstal”.

In the furnaces with walking beams and walking hearth at the preliminary stage the solution of nonlinear algebraic equations of heat balance is carried out taking into account the fact that the blank surface is a conventional plane, and at the next stage the calculation results are adjusted with regard to the reradiation between the blanks on the the furnace hearth with a gap [2, 3].

One of the most important aspects of improvement of technical and economic performance of the heating furnaces is the organization of metal hot charge. For this purpose an effective industrial heat technology of through (combined) process “steelmaking – pre-rolling heating” was developed [4] from the standpoint of reducing the specific fuel consumption and increase of furnace units productivity, as well as with regard to the thermal metal strength.

In this case the boundary conditions are of the form:

$$-\lambda\left(\frac{\partial T}{\partial n}\right)_{n=R_1,R_2}=\begin{cases}\sigma\left(T^4_{\text{surface}}-T^4_{\text{environment}}\right)+\alpha\left(T_{\text{surface}}-T_{\text{environment}}\right)-\text{during hardening and cooling}\\ \text{at CCM furnace;}\\ \sigma_{air}\left(T^4_{\text{surface}}-T^4_{\text{air}}\right)+\alpha_{air}\left(T_{\text{surface}}-T_{\text{air}}\right)-\text{during transportation to the furnace;}\\ \sigma_{\text{furnace}}\left(T^4_{\text{furnace}}-T^4_{\text{surface}}\right)+\alpha_{\text{furnace}}\left(T_{\text{furnace}}-T_{\text{surface}}\right)-\text{during heating in the furnace.}\end{cases}$$

In general, the thermo-physical model for calculating temperature fields is supplemented by the expressions to determine the thermal stresses in the growing crust and in fully hardened billet during heating with hot and cold charges. At the same we used different classical theories (of Norton, Maxwell, Boli-Weiner, Parkusa, etc.). During calculation of thermal stresses of the crystallized billet the “building effect” and the stresses relaxation at increased temperatures were taken into account.

Figure 5 shows as an example the results of calculation of the temperature field in the casting process, transportation and heating in the bar furnace applied to the conditions of BMP.

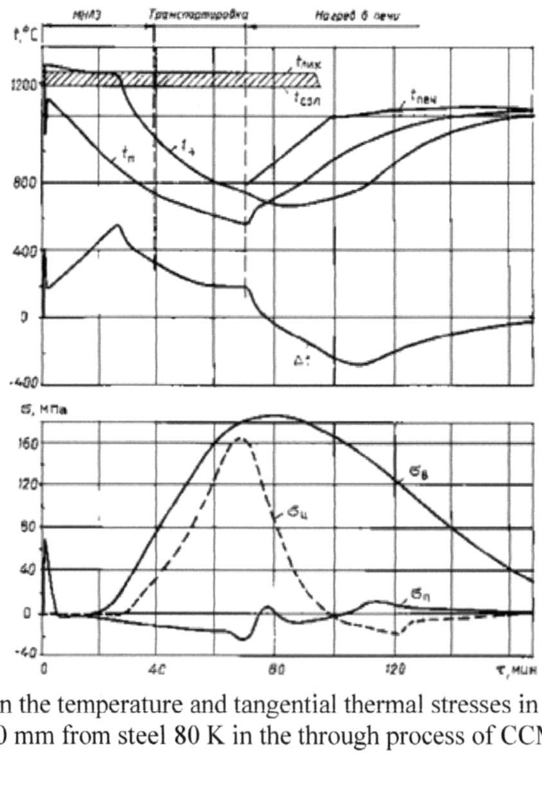

Figure 5. Changes in the temperature and tangential thermal stresses in the centre and on the billet surface 250 × 300 mm from steel 80 K in the through process of CCM – furnace of mill 850.

Much attention is paid to the complex research of thermal-physical processes of metal heating together with the processes of scale formation and decarbonization of blanks and special types of metal-roll [2, 4, 6, 9, 10].

In order to reflect the interconnected processes of scaling formation and decarbonization the deterministic mathematical model during metal heating [6, 9, 10], according to which the temperature change of the metal surface was approximated by the equation:

$$t = t_0 + A_1\left(1 - \exp\left(-B_1\frac{\tau_i}{\tau}\right)\right)$$

where t_0 – initial temperature of the metal;
A_1, B_1 – empirical coefficients depending on the heating mode;
τ_I – current point of time from the start of heating;
τ – total time of heating.
The thickness of the metal layer, which transformed into the scale, was determined by the ratio:

$$s = A\exp\left(-\frac{B}{T}\right)\tau^{\cdot}$$

where A, B, n – empirical coefficients depending on the chemical composition of the steel, the composition of the atmosphere and other factors.

Unsteady carbon concentration fields were found on the basis of differential diffusion equation

$$\frac{\partial C}{\partial \tau} = \frac{\partial}{\partial x}\left(D\frac{\partial C}{\partial x}\right)$$

where C – carbon concentration;
D – diffusion coefficient.
The diffusion coefficient was determined by the formula

$$D = (0.07 + 0.06C)\exp\left(-\frac{32000}{1.99T}\right)f_c^{alloy}$$

where C – carbon content in steel, %;
f – relative activity ratio of carbon in the alloy, taking into account the influence of alloying elements on the diffusion coefficient.

On the metal surface, contacting with the atmosphere, the boundary condition was of III type

$$-D\frac{\partial C}{\partial \tau} = \beta(C_{surf} - C_{atm})$$

where β – mass-transfer coefficient;
C_{surf} – the current concentration on the surface of the solid phase at the moment of time τ;
C_{atm} – carbon potential of the atmosphere.

Using the given mathematical model implemented by the numerical method, rational temperature regimes for heating the continuous furnaces were developed [9, 10].

The experimental studies of kinetics of scale formation and decarburization of the rail steel were performed that allows us to obtain dependences of oxidation rate of rail steel grades on the temperature [10].

As an example in Figure 6 the nomogram for determining the decarbonized layer of steel during rapid heating is given, which is of great practical value in terms of improvement of furnaces productivity and ensuring the needed heating quality.

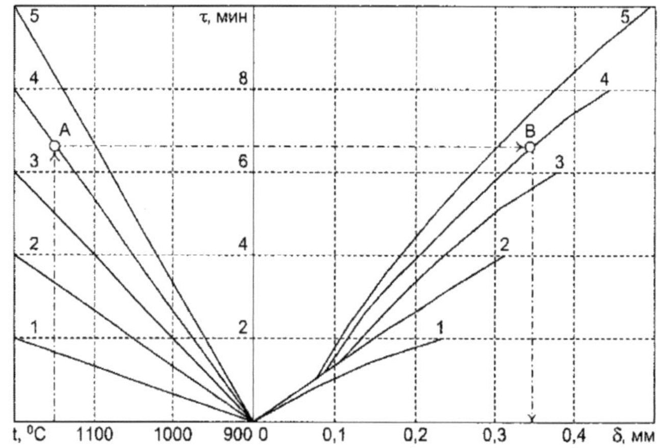

Figure 6. Nomogram for calculating the depth of the visible decarbonized layer during flash heat (heating rates: 1 – 150; 2 – 75; 3 – 50; 4 – 37.5; 5 – 30 deg/min).

The problem of optimal heating in the furnaces with walking beams with a minimal scale formation was formulated and solved [3, 4] using the method of main asymptotic optimization. The internal problem of prism heating is supplemented by the control function, in which the temperature of the furnace is given

$$A_1 \le T_{furnace} \le A_2, \quad \tau \in \left[0, \tau_{final}\right] \tag{1}$$

where A_1, A_2 – minimum and maximum temperature values of the furnace.

Since at the end of heating the blank should have a temperature distribution over the cross-section as close as possible to the desired uniform one, the restriction was

$$max\left|T(x, y, \tau) - T_{final}\right| \le \varepsilon \, . \tag{2}$$

The formation of scale on the surface (on the edge) was described by Evans law.

The optimal control problem consists in the choice of the furnace temperature in the class of piecewise continuous functions, which satisfies constraint (1) and during the solution of the temperature problem, satisfying condition (2), gives the minimum value to the scale.

A similar approach is used to solve the optimal control problem by heating the cylindrical billets under the terms of heat strength. In this case as a constraint was taken

$$\sigma_z \le \sigma_{per}; \ \sigma_r \le \sigma_{per}; \ \ \sigma_\theta \le \sigma_{per}$$

where σ_{per} – maximum permissible tensile stress;

σ_z, σ_r, σ_θ – axial, radial and tangential stresses in the cross-section of the billet.

3. Conclusions

In conclusion, it should be noted that the research team analyzed the thermal performance of continuous furnaces of rolling mills of various designs. The theoretical base of the international Ukrainian (Dnepropetrovsk) and Ural-Siberian scientific schools served as a foundation for the conducted research in relation to the existing equipment. The results of complex theoretical and experimental research allowed us to develop rational temperature-thermal heating modes, implementation of which ensured reduction in specific consumption of equivalent fuel, increase of heating furnaces productivity, improvement of heating quality due to reduction in scale formation and decarburization.

4. References

[1] Temlyantsev M V and Temlyantsev N V 2010 *Ferrous metals and thermal engineering. The history of science and technology development from ancient times to the present day* (M.: Teplotekhnik) p 171
[2] Timoshpolskiy V I 1995 *Technological Fundamentals of Metallurgical Processes and Aggregates of the Highest Technical Level* (Mn: Navuka I Tekhnika) p 256
[3] Timoshpolskiy V I et al 2001 *Heating* (Minsk: Belorusskaya Nauka) **3** p 879
[4] Timoshpolskiy V I et al 1992 *Thermal Technology of Metallurgical Mini-palnts* (Mn: Navuka I Tekhnika) p 158
[5] Timoshpolskiy V I 2002 *Steel* **10** pp 8–12
[6] Peretaytko V N, Temlyantsev M V et al 2008 *Heating of Steel Slabs* (M.: Teplotekhnik) p 192
[7] Gubinsky V I, Timoshpolskiy V I et al 2007 *Metallurgical Furnace: Textbook in Two Volumes* (Minsk: Belorusskaya Nauka)
[8] Timoshpolskiy V I, Postolnik et al 2005 *Theoretical Foundations of Thermal Physics and Thermal Mechanics in the Industry* (Minsk: Bel. Navuka) p 560
[9] Temlyantsev M V, Mikhaylenko Yu E 2006 *The Oxidation and Decarburization of Steel in the Process of Heating for Metal Forming* (M.: Teplotekhnik) p 200
[10] Pavlov B B, Temlyantsev M V et al 2007 *Advanced Technologies of Heat and Thermal Treatment in the Rails Production* (M.: Teplotekhnik) p 280

Development of converter operation technology using iron-bearing concentrates and clinker-steel cakes

E V Protopopov[1], S V Feyler[1], A V Amelin[2] and E P Chumov[2]

[1] Institute of Metallurgy and Material Science, Siberian State Industrial University, 42 Kirova Street, Novokuznetsk, 654007, Russia
[2] EVRAZ Consolidated West-Siberian Metallurgical Plant, 16 Kosmicheskoye Road, Novokuznetsk, 654042, Russia

E-mail: feyler@rdtc.ru

Abstract. Technological process of steel production in converters using clinker-steel cakes and iron-bearing concentrates of "EVRAZ ZSMK" JSC slag processing unit ensuring liquid-phase reduction of iron from oxides is designed. Developed technology allows implementation of iron-containing materials recycling, reduces consumption of scrap, increases liquid metal yield, improves environmental safety in the region.

1. Introduction

Steel slags comprise significant part in metallurgical wastes. Their yield in average is from 150 to 200kg/t of steel [1]. At current steel production level in the Russian Federation annually about 9 million tons of steel slags are produced. They contain 10% of pure metal, and 15 - 40% of iron oxides. Total amount of iron is 20 - 30% of slag mass [2-4].

For processing of generated steel slags, slag processing unit was put into operation at "EVRAZ ZSMK" JSC in the end of 2014, the processing line allows to extract gradually iron-bearing inclusions from converter slag using magnetic separation technology. Performance of the unit is 1400000 tons of slag per year, metal-bearing concentrates yield is 250th. t. per year. Resulting iron-bearing concentrate is ferrous material of the following fractions: 0-10mm (U4), 10-80mm (29ADUZh) and 80-250mm (29CSC).

2. Research and results

To assess possibility of iron-bearing concentrates recycling in steel production and develop relevant technology of converter melting laboratory tests were conducted for U4 and 29ADUZh concentrates which consisted of sequential abrasion of 1kg of material for 5 minutes and its screening through a sieve with 0.5mm mesh size. Oversize product was subjected to subsequent abrasion. After five rounds of abrasion oversize material appeared to be metallic particles. Results of the tests are shown in Table 1.

Table 1. Mass of minus in sequential abrasion and screening of iron-bearing concentrates, g.

Attrition round	Concentrate type	
	U4	29ADUZh
1	693.14	145.97
2	130.49	121.86
3	32.93	22.45
4	21.29	28.37
5	15.25	14.57
Residue (metal part)	106.90 (10.69 %)	666.77 (66.68 %)

Performed chemical analysis of minus (Table 2) has shown that in average mineral part of U4 and 29ADUZh contains 40.7 and 27.0% Fe_{total}, respectively. Thus, total content of iron in the tested concentrates is 47 and 76%, respectively. Average iron content in U4 concentrate in form FeO and Fe_2O_3 oxides comprises 25.6 and 11.06%, respectively, in 29ADUZh concentrate – 25.8 and 9.98 %.

Table 2. Chemical composition of minus.

	Sample №	Elements content, %														
		Fe_{total}	S	CaO	SiO_2	FeO	MgO	MnO	P_2O_5	Al_2O_3	Na_2O	K_2O	TiO_2	ZnO	Fe_{met}	Fe_2O_3
U4	1	21.8	0.095	31.5	15.0	14.7	9.55	3.96	0.37	3.0	0.13	0.099	1.28	0.031	-	14.9
	2	29.5	0.085	28.9	12.7	31.5	9.16	3.94	0.39	1.42	0.12	0.070	1.28	0.023	-	7.22
	3	37.9	0.080	25.1	11.2	41.7	8.25	3.58	0.37	0.94	0.12	0.066	1.18	0.022	5.58	-
	4	55.8	0.055	16.9	7.22	21.6	6.41	2.49	0.31	0.93	0.12	0.078	0.75	0.019	41.1	-
	5	58.5	0.103	16.3	7.54	18.7	4.86	2.22	0.31	2.20	0.13	0.14	0.64	0.023	44.7	-
	Average	40.7	0.084	23.7	10.73	25.6	7.65	3.24	0.35	1.70	0.12	0.091	1.03	0.024	30.46	11.06
29ADUZh	1	26.0	0.065	34.1	13.4	23.5	8.47	2.8	0.34	3.96	0.090	0.055	0.63	0.019	-	11.1
	2	28.0	0.081	32.4	12.9	26.2	8.06	2.94	0.33	3.26	0.070	0.054	0.59	0.017	-	11.0
	3	27.2	0.092	33.8	13.2	28.0	8.81	2.79	0.35	2.77	0.070	0.051	0.62	0.022	-	7.82
	4	27.8	0.094	31.5	13.8	26.2	9.04	3.27	0.31	2.90	0.080	0.057	0.61	0.026	-	10.7
	5	25.9	0.076	31.4	14.2	25.0	9.03	3.52	0.32	3.22	0.080	0.061	0.63	0.021	-	9.29
	Average	27.0	0.082	32.6	13.5	25.8	8.68	3.06	0.33	3.22	0.080	0.056	0.62	0.021	-	9.98

Designed converter melting technology with recovery of iron from oxides of slag component of the slag processing unit concentrates or clinker-steel cakes involves two-stage gas-oxygen refining of melt in converter with liquid-phase recovery elements. In accordance with proposed variation of technology melting is carried out in two stages (see Figure 1):

- on the first, reduction stage, high degree of reduction of iron from the slag component of concentrates is provided, for that purpose poured in iron ("supporting plate"), SPC concentrates and carbonaceous material (coke, anthracite) added to the aggregate, are mixed with nitrogen (at a rate of 400-500m^3/min) or gas-oxigen mixture (nitrogen at a rate of 300-350 m^3/min and oxygen at a rate of 50-150m^3/min) to intensify mass transfer processes during reduction;
- on the second stage oxidizing blasting in converter is implemented according to developed blasting and slag mode of melting.

To solve the assigned task preliminary thermodynamic analysis of processes accompanying steel production in oxygen converters using iron-bearing concentrates was performed. During BOF melting oxidative pattern of the processes is most clearly revealed, in this regard, iron reduction from oxides is the most difficult.

In determining the optimum parameters of technological mode it is advised to establish possible limits of iron reduction from oxide component, input of the accompanying processes in melting thermal balance, as well as the process rate.

Considering additional components of the BOF burdening, the following physical and chemical processes are possible:

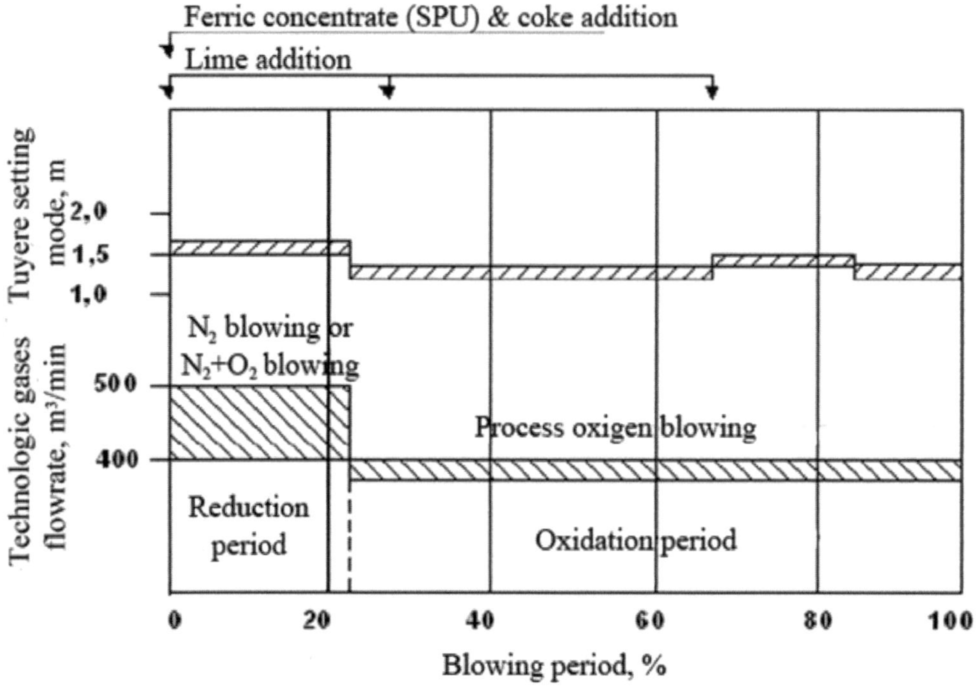

Figure 1. Processing mode of the two-stage converter melting with the elements of liquid-phase recovery.

- heating of solid oxidizers to the bath temperature (T_b) and their dissolution in oxide melt;
- heating of solid carbon to the bath temperature, and its dissolution in metal;
- reduction of iron from oxides with carbon, manganese, phosphorus, silicon dissolved in metal [5]:

$$(FeO)+[C]=[Fe]+\{CO\};$$
$$(FeO)+[Mn]=[Fe]+(MnO);$$
$$(FeO)+2/5[P]=[Fe]+1/5(P_2O_5);$$
$$(FeO)+1/2[Si]=[Fe]+1/2(SiO_2);$$

- reduction of iron from oxides with carbon monoxide [6]:

$$(FeO)+\{CO\}=[Fe]+\{CO_2\};$$

- oxidation of carbon dissolved in metal by gaseous oxygen [7]:

$$[C]+1/2\{O_2\}=\{CO\};$$

- iron oxidation with oxygen [8]:

$$[Fe]+1/2\{O_2\}=\{CO\};$$

- solid carbon burning [9]:

$$C_s+1/2\{O_2\}=\{CO\};\ C_T+\{O_2\}=\{CO_2\};$$

- gasification reaction [10]:

$$C_s+\{CO_2\}=2\{CO\};$$

- iron oxide dissolution in slag, [11]:

$$(FeO)_s=(FeO)_l;$$

- carbon dissolution in metal [12]:

$$C_s=[C].$$

Schematic diagram of the converter process with technological wastes liquid-phase recovery is presented on Figure 2.

Recovery reaction zones: Zone 1 - Surface of the up-floating ferric wastes particles (FW) contact with liquid cast iron; Zone 2 - Surface of up-floating slag drops (S) contact with liquid cast iron; Zone 3 - Surface of up-floating slag drops contact with carbonaceous material suspended in melt (CM); Zone 4 - Surface of slag contact with liquid iron; Zone 5 - Surface of slag contact with carbon material suspended in the melt.

Liquid iron I is the initial heat "capacity" and constant “storage” for reduced iron.

II - solid mixture of CSC (concentrate) and carbon (anthracite, steam coal, coal dust) in which the main reduction processes flow; the liquid phases, the oxide phase (slag) and the metal phase (iron-based) which transfers into I, subsequently allocate here.

III - gas phase, zone of conversion products, natural gas and carbon monoxide and hydrogen combustion products mixing. This zone is distributed throughout all inner volume of the unit with possibility of process gases supply through the upper primary and possibly auxiliary tuyere into gas mixing zone, into the oxide phase and, possibly through basal tuyeres into the metallic phase.

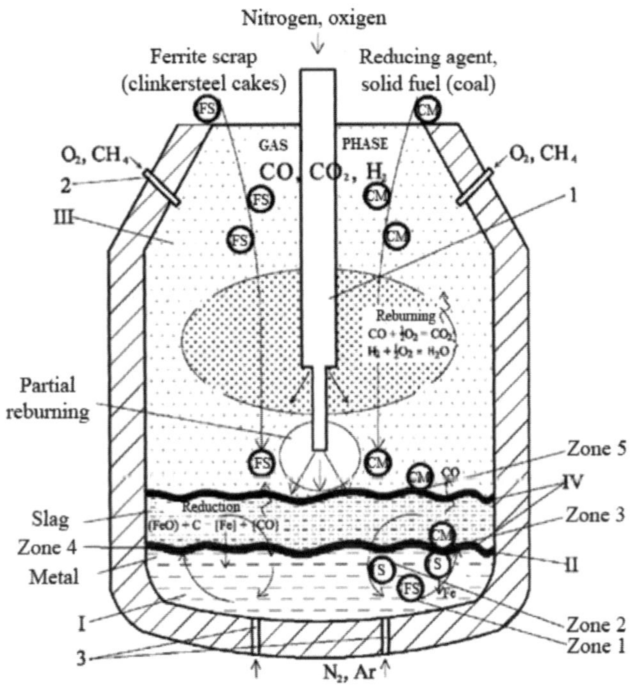

Figure 2. Scheme of BOF process with liquid-phase reduction of iron-containing materials.

IV – separation "auxiliary phase", solid carbon (reducing agent and fuel) consists of two conventional areas: in the lower part between the metallic (I) and the oxide (II) phases carbon concentration in metal at saturation level is kept, and iron is recovered from its monoxide; in the upper part between the gas mixing zone and oxides there is a conventional space separation boundary with predominance of deoxidizing (II) and oxidizing (III) properties; moreover, in IV a certain portion of carbon burns reacting with not consumed oxygen of process gases and serves as a source of heat.

3. Conclusions

To implement presented process scheme technically acceptable is the following distribution of process gases among tuyeres. Through the top tuyere 1 in certain periods technically pure oxygen, neutral gas or gas mixtures for oxidation, heating and bath mixing are supplied. Through tuyere 2 natural gas and oxygen for flue gases reburning are supplied. Mixture of natural gas and oxygen in 2:1 rate for metal conversion reaction. Basal tuyeres 3 are primarily intended for supply of neutral gas for additional mixing of the bath. To improve process thermal balance and more complete reduction reactions oxygen can be supplied through basal tuyeres mixed with neutral or natural gas.

Presented technological scheme of converter melting allows for better use of technogenic materials potential in steel production.

4. Acknowledgements

The work was performed at Siberian State Industrial University in the framework of the state task of the Ministry of Education and Science of the Russian Federation No. 2556.

5. References

[1] Yusfin Yu S et al 2002 *Industry and Environment* (Moscow: Akademkniga) p 469
[2] Butorina I V 2005 *Fundamentals of Iron and Steel Industry Stable Development* (Donetsk: Kashtan) p 332
[3] Das B, Prakash S et al 2007 *Resources, Conservation and Recycling* **50** 40–57
[4] Mihok L, Demeter P et al 2006 *Metalurgija* **45** 163-168
[5] Kudrin V A 2003 *Theory and Technology of Steel Production* (Moscow: Mir) p 528
[6] Bugayev K et al 2001 *Iron and Steel Production* (The Minerva Group, Inc) p 252
[7] Deo B and Boom R 1993 *Fundamentals of Steelmaking Metallurgy*
[8] Boichenko B M et al 2006 *Converter Steel Production: Theory, Technology, Steel Quality, Units Design, Materials Recirculation and Environment Protection* (Dnepropetrovsk: Dnepr-VAL) p 454
[9] Ryman C and Larsson M 2006 *ISIJ International* **46** 1752–58
[10] Baptizmanskiy V I et al 1988 *Thermal Work of Oxygen Converters* (Moscow: Metallurgiya) p 174
[11] Yavoyskiy V I 1967 *Theory of Steel Production Processes* (Moscow: Metallurgiya) p 467
[12] Baptizmanskiy V I 1960 *Mechanism and Kinetics of Converter Bath Processes* (Moscow: Metallurgizdat) p 286

Automated information system for analysis and prediction of production situations in blast furnace plant

V V Lavrov and N A Spirin

Department of Thermophysics and IT in Metallurgy, Ural Federal University, 19 Mira Street, Ekaterinburg, 620002, Russia

E-mail: spirin@urfu.ru

Abstract. Advances in modern science and technology are inherently connected with the development, implementation, and widespread use of computer systems based on mathematical modeling. Algorithms and computer systems are gaining practical significance solving a range of process tasks in metallurgy of MES-level (Manufacturing Execution Systems – systems controlling industrial process) of modern automated information systems at the largest iron and steel enterprises in Russia. This fact determines the necessity to develop information-modeling systems based on mathematical models that will take into account the physics of the process, the basics of heat and mass exchange, the laws of energy conservation, and also the peculiarities of the impact of technological and standard characteristics of raw materials on the manufacturing process data. Special attention in this set of operations for metallurgic production is devoted to blast-furnace production, as it consumes the greatest amount of energy, up to 50% of the fuel used in ferrous metallurgy. The paper deals with the requirements, structure and architecture of BF Process Engineer's Automated Workstation (AWS), a computer decision support system of MES Level implemented in the ICS of the Blast Furnace Plant at Magnitogorsk Iron and Steel Works. It presents a brief description of main model subsystems as well as assumptions made in the process of mathematical modelling. Application of the developed system allows the engineering and process staff to analyze online production situations in the blast furnace plant, to solve a number of process tasks related to control of heat, gas dynamics and slag conditions of blast-furnace smelting as well as to calculate the optimal composition of blast-furnace slag, which eventually results in increasing technical and economic performance of blast-furnace production.

1. Introduction

The modern trend of science and technology development is characterized by development, implementation and wide application of computer decision support systems in the Industrial Control System (ICS) based on the methods of mathematical modelling [1–5]. At present the role of algorithms and computer programs is becoming more and more evident for solving a complex of process tasks in metallurgy at MES Level (Manufacturing Execution Systems are industrial process control systems) of modern automated information systems at the largest iron and steel works in Russia [1, 2].

This determines the requirement for development of simulation systems based on a set of mathematical models considering the physics of the process, fundamentals of heat and mass transfer, energy conservation laws as well as special influence of process and standard characteristics of raw materials upon production process data. A special place in this complex of metallurgical production

IOP Conf. Series: Materials Science and Engineering **150** (2016) 012010 doi:10.1088/1757-899X/150/1/012010

processes is taken by blast furnace production as the most energy-intensive production with its share of the fuel used in ferrous metallurgy up to 50%.

The authors have developed and implemented an automated information system for analysis and prediction of production situations in the blast furnace plant (AIS APPS BFP) of Magnitogorsk Iron and Steel Works.

2. System structure

The structure of the automated information system for analysis and prediction of production situations in the blast furnace plant is shown in Figure 1. Its main subsystems are as follows:

- "Data Acquisition and Storage" which fills the database with actual reported performance parameters of blast furnace production;
- "Visualization of Shift and Daily Average Data of Blast Furnace Operation" which forms graphic trends as per parameters chosen by the user in any combination for the specified period of time;
- "Preparation of Technical Report on Operation of Blast Furnaces and Blast Furnace Plant" which is designed for making a technical report on operation of blast furnaces and blast furnace plant in whole for the specified periods of operation;
- "Comparison of Reported Performance Parameters of Blast Furnaces and Blast Furnace Plant" which gives an opportunity to compare the performance parameters of the plant or furnace by a set of chosen parameters for the specified period of time;
- "Model-based Decision Support, Prediction of Process Situations and Diagnostics of Blast Furnace Operation". Implementation of the subsystem is based on calculation algorithms with the use of mathematical models of the blast furnace process for the period of operation of individual blast furnaces or the plant in whole as chosen by the user. The analysis of the blast furnace process makes it possible to ascertain that this subsystem shall include the following interconnected subsystems of a lower stratum: heat-and-mass balance calculations of blast furnace smelting; slag conditions; gas dynamic conditions; heat conditions; optimal distribution of fuel and energy resources; optimal selection of the burden composition as well as diagnostics of blast furnace smelting operation. Each of these subsystems interacts with other parameter blocks, subsystems and environment;
- "Visualization of Performance Parameters of Blast Furnace Production for OLAP" which provides authorized users with access to all reported data on blast furnace production in the company's corporate network.

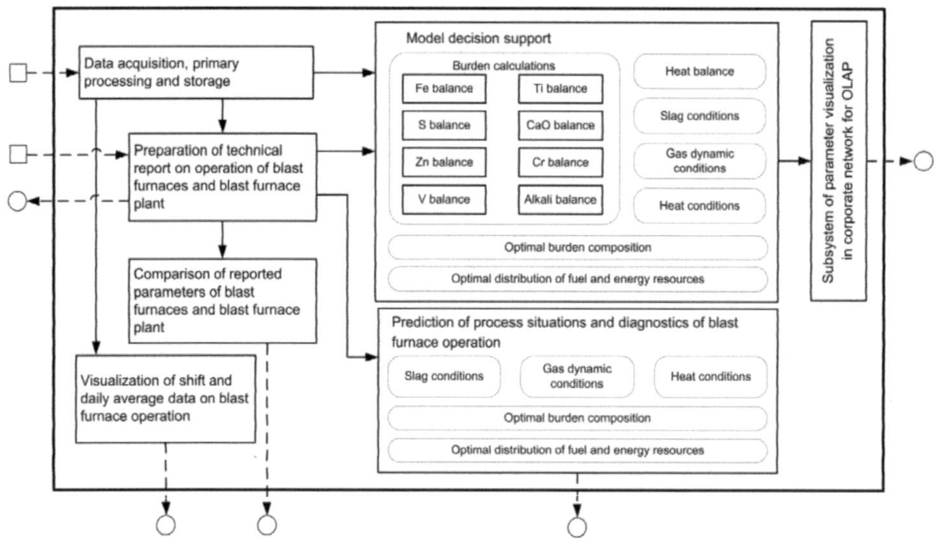

Figure 1. Structure of automated information system for analysis and prediction of production situations in blast furnace plant: – data source; – data presentation.

3. System engineering and implementation

In the course of system engineering a detailed development of functionality was required for individual subsystems. For this purpose the generalized functional model was developed on the basis of ideas and notations of IDEF0 (Integrated computer aided manufacturing DEFinition) Structured Analysis and Design Methodology [6]. It was implemented in AllFusion Process Modeler (BPwin). The use of this methodology has made it possible to create functional blocks of individual subsystems, to reveal actions performed by them and connections between these actions, controlling actions and execution mechanisms of each function. The total number of decomposed blocks of the functional model of AIS APPS BFP is 152. Decomposition of the model was performed up to and including the third stratum.

Analyzing the requests made by process staff, reference data of blast furnace production and functional modelling, mathematical and algorithmic software was developed and taken as a basis for software implementation of the subsystem displaying data on operation of individual blast furnaces and the plant in whole. The structured system analysis and engineering of mathematical model blocks were performed on the basis of the procedure-oriented approach. The basis of this approach is the use of data flow diagrams (DFD), an entity-relationship model where the main components are data flows transferring information from one module to another [6]. The notation of the DFD method involves breaking of the mathematical model into separate functional components (processes) and their presentation as a network interconnected by data flows.

4. Software components of the system

The software implementation of AIS APPS BFP – an integrated software complex for BF Process Engineer's Automated Workstation – has been made in Microsoft Visual Studio (C# Programming Language) [7]. Figure 2 shows the main components of the software architecture. The central link is the data storage structure [8] which is formed on the database server of the information computer center of blast furnace production. Its sources are hardware and software of the blast furnace plant ICS, corporate information system (CIS) and central control room (CCR) of the factory. Depending on the requirements of individual subsystems, it is possible to implement various periods of data averaging in the base using mechanisms of the database management system

The architecture shown in Figure 2 provides the required functionality, fulfilment of subject area requirements, relatively simple system extension and changes, possibility of free-running implementation of separate software modules and their independence from the data storage structure.

The list of the design software modules is as follows:

- "Heat Balance of Blast Furnace Smelting" makes automatic calculations and displays inputs and outputs of heat balance of blast furnace smelting on the basis of monthly average reported data on blast furnace plant operation for the period specified by the user;
- "Fe, S, Zn, Ti, CaO, Cr Mass Balances of Blast Furnaces" makes automatic calculations and displays mass balances of Fe, S, Zn, Ti, CaO, Cr in blast furnaces on the basis of monthly average reported data on blast furnace plant operation for the period specified by the user;
- "Gas Dynamic Conditions of Blast Furnace Smelting" makes automatic calculations and displays parameters of gas dynamic conditions of blast furnace smelting for the period specified by the user;
- "Slag Conditions of Blast Furnace Smelting" makes automatic calculations and displays parameters of slag conditions of blast furnace smelting on the basis of monthly average reported data on blast furnace plant operation for the period specified by the user;

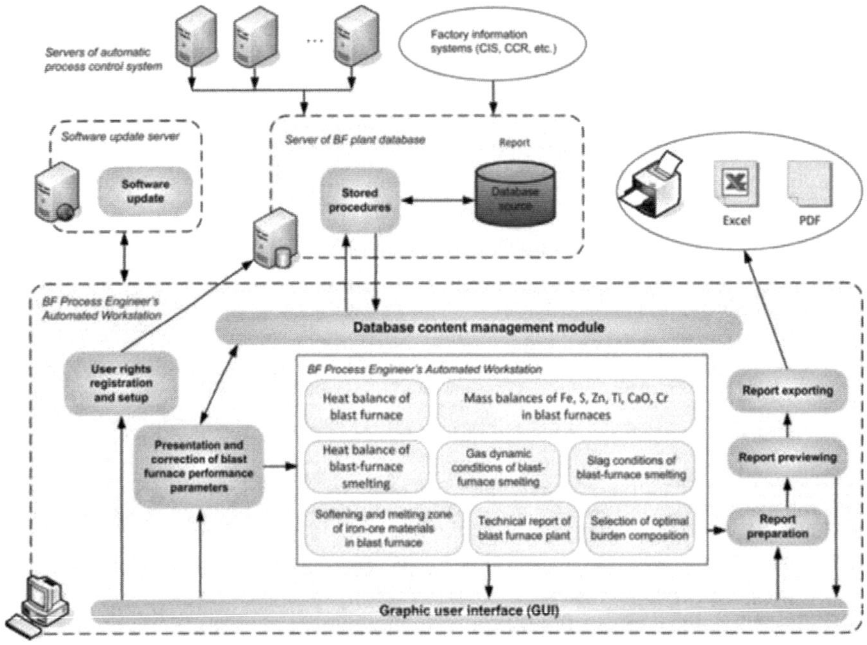

Figure 2. Architecture of software complex for BF Process Engineer's Automated Workstation.

- "Selection of Optimal Burden Composition" enables to solve process task related to selection of the optimal burden composition on the basis of monthly average reported data on blast furnace plant operation for the period specified by the user;
- "Softening and Melting Zone of Iron-Ore Materials in Blast Furnace" makes automatic calculations and displays parameters characterizing the shape and location of the softening and melting zone of iron-ore materials in the blast furnace on the basis of monthly average reported data on blast furnace plant operation for the period specified by the user;
- "Technical Report of Blast Furnace Plant" displays parameters of the technical report for a calendar month as well as from the beginning of the year to the month specified by the user.

The software complex is made as a separate client application installed on the user's computer by installation of the load file. After the user's registration in the software and loading of the main form the screen will display the reported data for the base period (actual reported calendar month) of blast furnace operation which are automatically retrieved from the database of MMK's ICS center for the calendar month (Figure 3). Data display in the main form is possible for two periods: base and comparative. The comparative period is used for comparison of actual reported performance parameters of blast furnace operation with analogous parameters in the base period for the calendar month. The complex has a wide selection of interactive settings for the user's improved readability and minimized routine actions. The results of the model reports are presented in the tabular and graphic user forms. Preparation of the report with the possibility of its previewing and exporting in other formats is provided.

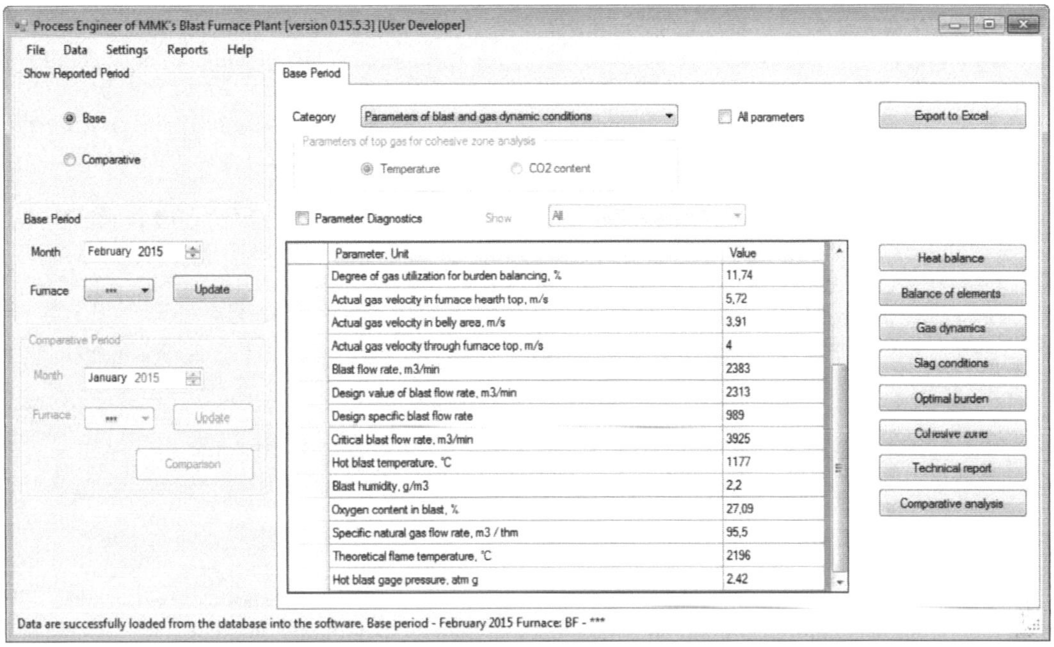

Figure 3. Main window of software complex for BF Process Engineer's Automated Workstation.

For easy and quick perception of process situations and blast furnace performance parameter trends in the base and comparative periods the user can see the data under separate categories: "Operating Parameters", "Furnace Structural Dimensions", "Coke Properties", "Burden Parameters and Consumption", "Blast Parameters', "Furnace Top Gas", "Liquid Molten Products: Iron", "Liquid Molten Products: Slag", "Heat Balance", "Fe Balance", "S Balance", "Zn Balance", "Ti Balance", "CaO Balance", "Cr Balance", "First Slag", "Final Slag"; "Gas Dynamics".

Apart from providing data for the current analysis of the reported parameters and evaluation of the production situations, the BF Process Engineer's Automated Workstation also enables to solve a number of process tasks related to control of heat, gas dynamic and slag conditions of blast furnace smelting as well as to calculate the optimal burden composition.

5. Conclusions

Using modern engineering technologies and software tools [6–8], a system for analysis and prediction of production situations in the blast furnace plant of Magnitogorsk Iron and Steel Works was developed and put into pilot industrial operation. This system provides the operators with a modern tool for online analysis of reported performance parameters [9, 10]. Application of the developed system will allow the operators to analyze online production situations in the blast furnace plant and to solve problems related to control of blast furnace smelting technology, which eventually results in increasing technical and economic performance parameters of blast furnace production.

6. Acknowledgements

The work was supported by Act 211 Government of the Russian Federation, contract № 02.A03.21.0006.

7. References

[1] Spirin N A, Lavrov V V, Rybolovlev V Yu et al 2011 *Model-Based Decision Support Systems in ICS of Blast Furnace Smelting* ed N A Spirin (Ekaterinburg: UrFU) p 426

[2] Onorin O P, Spirin N A, Terentyev V L et al 2005 *Computer Methods of Blast Furnace Process Modelling* ed N A Spirin (Ekaterinburg: USTU–UPI) p 301
[3] Feng Q and Wang L 2014 *WIT Transactions on Information and Communication Technologies* **46** (2) 1853–58
[4] Dimitrov B H, Nenov H B and Marinov A S 2013 *Conf. on Information and Communication Technology, Electronics and Microelectronics (Croatia)* (MIPRO) article No 6596383 pp 891–895
[5] Bartels R, Dohl Kw, Uhde H, Schroder A and Stumpe W 1994 *Stahl und Eisen* **114** (b) 75–79
[6] Odintsov I O 2004 *Professional Programming. System Approach* (SPb.: BHV-Peterburg) p 624
[7] Troelsen A 2013 *Pro C# 5.0 and the .NET 4.5 Framework* (M.: Williams) p 1311
[8] Date C J 2006 *An Introduction to Database Systems* (M.: Williams) p 1328
[9] Spirin N A , Lavrov V V, Burykin A A, Rybolovlev V Yu, Krasnobaev A V and Kosachenko I E 2011 *Metallurgist* **54** (9–10) 566–569
[10] Spirin N A, Lavrov V V, Rybolovlev V Yu et al 2014 *Mathematical Modelling of Metallurgical Processes in ISC: Study Guide* ed N A Spirin (Ekaterinburg: UrFU) p 558

Information modeling system for blast furnace control

N A Spirin, L Y Gileva and V V Lavrov

Department of Thermophysics and IT in Metallurgy, Ural Federal University, 19 Mira Street, Ekaterinburg, 620002, Russia

E-mail: spirin@urfu.ru

Abstract. Modern Iron & Steel Works as a rule are equipped with powerful distributed control systems (DCS) and databases. Implementation of DSC system solves the problem of storage, control, protection, entry, editing and retrieving of information as well as generation of required reporting data. The most advanced and promising approach is to use decision support information technologies based on a complex of mathematical models. The model decision support system for control of blast furnace smelting is designed and operated. The basis of the model system is a complex of mathematical models created using the principle of natural mathematical modeling. This principle provides for construction of mathematical models of two levels. The first level model is a basic state model which makes it possible to assess the vector of system parameters using field data and blast furnace operation results. It is also used to calculate the adjustment (adaptation) coefficients of the predictive block of the system. The second-level model is a predictive model designed to assess the design parameters of the blast furnace process when there are changes in melting conditions relative to its current state. Tasks for which software is developed are described. Characteristics of the main subsystems of the blast furnace process as an object of modeling and control – thermal state of the furnace, blast, gas dynamic and slag conditions of blast furnace smelting – are presented.

1. Introduction

Control of blast furnace operation is often classified as poorly structured problem. Because of this the base mathematical models have to be used at automated workstations of engineering and technological staff of blast furnace. The software for computer decision-making support system is a set of information-connected interactive (dialog) program modules integrated into a common environment. Integration of mathematical models is based on the application of both common initial data and the results of calculations performed by individual modules. Software modules implement mathematical models of heat-exchange, slag, blast and gas-dynamics regimes of blast furnace. This forms the basis of functionality of the decision making support system while the modular structure ensures its extensibility.

Solution of the problems of analysis and forecast of blast furnace operating parameters using available actual information on furnace performance requires decomposition of the problem into three main interrelated subsystems: the thermal condition of the furnace, blast and gas dynamics regimes of blast furnace, slag regime. The basic principle, which underlies the choice of specific mathematical equations to describe the individual subsystems, is to resolve the contradiction between the complexity of the modelled process and the need to resolve the technological problems in real time in course of the process and for a specified time interval using the actual available information.

2. Mathematical model

Modern advances in the theory and practice of the blast furnace process used as the basis of mathematical modelling [1–8]. Usage of the principle of mathematical modelling [2–5, 9–13] based on deviation of the actual operation from the base period of operation requires allocation of two levels of model: the basic state model and the predictive model. The basic state model allows estimation of the vector of system' parameters, This model is used to estimate adaptation coefficients of predictive model. The predictive model estimates the design parameters of blast furnace process when conditions of blast furnace operation deviate from the base state.

The initial data for the base state model is the data of the blast furnace operation in the period of time in which the dynamic measurement errors are negligible (base period). Conditions of the furnace operation in the base period should be stable and be close to the current state (current period), as the use of linear equations describing the nonlinear process is limited by the parameters of its nonlinearity. The current period is a period of furnace operation immediately preceding the time simulation. Data on furnace operation in the current period is required for the initial (baseline) data for projecting of blast furnace performance indicators in the case of change in raw materials or regime parameters. Additional data for predictive model are the vector of indicators of the state of the system, which is estimated in the base period model, and a set of data on changes in parameters of blast furnace operation relative to the current period (projected period).

A block diagram of computer decision support system is shown in Figure 1. The following subsystems are included in block diagram Figure 1.

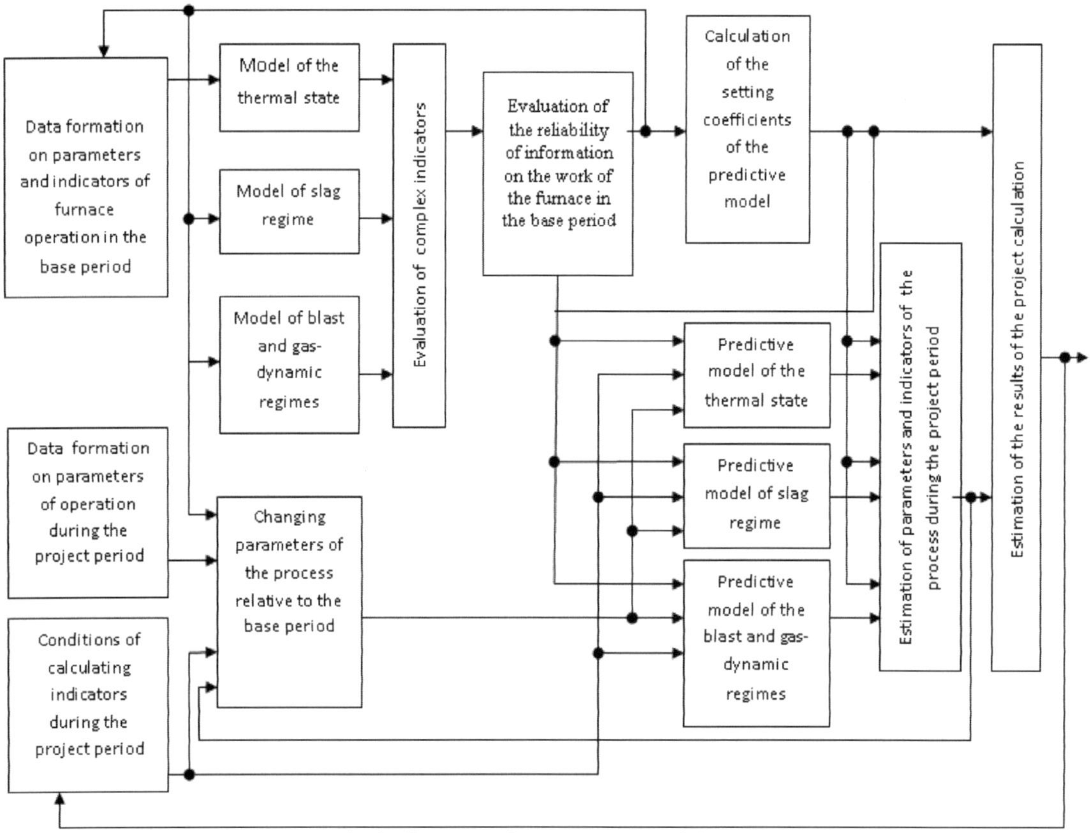

Figure 1. Block diagram of a computer decision support system.

Subsystem of blast furnace thermal state [1–5, 9]. Figure 2 illustrates the ideology of using this subsystem to estimate the required changes in coke rate in response to changes in burden quality. This subsystem has two major goals: assessment of the effect of regime parameters on the specific consumption of coke (ore load); forecast of chemical composition of smelting products at changes in parameters of burden and combined blast.

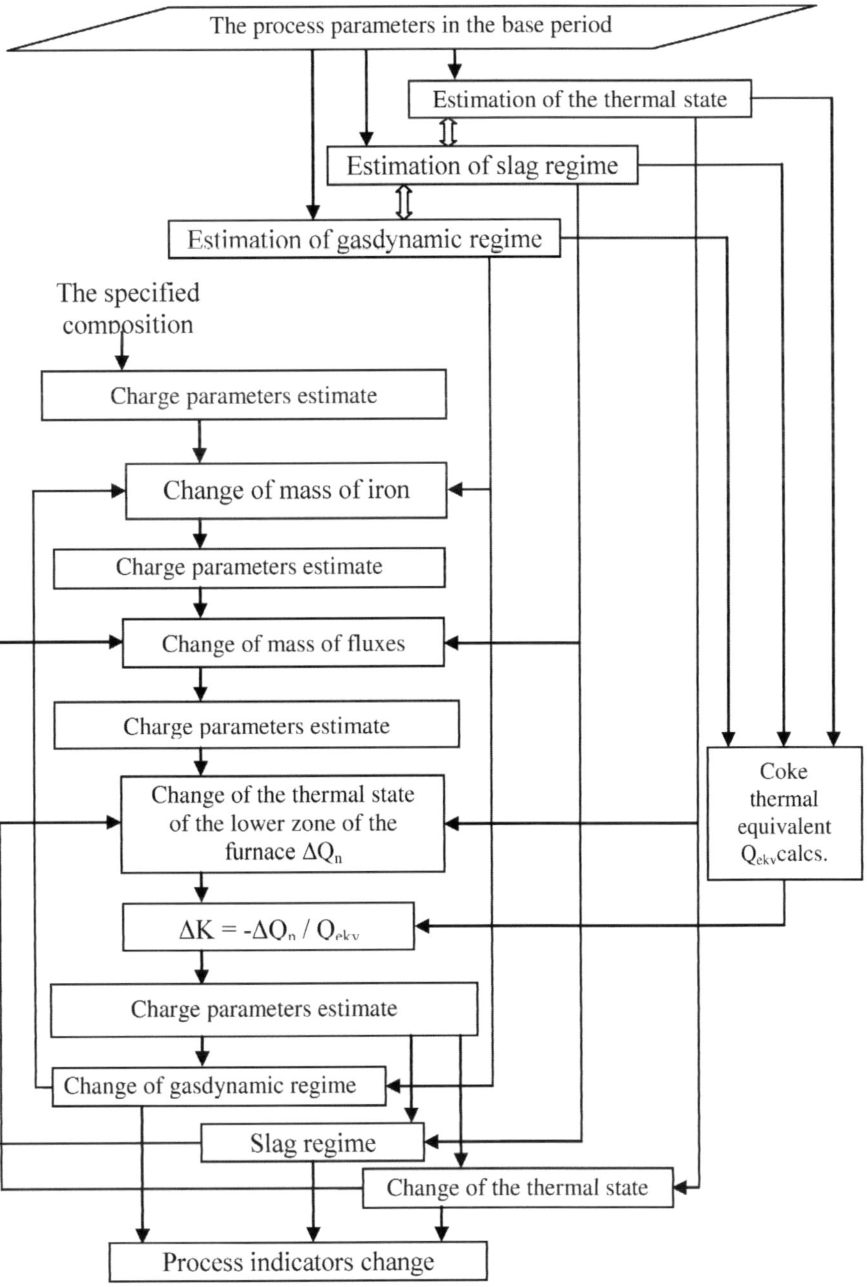

Figure 2. Algorithm to estimate the required changes in coke rate in response to changes in burden chemistry.

IOP Conf. Series: Materials Science and Engineering **150** (2016) 012011 doi:10.1088/1757-899X/150/1/012011

Regularities of heat transfer in modern blast furnaces form the basis of the thermal state model. These regularities confirm the possibility of independent evaluation of the upper and bottom heat exchange zones of blast furnace. The upper thermal zone determines the intensity of indirect reduction of iron. The lower thermal zone determines demand of enthalpy for the direct reduction of wustite and physical and chemical heating of slag and hot metal. The influence of the thermal state of the upper zone on the thermal state of the lower zone is developed through extend of direct and indirect reduction processes. Heat and mass balance relations of the model are supplemented by heat transfer equations to estimate the effect of gas flow performance and slag regime on heat conditions of the blast furnace.

Subsystem of blast and gas dynamics regimes [1–5, 10]. A block diagram of a gas dynamics modelling is shown in Figure 3. The gas dynamics modelling allows solution of the following set of problems:

- assess the impact of combined blast and burden parameters on raceway operation;
- estimate gas-dynamics parameters of materials' column;
- estimate the pressure drop along the furnace height;
- determine the reserves to increase intensity of blast furnace operation by parameters of blast.

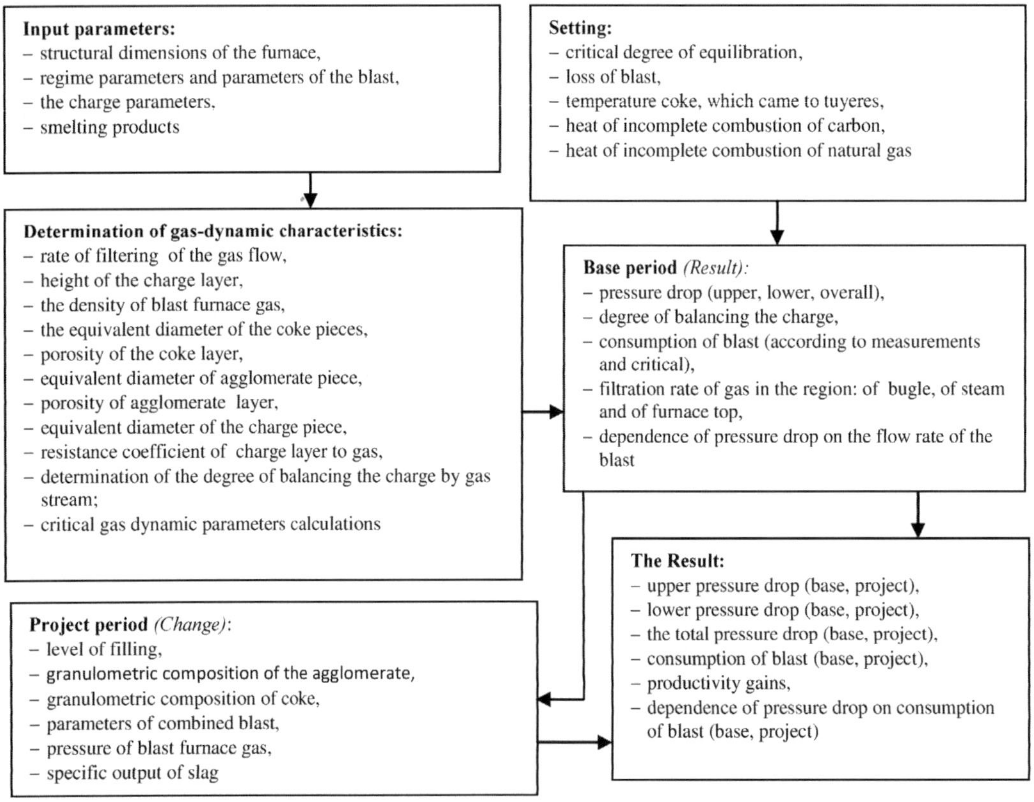

Figure 3. Block diagram of gas dynamics modelling.

Subsystem of slag regime [1–5, 11]. This subsystem evaluates the slag regime of blast furnace. Several problems have to be solved in the course of slag formation modeling at all stages of the process: materials fusion; filtration of primary slag through the coke bed; obtaining the final slag; obtaining the hot metal with desired chemical composition including minimum sulfur content. Block diagram of slag regime subsystem is presented in Figure 4.

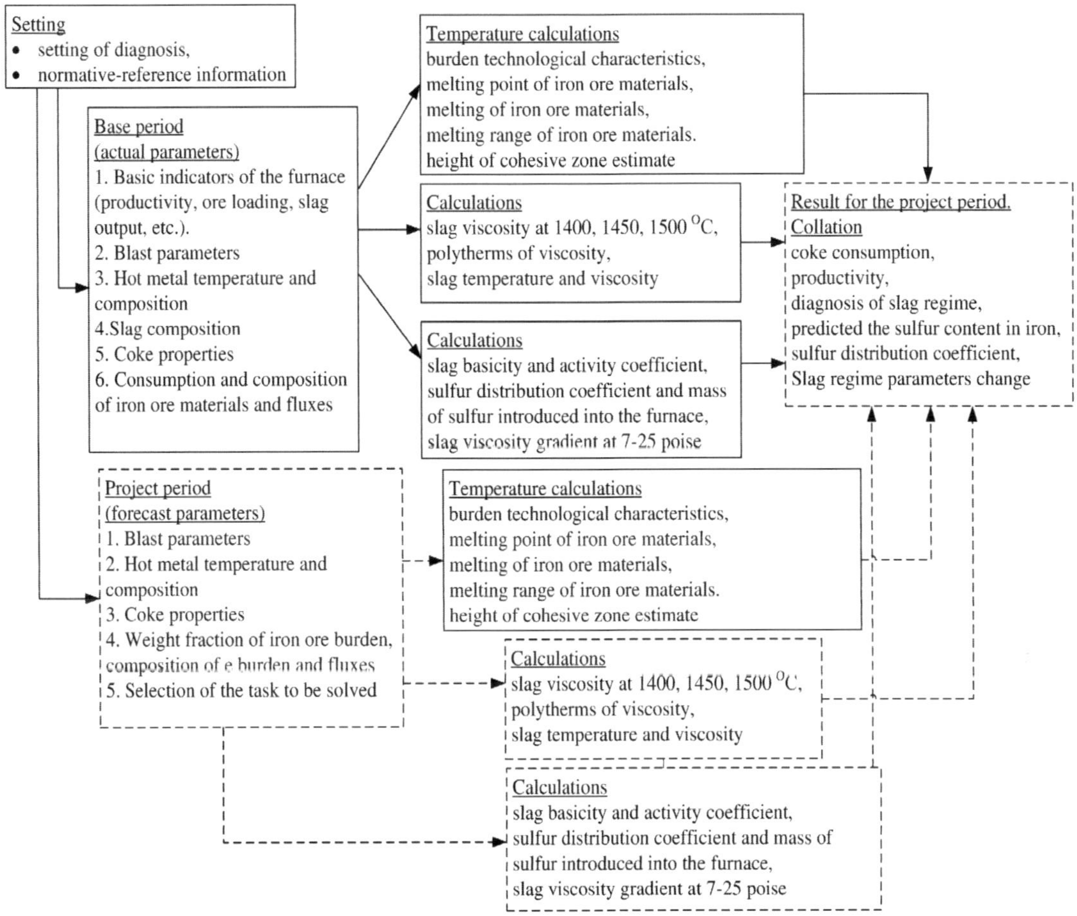

Figure 4. Block diagram of the slag regime modelling.

Four component slag system – CaO–SiO_2–MgO–Al_2O_3 in area of the real values of blast furnace slag plus known relationship between slag viscosity and its temperature are used for analytical estimation of slag viscosity for base and predictive periods. If significant quantities of other oxides are present in slag appropriate corrections are introduced taking into account the influence of these oxides on slag viscosity.

The dependence of blast furnace slag viscosity on temperature in the temperature range of 1350–1550 0C for the base period, as well as for the case of change in blast furnace operating conditions is estimated using the relationship between temperature and viscosity of homogeneous slag. The temperature of formation of the first portion of liquid phase (corresponds to the start of shrinkage – the temperature of start of fusion) and melting temperature (the temperature of the movable melt formation) are estimated by empirical equations for specific iron bearing material. Sulfur balance equation is used to predict sulfur content in hot metal. Detailed description of the subsystem of slag regime could be found in literature [3–5].

3. Results of modelling

The initial information on operating parameters and indicators of blast furnace is automatically extracted from the database to form the dataset of the furnace operation in the base period. Manual entry of the data and duration of the base period is envisaged. The period with duration of 24 hours (or one shift) closest to the current period of operation is set by default. The set of parameters which characterize heat, wind (blast), gas dynamics and slag regimes of the blast furnace is estimated during base period modelling (Table 1).

Table 1. Major indicators of blast furnace operation.

Indicators	Shift 1	Shift 2	Shift 3
Production, thm/period	1102	1215	1001
Theoretical productivity, thm/period	1116	1170	1179
Performance (actual/estimated), thm/day:	3340/3382	3681/3546	3033/3573
Coke consumption (Estimated production/minimum), kg /thm	430.5/390.7	428.3/373.0	431.4/381.3
Volumetric intensity index, tons of coke/(m3 per day)	1.05	1.10	1.11
Ore load (per wet coke) t/t	3.77	3.77	3.77
Degree of direct reduction, fraction.	0.30	0.34	0.33
Content of Fe in the iron part of burden,%	58.92	59.23	58.86
Hot metal temperature, 0C	1412	1405	1406
Slag volume (Actual/estimate), kg/ thm	310/298	312/289	334/296
Slag basicity CaO/SiO2 (actual/estimate), fraction	1.04/1.13	1.09/1.13	1.09/1.13
Slag viscosity at t=1500 0C (C-S-A)/ (C-S-A-M):	5.00/3.34	5.02/3.34	4.95/3.27
Sulfur distribution coefficient (Ls)	45	67	55
Temperature of melting/ temperature of slag good fluidity, 0C	1342/1494	1340/1494	1341/1491
Wind (blast) rate (actual/marginal), m3/min	3032/3267	3128/3401	3090/3242
Blast pressure (actual/marginal), atm.	2.71/2.91	2.76/3.00	2.76/2.90
Pressure drop along the burden height (actual/marginal), atm.	1.32/1.53	1.29/1.53	1.39/1.53
Equilibration degree of charge by gas flow (actual/marginal),%	47/55	47/55	50/55
Natural gas rate, % to wind rate	9	10	8
Raceway length, mm	1208	1222	1394
Oxidizing zone length, mm	1546	1564	1784
Oxidizing zone relative area,%	46	47	59
RAFT, 0C	1919	1933	2055
Top gas rate, m3/min	3912	4139	4540
Utilization degree of CO, fraction.	0.43	0.45	0.43
Thermal efficiency, %	66	67	62
Thermal state of the bottom zone indicator, mJ/thm	2870	2818	3342
Thermal equivalent of coke' carbon, kJ/kg Coke	9657	9638	11185

4. Conclusions

Experimental and industrial operation of the pilot expert system to control blast furnace operation and support management' decision making process demonstrates the correctness of the computational algorithm, the possibility of model adaptation and tuning to the specific conditions of individual blast furnace and the whole blast furnace shop

Developed software, with appropriate adaptation can be recommended for the following applications:

- control of blast furnace operation in real time;
- create automated workplaces for engineering staff of blast furnace shop;
- solution of the complex strategic problems for production planning, supply of iron ore and fuel, optimal energy management, etc.

5. Acknowledgements

The work was supported by Act 211 Government of the Russian Federation, contract No. 02.A03.21.0006.

6. References

[1] Kitaev B I, Yaroshenko Yu G, Sukhanov V D, Ovchinnikov Yu N and Shvidkii V S 1978 *Heat Operation of Blast Furnace Process* (Moscow: Metallurgiya) p 248
[2] Spirin N A, Ipatov Yu V and Lobanov V I 2001 *Information Systems in Metallurgy* (Ekaterinburg: USTU) p 617
[3] Onorin O P, Spirin N A, Terentev V L 2005 *Computer Techniques for Modeling Blast Furnace Process* (Ekaterinburg: USTU) p 301
[4] Spirin N A, Lavrov V V and Rybolovlev V Y 2011 *Model Systems of Support Decision Making in ASU TP of Blast Furnace* (Ekaterinburg: UrFU) p 462
[5] Spirin N A, Lavrov V V and Rybolovlev V Y 2014 *Mathematical Moderation of Metallurgical Processes in ASU TP* (Ekaterinburg: UrFU) p 562
[6] Fielden C J and Wood B I 1968 *J. Iron and Steel Institute* **206** 650–658
[7] Hatano M, Kurita K, Yamaoka H and Yokoi T 1982 *Tetsu-to-Hagane* **68** 2369–76
[8] Taguchi S, Kubo H, Tsuchiya N, Ichifuji K and Okabe K 1982 *Tetsu-to-Hagane* **68** 2303–10
[9] Zagainov S A, Onorin O P, Spirin N A and Yaroshenko Yu G 2003 *Steel in Translation* **33** (12) 1–5
[10] Onorin O P, Spirin N A and Lavrov V V 2005 *Steel in Translation* **35** (6) 3–6
[11] Spirin N A, Onorin O P and Rybolovlev V Y 2005 *Steel in Translation* **35** (8) 7–11
[12] Gileva L Y, Spirin N A, Rybolovlev V Y and Krasnobaev A V 2009 *Steel in Translation* **39** (12) 1060–63
[13] Spirin N A, Lavrov V V and Burikin A A 2011 *Metallurgist* **54** (9–1) 566–569

IOP Conf. Series: Materials Science and Engineering **150** (2016) 012012 doi:10.1088/1757-899X/150/1/012012

Mathematical model and software for control of commissioning blast furnace

N A Spirin, O P Onorin, K A Shchipanov and V V Lavrov

Department of Thermophysics and IT in Metallurgy, Ural Federal University, 19 Mira Street, Yekaterinburg, 620002, Russia

E-mail: spirin@urfu.ru

Abstract. Blowing-in is a starting period of blast furnace operation after construction or major repair. The current approximation methods of blowing-in burden analysis are based on blowing-in practice of previously commissioned blast furnaces. This area is theoretically underexplored; there are no common scientifically based methods for selection of the burden composition and blast parameters. The purpose of this paper is development and scientific substantiation of the methods for selection of the burden composition and blast parameters in the blast furnace during the blowing-in period. Research methods are based on physical regularities of main processes running in the blast furnace, system analysis, and application of modern principles for development and construction of mathematical models, algorithms and software designed for automated control of complex production processes in metallurgy. As consequence of the research made by the authors the following results have been achieved: 1. A set of mathematical models for analysis of burden arrangement throughout the height of the blast furnace and for selection of optimal blast and gas dynamic parameters has been developed. 2. General principles for selection of the blowing-in burden composition and blast and gas dynamic parameters have been set up. 3. The software for the engineering and process staff of the blast furnace has been developed and introduced in the industry.

1. Introduction

Blast furnace blowing-in and subsequent blowing period are critical operations and correct execution of these operations has an impact on normal performance of the blast furnace, its service life, quantity and duration of mid-life repairs. The analysis of literature data related to blast-furnace production has shown that there are approximation methods of blowing-in burden analysis based on blowing-in practice of previously commissioned blast furnaces [1–7]. Modelling of the one-dimensional non-steady state of temperature fields for blast furnace blowing-in was performed by foreign researchers [8–10]. However, this area is theoretically underexplored; there are no common scientifically based methods for selection of the burden composition and blast parameters.

In case of forced blowing-in (with controlled slag formation and reduction) the first portions of slag-forming and iron-ore materials are located in the furnace stack. In this case, there is no necessity to gradually increase the ore load [4–5, 7]. The structure of the mathematical model for calculation of blowing-in burden during forced blowing-in is shown in Figure1. The burden charged in the blast furnace is divided into three parts (zero burden, first burden, second burden).

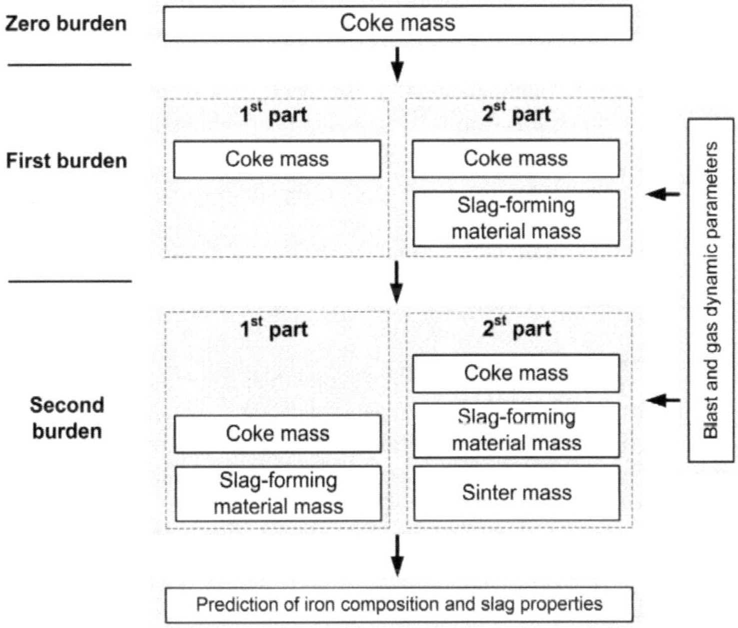

Figure 1. Structure of mathematical model for burden calculation during forced blowing-in.

The "zero burden" which consists only of coke does not take part in the combustion process and fulfils the role of drainage coke packing. The volume of the zero burden includes the following:

- well (hearth volume determined by the height of the fixed bed material);
- hearth volume from the iron tap-hole level to the blast tuyere level;
- coke volume in the cone of fixed materials from the blast tuyere level (dead man coke).

The mathematical model structure of the "first burden" analysis is shown in Figure 2.

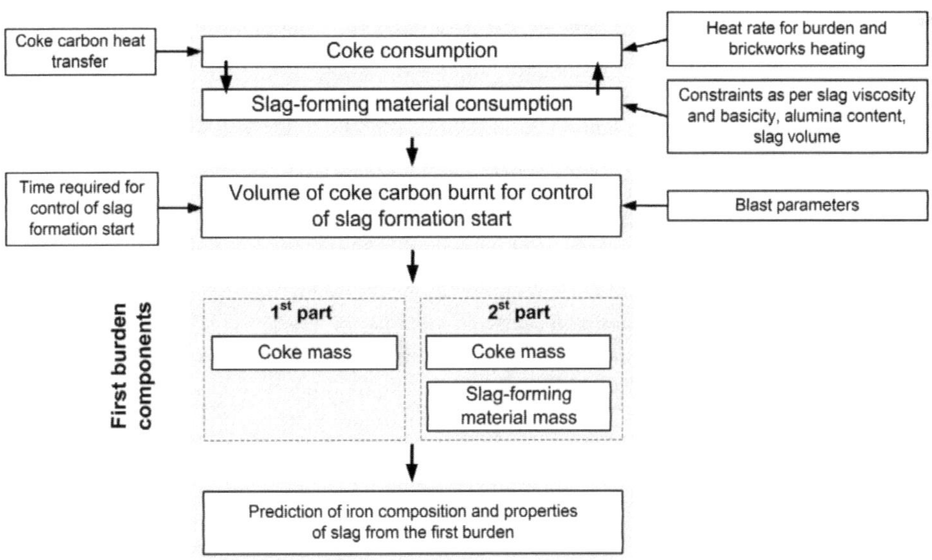

Figure 2. Structure of mathematical model for "first burden" calculation.

The first burden is divided into two parts; the first part consists of coke and the second part consists of coke and slag-forming materials. The sequence of calculations for determining parameters of the first burden is as follows:

- to determine the required coke consumption for heating of burden and refractory brickworks up to the operating temperatures on the basis of heat balance, i.e. the quantity of heat from burning coke of the first burden shall be sufficient for heating of refractory brickworks and materials of this burden up to the operating temperatures typical of the lower step of heat transfer;
- to calculate consumptions of slag-forming materials on the basis of the assigned slag composition (CaO/SiO_2, Al_2O_3) and assigned slag viscosity at 1400°C and 1500°C; the slag viscosity calculation is based on mathematical treatment of known slag diagrams at these temperatures;
- due to addition of slag-forming materials into the burden some additional coke consumption is required for heating and smelting of these materials; since the coke consumption changes, it is necessary to correct the burden composition as at the increased coke consumption there are changes in the slag composition and properties. The iterative procedure for calculation of coke and slag-forming material consumption is shown in Figure 3;
- to determine the location of the first burden along the furnace height; addition of slag-forming materials into the first part of the first burden is not provided; slag-forming materials coming from coke ash during combustion of first burden coke are distributed in fractional voids of the coke packing, they do not form free-running slag.

This method of furnace blowing-in means operation of the furnace hearth without slag, which gives an opportunity to remove hot gases through the pipes embedded in iron tap-holes. Otherwise, in case of early slag formation the pipes will be inevitably filled up with slag and gas flow in the hearth bottom will stop, which will not allow heating of the hearth bottom to the operating temperatures providing normal production of the first portions of molten slag and iron. Industrial blowing-in of the blast furnaces using the technology of the delayed slag formation shows that the furnace operation time before slag will appear in the hearth shall be more than 6 hours and reach up to 10 hours without any risk of significant breaking in gas dynamic properties of the burden column [4, 5]. The speed of burden movement in the furnace and arrival time of slag-forming materials to the tuyeres are determined by flow rates of blast, natural gas and oxygen as well as by blast humidity. After changes in these parameters are set in the course of blowing-in, it is necessary to calculate the coke quantity required for timing of the slag formation start.

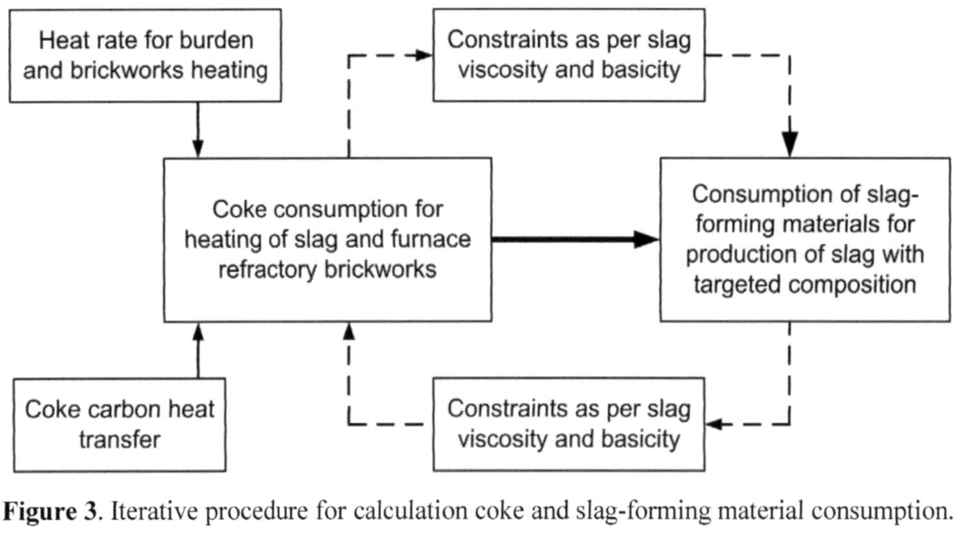

Figure 3. Iterative procedure for calculation coke and slag-forming material consumption.

The structure of the mathematical model for the "second burden" analysis is shown in Figure 4.

The same as the first burden, the second burden is divided in two parts. The first part consists of coke and slag-forming materials and the second part consists of coke, slag-forming and iron-ore materials.

In order to determine the consumption of burden components in total for the second burden, A.N. Ramm's calculation method [11] is applied. This method is used to determine the specific consumption of iron-ore materials and coke for one ton of iron as well as the average value of the ore load. Consumption of slag-forming materials and flux is calculated in the same way as for the first burden. The difference is that the presence of iron-ore materials is considered in the second part of the second burden. Arrangement of the first portions of ore materials of the second burden throughout the furnace height is of great importance. This is related to the necessity of good heating and complete reduction of iron oxides in the ore material.

In case iron-ore materials are located at a low height, it is possible that high-iron slag will go to the furnace bottom, which will result in chilling of the furnace bottom, inappropriate filtration of the melts in the hearth and problems with molten products in the course of tapping. This distance is related to the residence time of burden in the furnace. In this respect the permissible time of iron-ore material arrival to the blast tuyeres is used as a model setting and at the assigned rate of furnace blowing-in the level of ore material in the blowing-in burden is determined. The experience of blast furnace blowing-in gives evidence that the value of the theoretical flame temperature at the initial period of furnace blowing-in shall be as low as possible. This is explained by the fact that at this stage of blowing-in there are no heat consumers (ore and slag-forming components) in the furnace hearth. The permissible theoretical flame temperature is maintained by blast parameters, particularly by blast humidity.

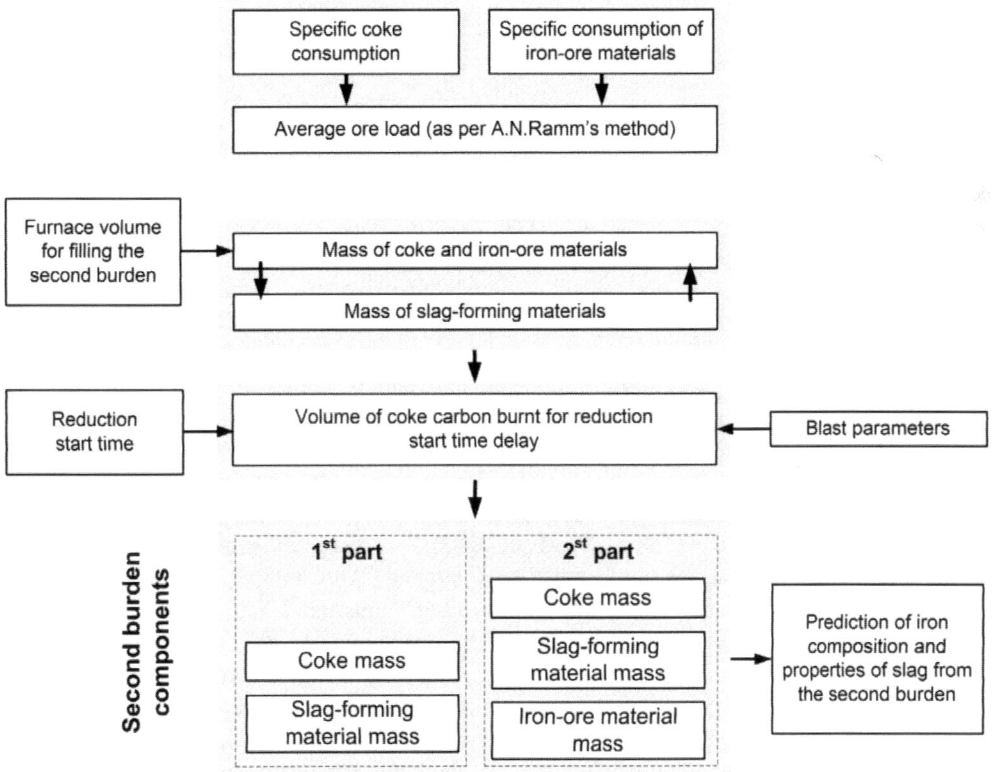

Figure 4. Structure of mathematical model for "second burden" calculation.

The predicted content of sulphur in cast iron is calculated as per sulphur balance which is made up of all sulphur-containing components of burden, slag and iron [12]. The coefficient of sulphur distribution between iron and slag LS is determined by many factors, particularly by slag temperature, activity of CaO in slag, CO partial pressure in the area of iron direct reduction and others.

Modelling of combustion zone parameters. With the small blast flow rate set at the initial stage the development of combustion zones is of the utmost importance as the small blast flow rate causes a small development of combustion zones, poor heating of the axial part of the burden column and leads to a peripheral gas flow with low utilization of the thermal and reducing capacity of the gas flow. The combustion zone performance analysis based on the use of empirical dependences makes it possible to evaluate the length of the raceway and oxidizing zones, the area of the oxidizing zones and their changes at varying smelting conditions. The experience of furnace blowing-in gives evidence that the value of the theoretical flame temperature at the initial period of furnace blowing-in shall be as low as possible – about 1850-1900°C. This is explained by the fact that during blowing-in there are no heat consumers (ore and slag-forming components) in the furnace hearth, and due to extreme heating of the furnace bottom during coke combustion hot suspensions of burden are inevitable at the tuyeres.

Modelling of gas dynamic melting conditions. It should be noted that the blowing-in burden layer is characterized by the decreased bulk density due to the large consumption of coke in burden and, therefore, burden suspension occurs at a lower differential than it happens in the normal melting conditions. The degree of gas utilization (DU) for burden balancing is taken as a criterion of burden layer stability in the furnace exposed to the gas flow. The limit differential pressure over which burden suspension in the furnace occurs is reached at the critical degree of gas utilization (DUcrit.) [12].

Optimization model of blowing-in burden. It is required to determine the burden composition and select the blast parameters on the basis of the assigned slag basicity (CaO/SiO_2). In this case the process constraints for heat, blast, gas dynamic and slag conditions shall be taken into account. These constraints are as follows:

- slag viscosity at 1400°C and 1500°C, poise;
- Al_2O_3 content in slag, %;
- slag yield, t/t;
- sulphur content in cast iron, %;
- silicon content in cast iron, %;
- theoretical flame temperature t at the tuyeres, °C;
- degree of gas utilization for burden balancing, %.

The task description for selection of the optimal burden composition, blast and gas dynamic parameters during furnace blowing-in as well as the sequence of optimization task solution are shown in Figure 5.

The characteristic feature of finding a solution for complex optimization tasks in mathematical programming is that there can be cases when the constraints put on heat, gas dynamic and slag conditions of blast furnace operation during blowing-in are contradictory, i.e. there is no region of feasibility. In this regard the solution algorithm provides a stage of task solution analysis. If there is no task solution and the conditions are contradictory, it is necessary to use the stage of correction, i.e. evaluate reasonability for the use of burden material data, constraints, reliability of the input data, etc.

The main functional capabilities of the software developed by the authors:

- maintenance of directories (blast furnace design characteristics; parameters of previous blowing-ins, properties of slag-forming materials, iron-ore materials and coke, etc.);
- calculation of blowing-in burden composition and iron / slag properties with the consumption of slag-forming materials and fluxes, blast parameters assigned by the user;
- calculation of the optimal composition of blowing-in burden and selection of blast parameters with account of the constraints for heat, slag and gas dynamic conditions and quality of the produced cast iron at any assigned combinations of input parameters;

- diagnostics and presentation of the calculation results in tables and graphics, storage in the database.

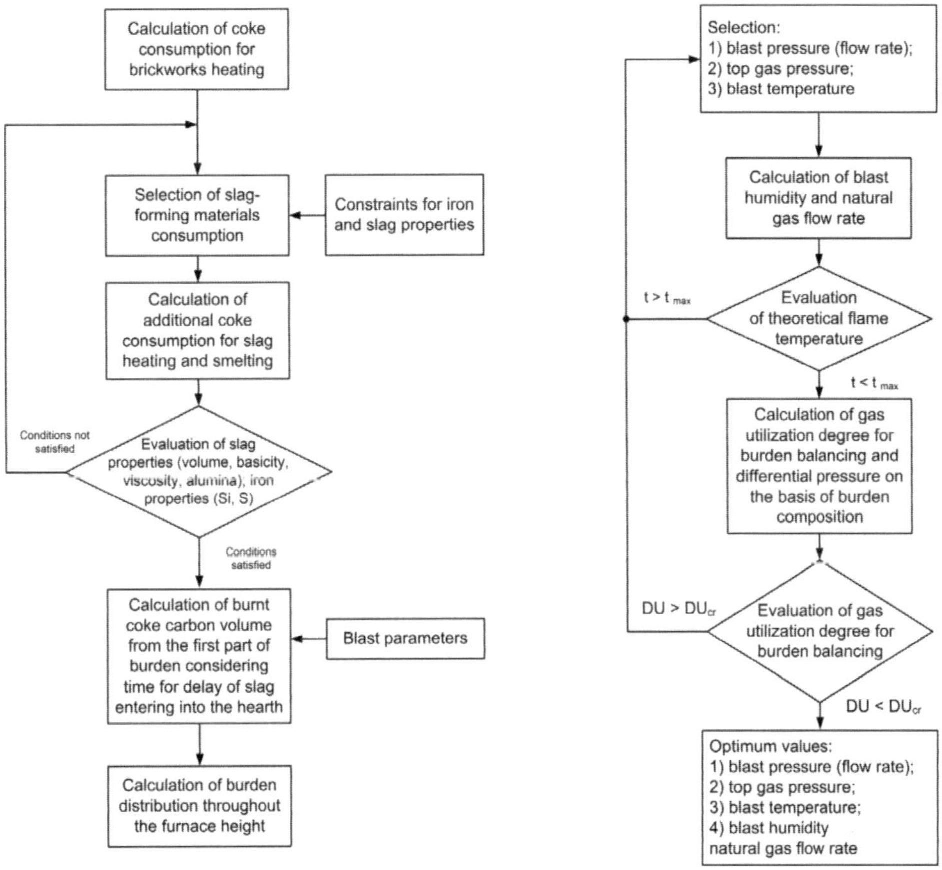

Figure 5. Task solution diagram for selection of the optimal composition of the first burden, blast and gas dynamic parameters.

The analysis of the calculation results has made it possible to set up general principles for selection of the blowing-in burden composition and blast and gas dynamics parameters:

- basicity of final slag as per CaO/SiO_2 ratio changes within the range of 0.9–1.0 depending on the silicon content in iron;
- average ore load without fluxes and slag-forming materials is 0.6–0.7 t/t;
- ore load in the iron-ore part of the blowing-in burden (the second part of the second burden) is 2.4–3.0 t/t;
- arrangement of blowing-in burden materials shall be in accordance with the following principles:
- slag-forming materials shall be charged at the height of minimum 7.0 m from the blast tuyere level (for furnaces of medium and large capacity);
- iron-ore materials shall be located at the height of minimum 11.5–12.0 m from the tuyere level (for furnaces of medium and large capacity);

- blowing-in burden is calculated for production of cast iron of medium grades with the silicon content of 1.5–3.0%;
- blast parameters shall be selected in such a way that the theoretical flame temperature at the tuyeres at the initial period of burden melting is 1850°C and the degree of gas utilization for burden balancing doesn't exceed 0.45.

2. Conclusions
The developed set of the mathematical models, algorithms and software is designed for analysis of furnace blowing-in burden, its arrangement throughout the furnace height, selection of the optimal blast and gas dynamic parameters, which can be used for blowing-in with controlled slag formation and reduction processes. The functional capabilities of the software make it possible to solve quickly the optimization tasks related to selection of the blowing-in burden composition, blast, gas dynamic and slag conditions, to study the influence of different input factors on selection of the blowing-in burden composition and blast parameters.

3. Acknowledgements
The work was supported by Act 211 Government of the Russian Federation, contract № 02.A03.21.0006.

4. References
[1] Zherebin B N 1980 *Practice of Blast Furnace Operating* (M.: Metallurgy) p 248
[2] Volkov Yu P, Shparber L Ya and Gusarov A K 1986 *Blast Furnace Process Engineer* (M.: Metallurgy) p 26
[3] Ostroukhov M Ya and Shparber L Ya 1976 *Reference Guide of Blast Furnace Foreman* (M.: Metallurgy) p 304
[4] Chernobrivets B F, Kaporulin V V and Zavidonskiy V A 1992 *Practice of Blast-Furnace Production* (M.: Metallurgy) p 111
[5] Kaporulin V V 1989 Steel **6** 17–20
[6] Ulakhovich V A, Mozharenko N M, Netronin V I, Ikkonen A K and Kotov A P 1988 *Steel* **1** 12–18
[7] Spirin N A, Gibadulin M F, Kosachenko I E et al 2006 *Steel* **3** 7–11
[8] Fielden C J and Wood B I 1968 *J. Iron and Steel Inst.* **206** 650–658
[9] Hatano M, Kurita K, Yamaoka H and Yokoi T 1982 *Tetsu-to-Hagane* **68** 2369–76
[10] Taguchi S, Kubo H, Tsuchiya N, Ichifuji K and Okabe K 1982 *Tetsu-to-Hagane* **68** 2303–10
[11] Ramm A N 1980 *Modern Blast-Furnace Process* (M.: Metallurgy) p 304
[12] Spirin N A, Lavrov V V, Rybolovlev V Yu et al 2011 *Model-Based Decision Support Systems in ICS of Blast-Furnace Smelting* (Ekaterinburg: UrFU) p 426

Mathematical model of layered metallurgical furnaces and units

V S Shvydkiy, N A Spirin and V V Lavrov

Department of Thermophysics and IT in Metallurgy, Ural Federal University, 19 Mira Street, Yekaterinburg, 620002, Russia

E-mail: vshvit@isnet.ru

Abstract. The basic approaches to mathematical modeling of the layered steel furnaces and units are considered. It is noted that the particular importance have the knowledge about the mechanisms and physical nature of processes of the charge column movement and the gas flow in the moving layer, as well as regularities of development of heat- and mass-transfer in them. The statement and mathematical description of the problem solution targeting the potential gas flow in the layered unit of an arbitrary profile are presented. On the basis of the proposed mathematical model the software implementation of information-modeling system of BF gas dynamics is carried out. The results of the computer modeling of BF non-isothermal gas dynamics with regard to the cohesion zone, gas dynamics of the combustion zone and calculation of hot-blast stoves are provided

1. Introduction

One of the main problems in the field of development of the theory of layered metallurgical processes is the maximal use of the units capacities, thermal and reducing potentials of gases and, ultimately, achievement of highest furnaces productivity and minimal fuel consumption.

From the point of view of this problem of utmost importance has the mathematical description of the physical, mechanical, thermal and chemical processes that occur in the layered metallurgical furnaces and units, creation of mathematical models of machines of various application.

It is to be recalled that the mathematical model of the object is usually understood as the simplified representation of processes in the unit, which preserves the most essential properties of the real object and transfers them in the mathematical form. In this sense we can speak about the homomorphism of the object (of the unit or its components) in its model. Depending on the task the mathematical model takes into account a different number of properties and attributes of the object, and therefore can be wide or narrow. The wide model more fully and more accurately than the narrow one reflects the properties, but, as mentioned above, it is only an idealized reflection of the real object. In a certain area of the model these model are homomorphic, i.e. may provide similar or identical results.

Mathematical model of the layered metallurgical furnace or unit must be, on the one hand, easy enough to visually and distinctly transmit all the qualitative aspects of the phenomenon we are interested in (only in this case it is possible to maintain "physical control" over the used model), and, on the other, precisely enough to pass the quantitative patterns of process flow. These requirements are in conflict, because without a thorough study of the system properties it is not clear which factors are important. The only rational compromise can be found only within a hierarchical approach, suggested by the very nature of a layered unit.

Layered metallurgical furnaces and units, being complex objects, have a multi-stage structure, and their mathematical models should be built consistently on the basis of preliminary building of components models. The research of a complex process by parts makes it possible to move to a higher level model, including into it as constituent a narrow pattern of a lower level. However, this transition will be correct unless a number of conditions are fulfilled, namely:

- the whole range of change in the dependent and independent variables, we are interested in, are included into a broader model of lower-level models, is known (including the boundary conditions);
- the narrow model in this entire range is analyzed;
- it is shown that the narrow model is adequately describes the process at a lower level.

The indicated method of constructing mathematical models of a layered unit on levels suggests that during the model building of this level all essential physical and chemical laws, which determine the properties of this level, are deeply studied and experimentally verified. In this case, the patterns acquire a predictive power of physical laws; they are invariant in space and autonomous over time. Hence follows the principle of invariance of the model components to the geometric dimensions of the system and the autonomy of these parts over time [1]. It can be formulated as follows: Regularities of processes in the component parts of the models of this level, as well as regularities of interaction between these parts can be expressed in a form that does not depend on the scale of the level in question and the point of time. It should be noted that in this definition we are talking about invariance and autonomy of the processes regularities at the lowest level to the processes at the highest level, but not about the invariance and autonomy of quantitative dependences.

For a correct transition from the lowest level to the highest one invariant relative to the scale and autonomous over time should remain not only the regularities, but also the parameters in the mathematical description of the components of the highest level and in the boundary conditions. This independence of regularities and parameters on the scale of level and time indicates their objectivity and allows the process in units of almost any type and size to be predicted since these parameters determine the value of similarity numbers of processes in units of various sizes.

The successive building of a mathematical model from the lowest to the highest level on the basis of invariance and autonomy principles of the model constituent parts opens the possibility of independent experimental determination of the unknown parameters of this model in the conditions, when, firstly, there is a greater sensitivity of the process to these parameters (and therefore ensures a high precision of the experiments), and, secondly, the influence on the process by the most part of other factors is practically excluded.

The building of the mathematical models on the basis of these principles does not exclude, but presupposes the mathematical procedure of enlargement, “compression” of the information about any part of the level, presenting it in a compact, convenient for the subsequent analysis form. In this from the mathematical model may lose its physical visualization, may appear and appear the “signs of empiricism”, and the coefficients become “effective”.

Thus, the division of a complex process in layered metallurgical furnace or unit into its constituent parts must be preceded by a thorough study of the nature of physical and chemical phenomena, which, as a result, further leads to the work not only with mathematical expressions, but with independent from the scale and time physical concepts. Reliability and accuracy of the mathematical models and the results obtained on their basis depend on the level of development of these ideas. The knowledge about the mechanisms and physical nature of the charge column movement and the gas flow in the moving layer, as well as the patterns of development of heat- and mass-transfer in them, has the particular importance when it is applied to the layered metallurgical furnaces and units.

A detailed overview of the current state of the theory of thermal processes in the metallurgical units with a dense layer with an extensive bibliography is presented in monographs [2-4]. Therefore, here we formulate only the main conclusions resulting from the indicated review and largely determining the content of this paper.

1. Despite the abundant experimental data and physical substantiation of the mechanism of granular media flow through the hole, up to now there is no sufficient logical and reliable mathematical model describing the material movement in the shaft furnaces.

2. The experimental studies of the layer gas mechanics continue to develop rapidly. New methods of measurement of velocities and gas pressure in the layer are offered, which allowed the processes of gas motion in real furnaces to be studied, and numerous experiments, carried out on the ovens and models, made it possible to clarify the dependence of pressure drop on gas velocity and to identify the influence of various parameters of shaft furnaces on gas distribution in them, and also offer a variety of methods for gas flow control. However, the experimental methods are characterized by the well-known disadvantages (great labour intensity, complexity of the conditions existing in the shaft furnaces and complicating the research, difficulties in building of a large variety of physical models of furnaces, etc.) that make use the calculation and theoretical investigations of the layer gas mechanics.

3. Currently, the calculation and theoretical investigations of gas movement in the in general and in the blast furnaces in particular are developing very intensively; in the world several metallurgical centers are engaged in the mathematical description of gas mechanics of the lump materials layer. The model of the gases movement in the layer, offered by the research team of the Department of Thermophysics and IT in Metallurgy, Ural Federal University, Russia (see the indicated above references) are characterized by its completeness.

4. The theory of heat-transfer in the layer achieved a great success. Today it can be stated [2–4, 5], that the one-dimensional problem of heat-transfer in the layer is almost exhausted. At the same time it should be taken into account that in real layered metallurgical furnaces and units the gases movement is usually multidimensional. The uneven distribution of gas velocity in the layer volume and along the cross-section of units is quite substantial.

The problem of the development of complex mathematical models of layered metallurgical furnaces and units consists in derivation of the gas flow equation for moving layer, connecting the average velocity in the inter-lump area (true speed), its actual pressure and temperature, as these values influence the intensity of flow of physical and chemical transformations. At the same time the laws of conservation of mass, quantity of motion and energy should be fulfilled. Apparently, the solution to this problem should be based on the achievements in the modern theory of continuous media [6] using the parallels between the gas flow in the layer and the movement of a hypothetical fluid, which occupies the entire volume of the unit, including the volume of lump materials.

The consistent satisfaction of the requirements of dynamic properties equality of this liquid to the local averaged parameters of real heterogeneous medium (aggregate of lump materials and gases) makes it possible to reduce a complicated problem of gases movement in the layer of variable structure to a relatively simple problem of the homogeneous liquid movement, overcoming, however, some additional resistance.

2. Averaged continuity equations of motion and gas energy in the layer

A detailed derivation of all equations of layer gas mechanics is given in [2-4], so here we shall only illustrate the realization of the gas flow analogy in the real layer and a hypothetical homogeneous liquid on the example of continuity equation. The remaining equations we shall provide without conclusion.

So, let the shaft furnace of an arbitrary profile be filled with lump material of porosity ε and cross-sectional pore aperture ε_p. In general ε and ε are the functions of the form of the charged lumps and coordinates. In the bottom the gas is injected through the tuyere or otherwise. The medium thermal properties are the functions of pressure, temperature, and coordinates. The gas gives the material its heat and can chemically interact with it. It is required to find the distribution of velocities, pressures, temperatures and concentrations of gas components along the height and cross-section of the furnace.

Let us distinguish in the layer an elementary volume V (m^3). In this volume the gas occupies V' (m^3). If its density ρ (kg/m^3), the mass of the gas in the indicated volume will be $\rho V'$ (kg).

Let us consider the mass balance in the element of volume V. Change in the gas mass in this amount equals to

$$\frac{d}{d\tau}\int_{V'}\rho dV' = \int_V \frac{\partial}{\partial\tau}\left(\rho\frac{dV'}{dV}\right)dV = \int_V \frac{\partial}{\partial\tau}(\rho\varepsilon)dV, \tag{1}$$

where ε– average porosity of layer in the volume V:

$$\varepsilon = \frac{dV'}{dV} = \frac{1}{dV}\int_V \varepsilon_л dV, \tag{2}$$

where ε_l– local value of porosity.

Mass change equals to its flow through the closed surface S' (contour) of the volume V', and the full flow of mass in this case is

$$\oint_{S'}(\rho\vec{v})d\vec{S}' = \oint_S(\rho\vec{v})\left(\frac{dS'}{dS}\right)d\vec{S} = \oint_S(\rho\varepsilon_n\vec{v})d\vec{S} = \oint_S(\rho\varepsilon_n\vec{v})\cdot\vec{n}dS. \tag{3}$$

Here ε_p– average cross-sectional pore aperture of the layer in the volume V, and n– normal to a surface of contour S'.

Taking into account the effect of inner sources (sinks) of mass with capacity q_V (the mass formation due to the physical and chemical transformations), after conversion of a surface integral into the volumetric one according to the divergence theorem [7-9], we obtain

$$\int_V\left[\frac{\partial}{\partial\tau}(\rho\varepsilon) + \nabla\cdot(\rho\varepsilon_n\vec{v}) - q_V\right]dV = 0, \tag{4}$$

from it because of the arbitrariness of volume V we find

$$\frac{\partial}{\partial\tau}(\rho\varepsilon) + \nabla\cdot(\rho\varepsilon_n\vec{v}) = q_V. \tag{5}$$

Comparing equation (5) with continuity of hypothetical continuous flow [10-12]

$$\frac{\partial\rho_*}{\partial\tau} + \nabla\cdot(\rho_*\vec{v}_*) = q_V, \tag{6}$$

we find, that relations (5) and (6) become equivalent, if the equalities are satisfied

$$\rho_* = \varepsilon\rho; \quad \mathbf{v}_* = (\varepsilon_p/\varepsilon)\,\mathbf{v}. \tag{7}$$

Thus, as it is shown in [13, 14], the physical meaning of equations content (7) is as follows. To describe the processes occurring in the layer, it possible to use the equations derived for a continuous medium, if we know the effective values of its thermal parameters. It is obvious that these values are not equal to the values of real gas parameters. We should emphasize the fact that any mathematical model of gas motion in the layer describes the patterns of change in average characteristics of the flow field.

During derivation of motion equations Cauchy stress principle is usually taken as a postulate of [5, 6, 10], stating that for any closed surface Ω there is a distribution of voltages vector **t** with the resultant and the moment equivalent to the forces field influencing on the continuous medium, inside Ω, from the medium side located outside this surface. It is assumed that at a given moment of time vector **t** depends only on the position and orientation of the surface element *dS*, i.e. if we denote by **n** the outer normal to the surface Ω, then $\mathbf{t} = t(\mathbf{x},\tau,\mathbf{n})$.

Let us now recall the basic dynamics principle of fluid motion, called the principle of momentum conservation [5, 6, 10]: rate of change in fluid movement quantity, enclosed in the moving volume Σ equals to the resultant of the forces acting on the fluid.

The analytical expression of this principle is the equation

$$\frac{d}{d\tau}\int_{\Sigma}\rho_*\vec{v}_* dV = \int_{\Sigma}\rho_*\vec{f}\,dV + \int_{\Omega}\vec{t}_* dS, \qquad (8)$$

where **f**– field of external forces attributed to the unit mass.

Averaging this equation for the volume of *V*, as we did it for the mass balance equation (for details see [4]), we arrive at an intermediate form of the motion equations of

$$\varepsilon\rho\frac{d}{d\tau}(\varepsilon_n\vec{v}/\varepsilon) = \varepsilon\rho\vec{g} - grad(\varepsilon_n p) - \frac{2}{3}grad\left[\varepsilon_n\mu\, div(\varepsilon_n\vec{v}/\varepsilon)\right] + $$
$$+ 2\cdot \mathrm{div}(\varepsilon_p\mu\mathbf{D}) + \mathbf{R}. \qquad (9)$$

Here *p*– gas pressure, μ – coefficient of dynamic viscosity of gas, **R** – additional resistance that occurs during the transition from real gas to homogeneous fluid, and **D** – strain velocity tensor with components

$$D_{ij} = 0.5(\partial v_i/\partial x_j + \partial v_j/\partial x_i), \qquad (10)$$

where $i,j = 1, 2, 3$; $x_1 = x, x_2 = y, x_3 = z$; $v_1 = u$, $v_2 = v$, $v_3 = w$. Note, that in equation (9) $d/d\tau$ characterizes the so-called *full* or *substantial* derivative, i.e. $d/d\tau = \partial/\partial\tau + u(\partial/\partial x) + v(\partial/\partial y) + w(\partial/\partial z)$.

The last summand of the right-hand side (9), which represents additional resistance to the gas movement through the layer, at the present level of knowledge can not be expressed in quadratures. The only way to determine the value of R is an experiment. It is clear, however, that the resistance **R** should be a unique function of layer porosity, i.e. $\mathbf{R} = \mathbf{R}(1-\varepsilon)$, since in the absence of solid particles ($\varepsilon = 1$) the value $\mathbf{R} \equiv 0$. In [2-4, 15-17] various approximations of the results of experimental determination of R are discussed. In [4] it is stated that the most reliable dependence for operation of shaft metallurgical furnaces is the following:

$$\mathbf{R} = -2\,\mathrm{div}(\varepsilon_p\mu\mathbf{D}) + (2/3)\mathrm{grad}[\varepsilon_p\mu\mathrm{div}(\varepsilon_p\mathbf{v}/\varepsilon)] - 150\left(\frac{1-\varepsilon}{\varepsilon d_\text{к}\Phi}\right)^2\mu\frac{\varepsilon_p^2}{\varepsilon}\mathbf{v} -$$
$$- 1.75\left(\frac{1-\varepsilon}{\varepsilon d_\text{к}\Phi}\right)\rho\frac{\varepsilon_p^2}{\varepsilon}\left(\frac{\varepsilon_p}{\varepsilon}|\mathbf{v}|\right)\mathbf{v} \qquad (11)$$

Here Φ is the particle shape factor of the material (1 for a sphere); $d_к$– particle diameter.

The last two summands of relation (11) for **R** define the so-called law of resistance of S. Ergan.

Using (11) the equations of motion (9) are simplified and have the from

$$\varepsilon\rho\frac{d}{d\tau}\left(\frac{\varepsilon_{\mathrm{p}}}{\varepsilon}\mathbf{v}\right)=\varepsilon\rho\mathbf{g}-\mathrm{grad}(\varepsilon_{\mathrm{p}}p)-\left[150\left(\frac{1-\varepsilon}{\varepsilon d_{\kappa}\Phi}\right)\mu+1.75\rho\frac{\varepsilon_{\mathrm{p}}}{\varepsilon}|\mathbf{v}|\right]\frac{1-\varepsilon}{\varepsilon d_{\kappa}\Phi}\frac{\varepsilon_{\mathrm{p}}^{2}}{\varepsilon}\mathbf{v}. \quad (12)$$

Instead of S. Ergan equation any experimental dependence can be used [2-4, 15-17].

The energy equation is derived by the similar method described in [4]. In this case, two circumstances must also be taken into account. Firstly, the heat-transfer by thermal conductivity is carried out not only in the gas, but from the gas to the surface of material particles. As the evaluation of the last component of heat-transfer requires data on identification of patterns of development of thermal and dynamic boundary layers, it is advisable to replace this summand by Newton's law of heat-transfer by convection. Secondly, the presence of additional force of resistance in the equations of motion leads to the appearance in the energy equation of a summand, characterizing the transition of mechanical energy into thermal energy. Omitting the intermediate calculations (for details see [2-4]), we shall write the final result

$$\varepsilon\rho C_v\frac{dT}{d\tau}=\varepsilon\rho q_R+\mathrm{div}(\varepsilon_p\lambda\,\mathrm{grad}T)-\alpha_v(T-t)-\varepsilon_p p\,\mathrm{div}(\varepsilon_p\mathbf{v}/\varepsilon)-$$
$$-(2/3)\,\varepsilon_p\mu[\mathrm{div}(\varepsilon_p\mathbf{v}/\varepsilon)]^2+2\varepsilon_p\mu\sum_{i,j}D_{i,j}^2+q_g', \quad (13)$$

where C_v – is gas heat capacity (kJ/(kg·K)), λ – coefficient of gas thermal conductivity (W/(m·K)), α_v – volumetric heat-transfer coefficient (W/(m^3·K)), q_R – velocity of heat inflow due to radiation (W/kg), q_g' –source (sink) of heat acting in the gas (W/m^3). This summand, as noted before, also includes the dissipation of mechanical energy occurring on the surface of material particles.

The first three summands of the equation right-hand member (13) characterizes three types of heat-transfer: radiation, conductivity and convection. For the operation conditions of metallurgical shaft furnaces the thermal conductivity of gases can be ignored [2]. Typically the gas thermal conductivity is taken equal to zero in the direction of its motion and infinity – across the flow. Since the effective beam length in the inter-lump space is insignificant, and the weakening capacity of dust-laden gas flows is great, the radiative transfer cannot also be taken into account. This assumption is especially justified by the fact that the heat-transfer coefficient α_v is usually determined experimentally. For this reason, the radiant component, which is difficult to calculate, is included into α_v through the dependence of the latter on temperature. However, it should be noted that this simplification does not lead to noticeable errors only for areas of the layer in which the coolant temperature does not exceed 500 °C [9].

The following three summands of the equation right-hand member (13) described the work of the pressure forces and the dissipation of mechanical energy (gas dynamic heat). These values become comparable with the convective heat- transfer only at the sub- and supersonic gas velocities. The shaft metallurgical furnaces do not have such velocities, then without any losses for the of calculations accuracy these summands can be neglected.

Thus, for the conditions of heat and gas dynamic operation of shaft metallurgical furnaces the energy equation transforms into the equation of heat- transfer

$$\varepsilon\rho C_{\mathrm{v}}\frac{dT}{d\tau}=(\varepsilon\rho q_R)-\alpha_{\mathrm{v}}(T-t)+q_{\mathrm{g}}, \quad (14)$$

where q_g sources (sink) of heat conditioned by the physical and chemical reactions proceeding in the gas. The first summand in the right-hand member is placed in brackets in order to emphasize the need for use of numerical methods of calculation when taking into account the radiation heat-transfer.

For system closure it is necessary to add the equation:

– gas condition

$$\rho = f(T, p); \tag{15}$$

– heating of material lumps

$$(1-\varepsilon)\rho_m C_m \frac{dT}{d\tau} = \alpha_v(T-t) + q_m; \tag{16}$$

– regularities of change in gas viscosity and het-conductivity

$$\mu = \mu(T, p); \quad \lambda = \lambda(T, p), \tag{17}$$

–and mass-transfer

$$\frac{dC}{d\tau} = -K_\Sigma f[C(1+\overline{K}) - \overline{K}\cdot\Sigma] \tag{18}$$

and

$$\frac{d\varphi}{d\tau} = n\frac{dC}{d\tau}. \tag{19}$$

In these expressions ρ_m – material density (kg/m^3), C_m – specific heat capacity of the material (kJ/(kg·K)), q_m – sources (sinks) of heat acting in the material (W/m^3); C –concentration of a reductant in the gas (m^3/m^3), $\overline{K}$ – equilibrium constant, K_Σ – effective constant of reaction rate (m/s), f – value of the reaction surface (m^2/m^3), Σ – equilibrium concentration, φ – degree of oxide reduction (fractions of a unit), n – ratio of gas flow mass-capacity W_g^* to material flow W_m^*.

Due to the substantial nonlinearity of equations (5), (12), (14) – (19) their joint solution can be done only numerically. For this reason, to obtain a sufficiently reliable result it is advisable to use the iterative method. First, for the arbitrarily given initial temperature field, the problem of layer gas mechanics is solved, then the heat- and mass-transfer equations. The iterations are carried out up the specified accuracy of calculations of temperature fields in the layer. From a practical point of view it is better not to use the equation of motion (12), but more convenient simplified expressions that take into account peculiarities of gas movement in the layer for a specific unit.

3. Possible simplifications of a gas mechanics model of the moving layer

Potential flow. One of the possible simplifications of the model is the assumption about the potential (non-vortex) nature of the gas flow. The mathematical condition of the vorticity absence is the ratio of rot $\mathbf{v} = 0$. From the field theory [7, 8] it is known that if this condition is satisfied, then there is a scalar function $\varphi(\mathbf{x})$, called the *potential function*, gradient of which equals to the velocity $\mathbf{v}$, i.e.

$$\mathbf{v} = \text{grad}\varphi. \tag{20}$$

The minus mark means that the flow moves from a larger to a smaller potential. Inserting the expression (30) into the equation (5) we obtain the equation for finding the potential function φ:

$$\mathrm{div}(\varepsilon_p \rho\, \mathrm{grad}\varphi) = 0. \quad (21)$$

The components of gas velocity are calculated according the known field φ using relation (20). Knowing the velocities values on the basis of equations the motion of gas pressure can be determined. For non-vortex flows equation (12) takes the form:

$$-\mathrm{grad}(\varepsilon_p p) = \rho\varepsilon_p\, \mathrm{grad}(q^2/2) - \rho\varepsilon \mathbf{v}[\mathbf{v}\, \mathrm{grad}(\varepsilon_p/\varepsilon)] + (A_1 + A_2 q)\,\mathbf{v} - \rho\varepsilon \mathbf{g}. \quad (22)$$

Here

$$A_1 = 150.0\left[\frac{(1-\varepsilon)\varepsilon_p}{\varepsilon d_э \Phi}\right]^2 \frac{1}{\varepsilon}\mu; \quad A_2 = 1.75\frac{(1-\varepsilon)\varepsilon_p^2}{\varepsilon^2 d_э \Phi}\rho; \quad q = |\mathbf{v}|.$$

$d_э$ – the equivalent diameter of the particle.

The complexity of calculations according to equations (20) – (22) lies in the fact that in order to calculate the potential φ it is required to know the density ρ, which depends on the pressure, cross-sectional pore aperture and porosity of the layer. The gas pressure can be determined by the velocity fields **v** and density ρ. The natural solution to this situation is the use of iterations.

In the case of a plane or axial-symmetric (two-dimensional) flow it is convenient to use the current function ψ, inserted by the relations:

$$\rho\varepsilon_p u = \partial\psi/\partial y; \quad \rho\varepsilon_p v = -\partial\psi/\partial x \quad (23)$$

and representing the mass consumption of gas.

With the introduction of this function, the continuity equation is satisfied identically, and the condition of the vorticity absence $\partial u/\partial y - \partial v/\partial x = 0$ leads to the equation for determining ψ

$$\frac{\partial}{\partial x}\left(\frac{1}{\rho\varepsilon_p}\frac{\partial\psi}{\partial x}\right) + \frac{\partial}{\partial y}\left(\frac{1}{\rho\varepsilon_p}\frac{\partial\psi}{\partial y}\right) = 0. \quad (24)$$

The boundary conditions of equations (21), (22) and (24) are defined by the profile of the unit and geometry of devices inputting and outputting the gas.

Illustration for using the above described equations of non-vortex flows, comparison of the results of their solution with the data obtained on the basis of the available in the literature of mathematical models of layer gas mechanics, as well as verification of the adequacy of the mathematical model of the potential motion to the real picture of gas flow are described in [2-4, 14-17] and in the works from an extensive bibliography in [3].

It is shown that in terms of uniformity of gas distribution the mathematical model of the potential gas flow in the layer can be effectively used for the selection of optimal geometry parameters of the tuyere device and the blast furnace (furnace profiles). As for the absolute values of velocities and pressures, the exact values of which are necessary for the calculation of heat- and mass- transfers occurring in the layered metallurgical furnaces and units, then the accuracy of their determination characteristic for the potential model should be specifically identified by comparing the solutions of the problems of the vortex and potential movements.

Vortex flows. Basically, the calculation of vortex flows is possible on the basis of equations (5) and (12) if we introduce the vector potential of velocity ψ by the relation $\rho\varepsilon_p\mathbf{v}$=rot ψ (details of this approach see in [3] pp. 18–19). Practice of calculations shows, however, that this method is ineffective, because the numerical schemes of finite differences converge very slowly. Therefore, it is appropriate to transform equations (5) and (12), so that they would directly include vorticity of the flow ω = rot **v**.

In studying the gases movement through the layer of lump materials we have to deal with two manifestations of the vortex nature of a vortex flow:

- *micro-vorticity*, i.e. with vortices, the scale of which is commensurable with the average sizes of inter-lump space (diameter of a layer particle); the appearance of these vortices is entirely due to gas friction on the surface of particles;
- *macro-vorticity*, i.e. with vortices, the scale of which is commensurable with the furnace working volume; these vortices are conditioned by the geometry of the unit workspace (furnace profile), the presence of combustion zones (circulation zones), etc.

First of all, we shall note that, according to Lagrange's theorem [10] the vortices can not be formed within the area of movement: they are either introduced into it, or formed on its boundaries. The vortices formation on the boundary of flow is due to the sticking phenomenon, i.e., the presence of tangential stresses. On the other hand, setting the condition of sticking as a boundary one for solving the motion equations is possible, if only the latter has the second order, whereas equation (12) is a differential equation of the 1st order.

Thus, the introduction of vortices into the area of movement is due to the vortex nature of the gas flow through the tuyeres. As the gas inflows into the developed area of circulation, the vorticity of flow can be enhanced. However, if the gas from the tuyere outflows into the dense layer, the initial vortices are likely to by quickly dampened.

However, this situation requires a special consideration, since in this case the interest is arisen, basically, by the principle problem of the possibility of ignoring the vortex nature of the gas flow during the analysis of gas distribution. In this case it is advisable to simplify the equations of gas motion in the layer (5) and (12) to the model of two-dimensional motion of the incompressible gas in the layer of constant porosity. Note, that these simplifications (incompressible gas and the layer of constant porosity) are not fundamental, since using the iterative method for solution of determinative equations, it is not difficult to take into account the changes in these characteristics.

For generalization of the results let us introduce the dimensionless coordinates

$$\tilde{x} = x/R, \quad \tilde{z} = z/R \tag{25}$$

and velocities

$$\tilde{v}_x = v_x/v_0; \quad \tilde{v}_z = v_z/v_0, \tag{26}$$

where R – furnace radius; $v_0 = Q_v/R$ – average consumption rate (velocity in the empty shaft); Q_V – blast consumption. Here the coordinate along the furnace height we denote by z, to enable solution of three-dimensional problems. Taking into consideration that for the two-dimensional problem of flow vorticity ω = rot **v** is a scalar $\tilde{\omega} = \partial\tilde{v}_x/\partial\tilde{z} - \partial\tilde{v}_z/\partial\tilde{x}$, from (5) we obtain the first equation of the model:

$$\frac{\partial^2\tilde{\psi}}{\partial\tilde{x}^2} + \frac{\partial^2\tilde{\psi}}{\partial\tilde{z}^2} = \tilde{\omega}, \tag{27}$$

where $\tilde{\omega} = \omega R / v_0 = \omega R^2 / Q_V$.

Equation (27) contains two unknown functions, therefore, to solve the problem one more relation is required, namely, the equation of motion (12), which, after simple transformations can be written as Lamb-Gromek equation [10], supplemented by resistance forces

$$\frac{\partial}{\partial \tilde{x}}\left(\tilde{\omega}\frac{\partial \tilde{\psi}}{\partial \tilde{z}}\right) - \frac{\partial}{\partial \tilde{z}}\left(\tilde{\omega}\frac{\partial \tilde{\psi}}{\partial \tilde{x}}\right) + M(\tilde{q})\tilde{\omega} = -1.75\left(\frac{\partial \tilde{\psi}}{\partial \tilde{x}}\cdot\frac{\partial \tilde{q}}{\partial \tilde{x}} + \frac{\partial \tilde{\psi}}{\partial \tilde{z}}\cdot\frac{\partial \tilde{q}}{\partial \tilde{z}}\right), \tag{28}$$

where $M(q) = A_1 + A_2 q$.

The system of differential equations (27), (28) describes a vortex motion of the gas in the layered unit of any profile. Therefore, we specify the task and choose the furnace profile shown in Figure 1 (a typical two-dimensional profile of a shaft furnace).

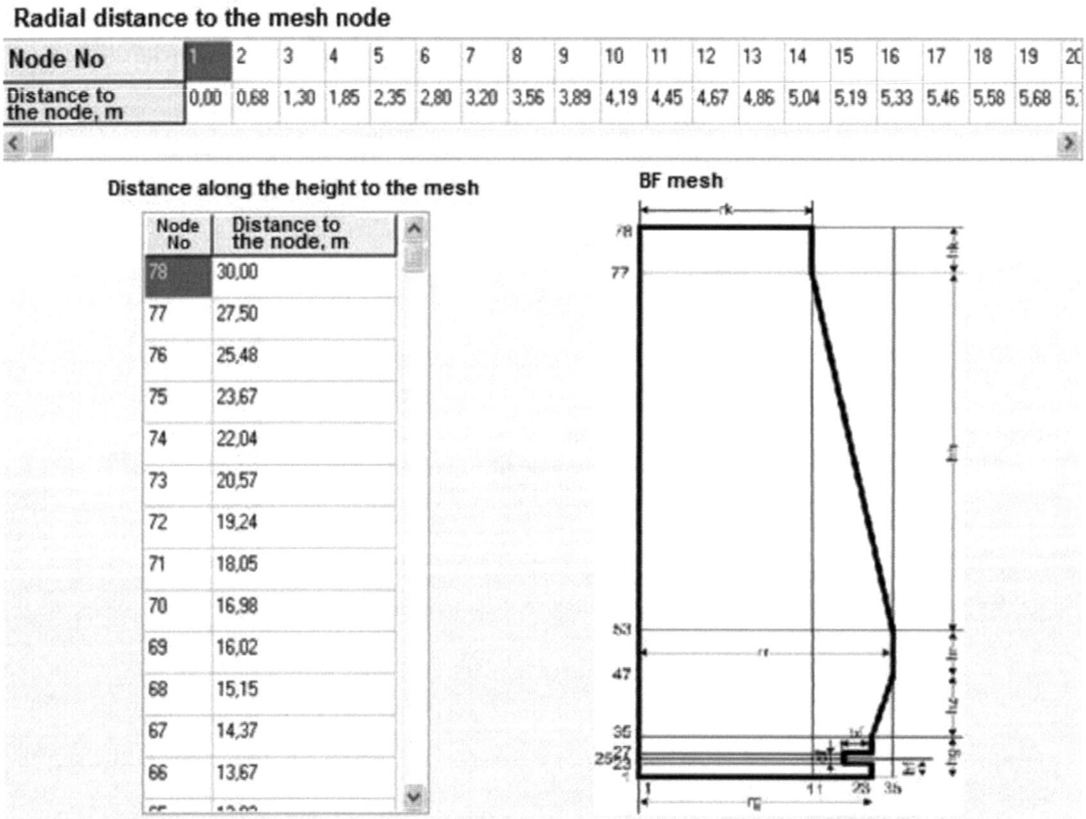

Radial distance to the mesh node

Node No	1	2	3	4	5	6	7	8	9	10	11	12	13	14	15	16	17	18	19	2[illegible]
Distance to the node, m	0,00	0,68	1,30	1,85	2,35	2,80	3,20	3,56	3,89	4,19	4,45	4,67	4,86	5,04	5,19	5,33	5,46	5,58	5,68	5,[illegible]

Distance along the height to the mesh

Node No	Distance to the node, m
78	30,00
77	27,50
76	25,48
75	23,67
74	22,04
73	20,57
72	19,24
71	18,05
70	16,98
69	16,02
68	15,15
67	14,37
66	13,67

Figure 1. Shaft furnace profile and the nodes coordinates of the finite-difference mesh.

For scheme in Figure 1 the boundary conditions for $\tilde{\psi}$ will be the following:

$$\tilde{\psi} = 0 \quad if \quad \begin{cases} \tilde{x} = 0, & 0 \le \tilde{z} \le H; \\ 0 \le \tilde{x} \le \tilde{x}_{\max}, & \tilde{z} = 0; \\ \tilde{x} = \tilde{x}_{\max}, & 0 \le \tilde{z} \le \tilde{L} - \tilde{r}_t; \end{cases} \tag{29}$$

$$\tilde{\psi} = 1 \quad if \quad \tilde{x} = \tilde{x}_{\max}, \quad \tilde{L} + \tilde{r}_t \le \tilde{z} \le H; \tag{30}$$

Here the index t means the lance blast furnace.

$$\frac{\partial \tilde{\psi}}{\partial \tilde{z}} = 0 \; if \quad 0 \le \tilde{x} \le \tilde{x}_{\max}, \quad \tilde{z} = H. \tag{31}$$

The value $\tilde{\psi}$ on the cross-section of the tuyere, i.e. if $\tilde{x} = \tilde{x}_{\max}$, $\tilde{L} - \tilde{r}_t \le \tilde{z} \le L + \tilde{r}_t$ is determined by the mode of blast motion in the tuyere. For laminar mode the velocity profile is described by a quadratic parabola, and the ratio of the average velocity to the maximal equals to 0.5. For the conditions Figure 1

$$\tilde{v}_x = -\frac{3}{4}\tilde{r}_t^3\left[\tilde{L}^2 - \tilde{r}_t^2 - \tilde{z}(2\tilde{L} - \tilde{z})\right].$$

Since $\tilde{v}_x = \partial\tilde{\psi} / \partial\tilde{z}$, then taking into account the fact that the below the tuyere $\tilde{\psi}$ = 0, after integration at $\tilde{x} = \tilde{x}_{\max}$, $\tilde{L} - \tilde{r}_t \le \tilde{z} \le L + \tilde{r}_t$ we find

$$\tilde{\psi} = 0.75\tilde{r}_t^3\left[(\tilde{L} + 2\tilde{r}_t)(\tilde{L} - \tilde{r}_t)^2 - (\tilde{L}^2 - \tilde{r}_t^2)\tilde{z} + \tilde{L}\tilde{z}^2 - \frac{1}{3}\tilde{z}^3\right],$$

i.e. the $\tilde{\psi}$ profile on the tuyere is a cubic parabola.

In case of turbulent gas flow in the tube the following equation best agrees with the experimental data throughout the turbulent core of the flow up to the pipe axis [21]:

$$\tilde{v}_x = 0.2\frac{\tilde{d}_t}{\mathrm{Re}^{1/8}}\left\{5.5 + 2.5\ln\left[0.2\mathrm{Re}^{7/8}\frac{\tilde{z} - \tilde{L} + \tilde{d}_t/2}{\tilde{d}_t} \times \frac{1.5[1 + (\tilde{L} - \tilde{z})/\tilde{r}_t]}{1 + 2[(\tilde{L} - \tilde{z})/\tilde{r}_t]}\right]\right\}. \tag{32}$$

Here $\tilde{d}_t = 2\tilde{r}_t$. Then if $\tilde{x} = \tilde{x}_{\max}$, $\tilde{L} - \tilde{r}_t \le \tilde{z} \le L + \tilde{r}_t$ we get

$$\tilde{\psi} = \int_{\tilde{L}-\tilde{r}_t}^{\tilde{L}+\tilde{r}_t} \tilde{v}_x(\tilde{x}_{\max}, \tilde{z})\, d\tilde{z}.$$

Within the laminar sublayer the relations are valid:

- if $\tilde{z} < \tilde{L} - \tilde{r}_t + [\tilde{d}_t/(0.04\mathrm{Re}^{7/8})]$ $\tilde{v}_x = 0.04\mathrm{Re}^{3/4}(\tilde{z} - \tilde{L} + \tilde{r}_t)/\tilde{d}_t^2$;

- if $\tilde{z} > \tilde{L} - \tilde{r}_t - [\tilde{d}_t /(0.04\mathrm{Re}^{7/8})]$ $\tilde{v}_x = 0.04\mathrm{Re}^{3/4}(\tilde{L} + \tilde{r}_t - \tilde{z})/\tilde{d}_t^2$.

Let us move to the boundary conditions on vorticityω. If $\tilde{x} = \tilde{x}_{max}$ and $\tilde{L} - \tilde{r}_t \leq \tilde{z} \leq L + \tilde{r}_t$, i.e. at the tuyere cross-section $\tilde{v}_z = \partial\tilde{v}_z / \partial\tilde{x} = 0$ and $\tilde{\omega} = \partial\tilde{v}_x / \partial\tilde{z}$. For laminar mode $\tilde{\omega} = 1.25(\tilde{L} - \tilde{z})/r_t^3$.

In case of turbulence mode at the lower tuyerestage $\tilde{\omega} = 0{,}04\mathrm{Re}^{3/4}/\tilde{d}_t^2$, and at the upper $\tilde{\omega} = -\ 0{,}04\mathrm{Re}^{3/4}/\tilde{d}_t^2$. For the flow core the vorticity is determined by the equation differentiation (32) $\tilde{z}$:

$$\tilde{\omega} = \partial\tilde{v}_x / \partial\tilde{z} = 2.24(\tilde{L} - \tilde{z})\tilde{r}_t / \{\mathrm{Re}^{1/8}[\tilde{r}_t^2 - (\tilde{L} - \tilde{z})^2][\tilde{r}_t^2 + \\ + 2(\tilde{L} - \tilde{z})^2]\}.$$

On the furnace axis if $\tilde{x} = 0$ and $0 \leq \tilde{z} \leq \tilde{H}$ $\tilde{v}_x = \partial\tilde{v}_x / \partial\tilde{z} = 0$, i.e. $\tilde{\omega} = -\partial\tilde{v}_z / \partial\tilde{x}$. However, we consider a symmetric problem, and in this case the profile of vertical component of velocity is symmetrical. But then $\partial\tilde{v}_z / \partial\tilde{x} = 0$ and $\tilde{\omega}(0, \tilde{z}) = 0$.

Expression (28) with regard to the continuity equation $\partial\tilde{v}_x / \partial\tilde{x} + \partial\tilde{v}_z / \partial\tilde{z} = 0$ can be written as:

$$\tilde{v}_x \frac{\partial\tilde{\omega}}{\partial\tilde{x}} + \tilde{v}_z \frac{\partial\tilde{\omega}}{\partial\tilde{z}} + (A_1 + A_2\tilde{q})\tilde{\omega} = A_2\left(\tilde{v}_z \frac{\partial\tilde{q}}{\partial\tilde{x}} - \tilde{v}_x \frac{\partial\tilde{q}}{\partial\tilde{z}}\right).$$

On the furnace wall if $\tilde{x} = \tilde{x}_{max}$ and $0 \leq \tilde{z} \leq \tilde{L} - \tilde{r}_t$ or $\tilde{L} + \tilde{r}_t \leq \tilde{z} \leq \tilde{H}$ $\tilde{v}_x = 0$ and $\tilde{q} = |\tilde{v}_z|$. Thus,

$$\tilde{v}_z \frac{\partial\tilde{\omega}}{\partial\tilde{z}} + (A_1 + A_2|\tilde{v}_z|)\tilde{\omega} = A_2\tilde{v}_z \frac{\partial|\tilde{v}_z|}{\partial\tilde{x}}. \quad (33)$$

With values $\tilde{\omega}$ on the tuyere cross-section, solving equation (33), we can determine the vorticity along the furnace wall upward from the tuyere to the charge level and downward from the tuyere to the bottom. At the bottom if $\tilde{z} = 0$ and $0 \leq \tilde{x} \leq \tilde{x}_{max}$ $\tilde{v}_z = 0$ and $\tilde{q} = |\tilde{v}_x|$. Then

$$\tilde{v}_x \frac{\partial\tilde{\omega}}{\partial\tilde{x}} + (A_1 + A_2|\tilde{v}_x|)\tilde{\omega} = -A_2\tilde{v}_x \frac{\partial|\tilde{v}_x|}{\partial\tilde{z}}. \quad (34)$$

At the charge level the relation (34) is also valid.

The finite-difference approximation of the above stated equations in the variables "vorticity - flow function" is a numerical model of the gas vortex flow in the layer of the lump material. Computing practice has shown that it is more rational to solve the equations of this model by the method of successive upper relaxation (SUR) [2-4, 11, 18-24].

4. The example of solution to the equations of a model of vortex problem for the blast furnace

For the operational conditions of blast furnace with height (along the layer) 30 m and a belly diameter 6.55 m the information-modeling system of gas dynamics of the BF process was developed. The system included a solution to the problem of non-isothermal vortex flow in the blast furnace at a given porosity field taking into account the cohesion area, the problem of gas dynamics of the combustion zone and a program for calculating the hot blast stove. The software implementation of the system was carried out in Delphi 7 and Compaq Visual Fortran. Unfortunately, the second software package does not function in Windows 7; therefore, it is impossible to illustrate the operation of the system in it. However, the application contains all program modules of the system in Compaq Visual Fortran, since, for example, in Windows XP this package functions perfectly.

In Delphi the system is started by command **Gazodin_DP**.exe, which opens the initial form (Figure 2). When choosing a problem to be solved, for example, "Calculation of non-isothermal problem of gas dynamics with regard to cohesion zone", the form with geometrical and regime parameters of the analyzed BF is opened (Figure 3).

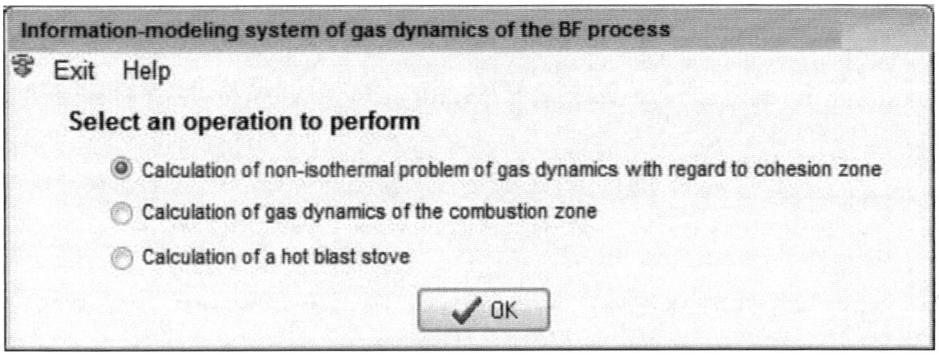

Figure 2. The form for system start-up.

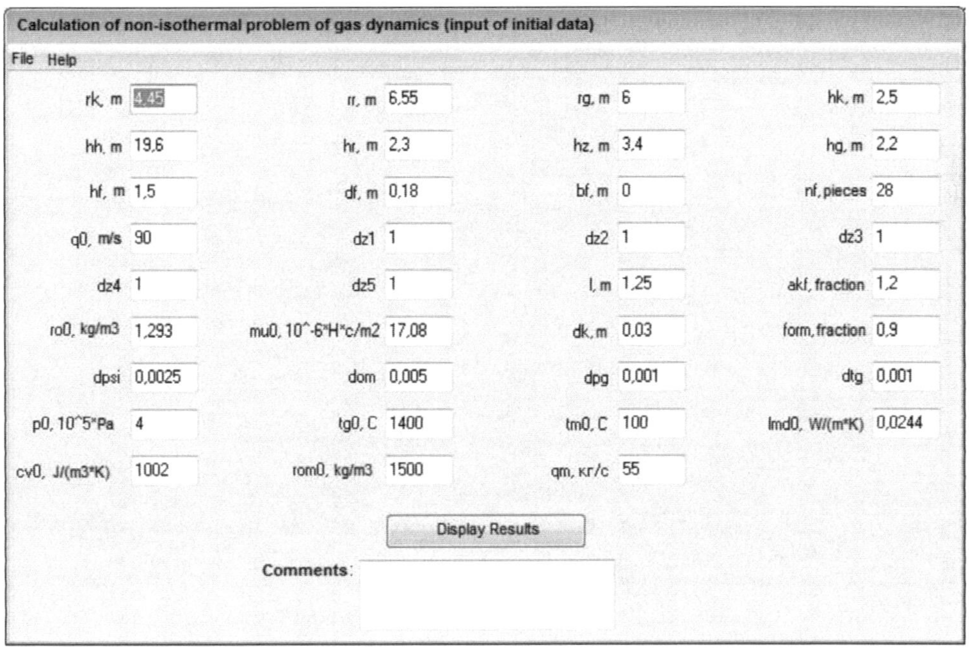

Figure 3. Setting BF operational parameters.

IOP Conf. Series: Materials Science and Engineering **150** (2016) 012013 doi:10.1088/1757-899X/150/1/012013

In this form rk, rr, rg – are radii of the furnace mouth, belly and hearth; hk, hh, hr, hz, hg, hf – are heights of the furnace mouth, shaft, belly, boshes and the tuyeres arrangement horizon. Similarly dk – diameter of charge lumps; df – the tuyres diameter; q0 – velocity of the blast outflow from the tuyere; dpsi, dom, dpg, dtg – precision of the calculations (iterations convergence criteria) of the flow function, vorticity, gas pressure and gas temperature. The remaining values are standard and do not require explanation.

As it can be seen from the data in Figure 3, the form offers *two* options. Clicking "File" you can enter the same data for another furnace or enter other values for the same furnace. By clicking "Display Results" the calculation according to the equations of numerical model is performed and the results are shown in the tabular form (Figure 1). The first form contains the nodes coordinates of finite difference mesh, which shows that along the furnace height (up to 30 m) 78 nodes are distinguished and along the furnace radius in the belly zone (up to 6.55 m) – 35 nodes. Note that the mesh cell dimensions are not the same – they are minimal near the combustion zone increase on approaching to the charge level. This form is shown in Figure 1 and is not given here.

The form with the calculation results reveals the distribution of nine gas-dynamic parameters in the layer volume (Figure 4).

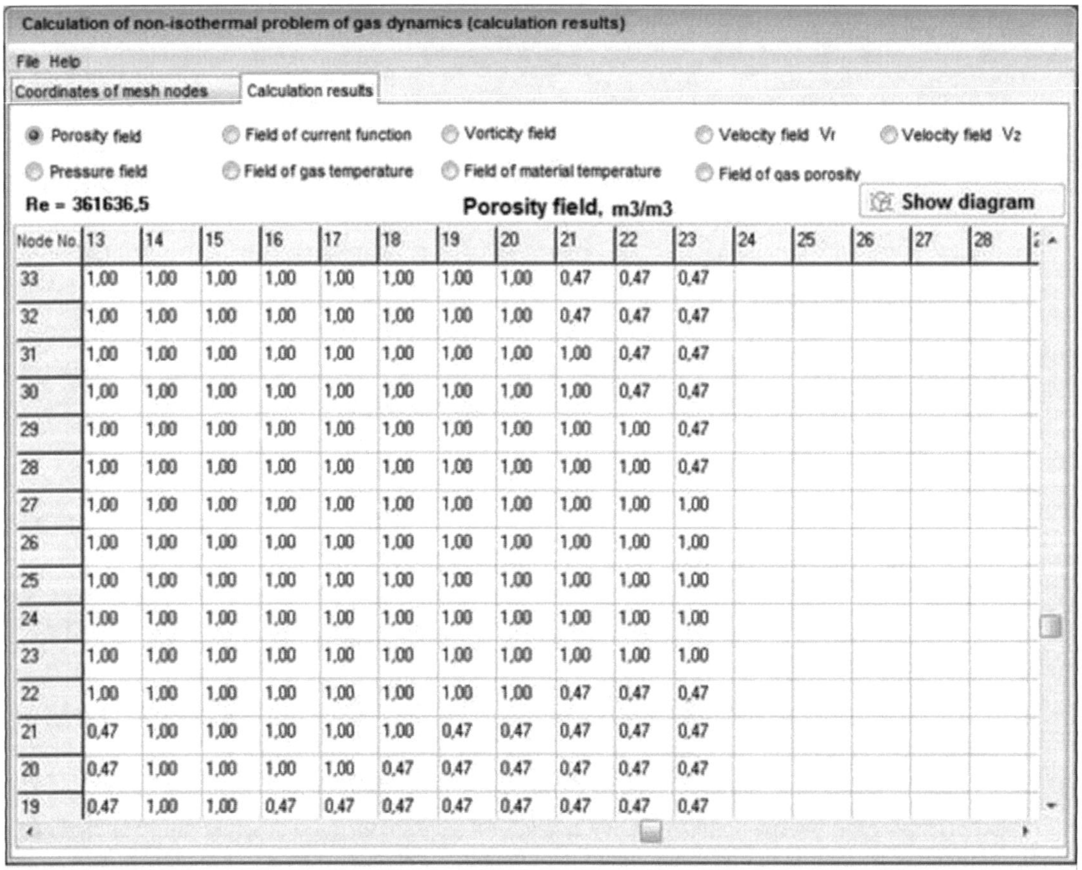

Node No.	13	14	15	16	17	18	19	20	21	22	23	24	25	26	27	28
33	1,00	1,00	1,00	1,00	1,00	1,00	1,00	1,00	0,47	0,47	0,47					
32	1,00	1,00	1,00	1,00	1,00	1,00	1,00	1,00	0,47	0,47	0,47					
31	1,00	1,00	1,00	1,00	1,00	1,00	1,00	1,00	1,00	0,47	0,47					
30	1,00	1,00	1,00	1,00	1,00	1,00	1,00	1,00	1,00	0,47	0,47					
29	1,00	1,00	1,00	1,00	1,00	1,00	1,00	1,00	1,00	1,00	0,47					
28	1,00	1,00	1,00	1,00	1,00	1,00	1,00	1,00	1,00	1,00	0,47					
27	1,00	1,00	1,00	1,00	1,00	1,00	1,00	1,00	1,00	1,00	1,00					
26	1,00	1,00	1,00	1,00	1,00	1,00	1,00	1,00	1,00	1,00	1,00					
25	1,00	1,00	1,00	1,00	1,00	1,00	1,00	1,00	1,00	1,00	1,00					
24	1,00	1,00	1,00	1,00	1,00	1,00	1,00	1,00	1,00	1,00	1,00					
23	1,00	1,00	1,00	1,00	1,00	1,00	1,00	1,00	1,00	1,00	1,00					
22	1,00	1,00	1,00	1,00	1,00	1,00	1,00	1,00	0,47	0,47	0,47					
21	0,47	1,00	1,00	1,00	1,00	1,00	0,47	0,47	0,47	0,47	0,47					
20	0,47	1,00	1,00	1,00	1,00	0,47	0,47	0,47	0,47	0,47	0,47					
19	0,47	1,00	1,00	0,47	0,47	0,47	0,47	0,47	0,47	0,47	0,47					

Figure 4. The form contains the results of the solution to the non-isothermal problem of gas dynamics.

Three parameters (flow function, gas temperature and material temperature) can be displayed graphically after clicking "Show diagram". After clicking in the start farm "Calculation of gas dynamics of the combustion zone" the form with combustion zone parameters is opened (Figure 6). Note that the theory of operation of the BF combustion zone and the detailed output of determining equations of gas dynamics and heat-transfer are presented in our works [25, 26]. As for the notations in Figure 6, then, in our opinion, should be explained only parameter 1 tuyere height, i.e. the distance of the tuyere cross-section to the inner wall of hearth, t_rec – the recycle temperature or, in other words, the gas leaving the tuyere, but returning to the combustion zone due to the flow vorticity, and PartRec – is the recycle portion from the total amount of the tuyere gas.

Clicking the button "Display results" the form with the tables of all combustion zone characteristics is opened. Perhaps the option of "File" is of particular interest, where the graph of changes in the combustion zone parameters can be obtained (Figure 8 and 9).

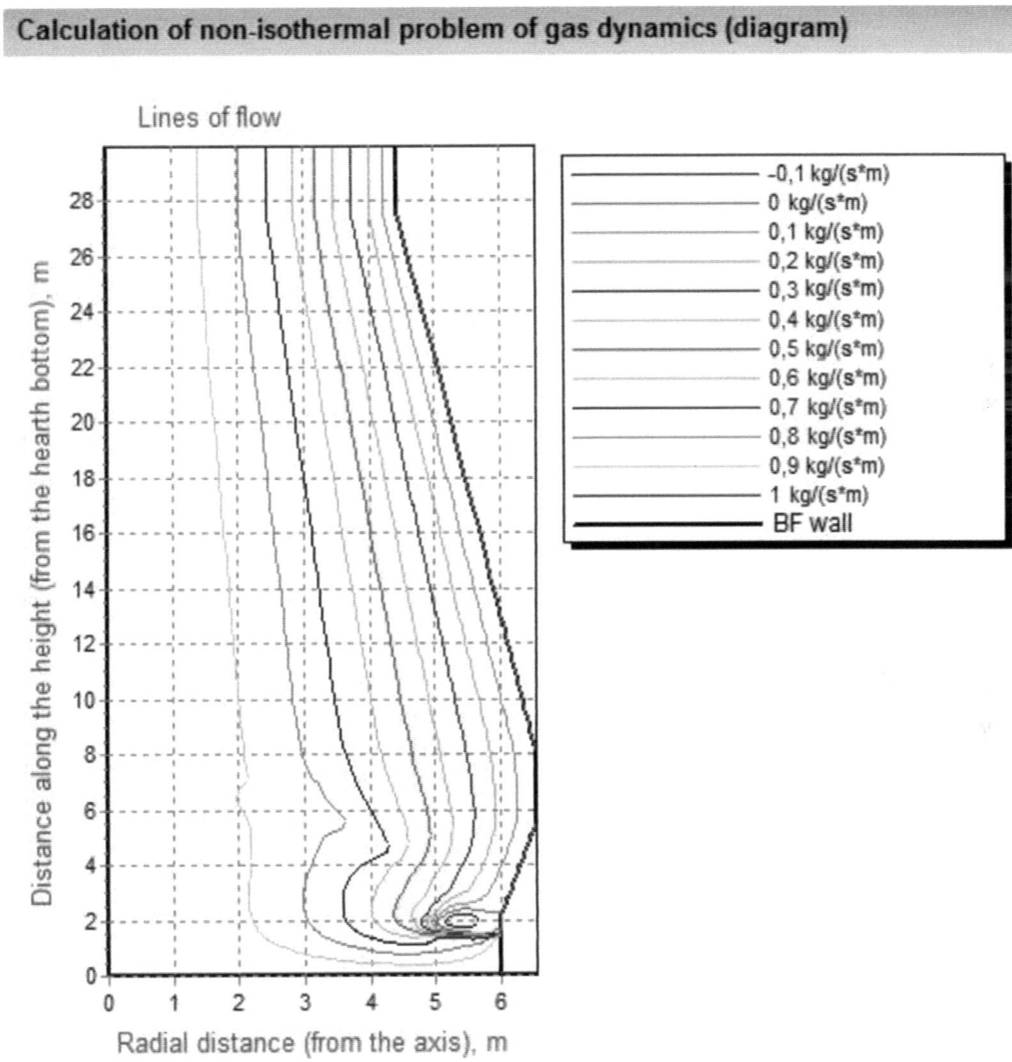

Figure 5. Graph of the gases flow in the blast furnace.

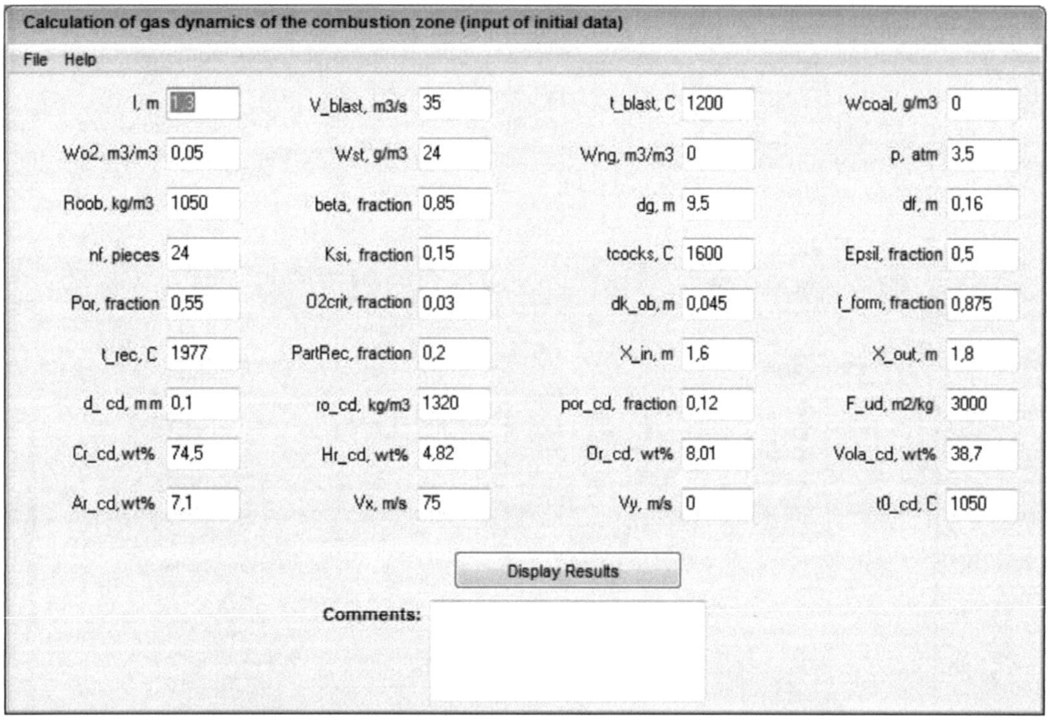

Figure 6. Operational parameters of BF combustion zone.

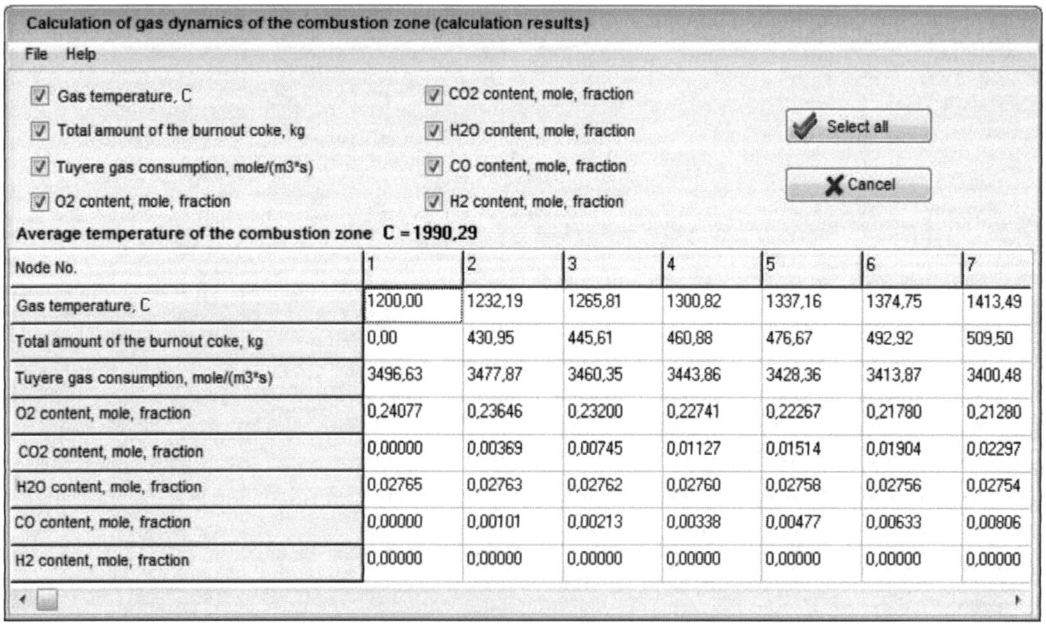

Node No.	1	2	3	4	5	6	7
Gas temperature, C	1200,00	1232,19	1265,81	1300,82	1337,16	1374,75	1413,49
Total amount of the burnout coke, kg	0,00	430,95	445,61	460,88	476,67	492,92	509,50
Tuyere gas consumption, mole/(m3*s)	3496,63	3477,87	3460,35	3443,86	3428,36	3413,87	3400,48
O2 content, mole, fraction	0,24077	0,23646	0,23200	0,22741	0,22267	0,21780	0,21280
CO2 content, mole, fraction	0,00000	0,00369	0,00745	0,01127	0,01514	0,01904	0,02297
H2O content, mole, fraction	0,02765	0,02763	0,02762	0,02760	0,02758	0,02756	0,02754
CO content, mole, fraction	0,00000	0,00101	0,00213	0,00338	0,00477	0,00633	0,00806
H2 content, mole, fraction	0,00000	0,00000	0,00000	0,00000	0,00000	0,00000	0,00000

Figure 7. Characteristics of BF combustion zone.

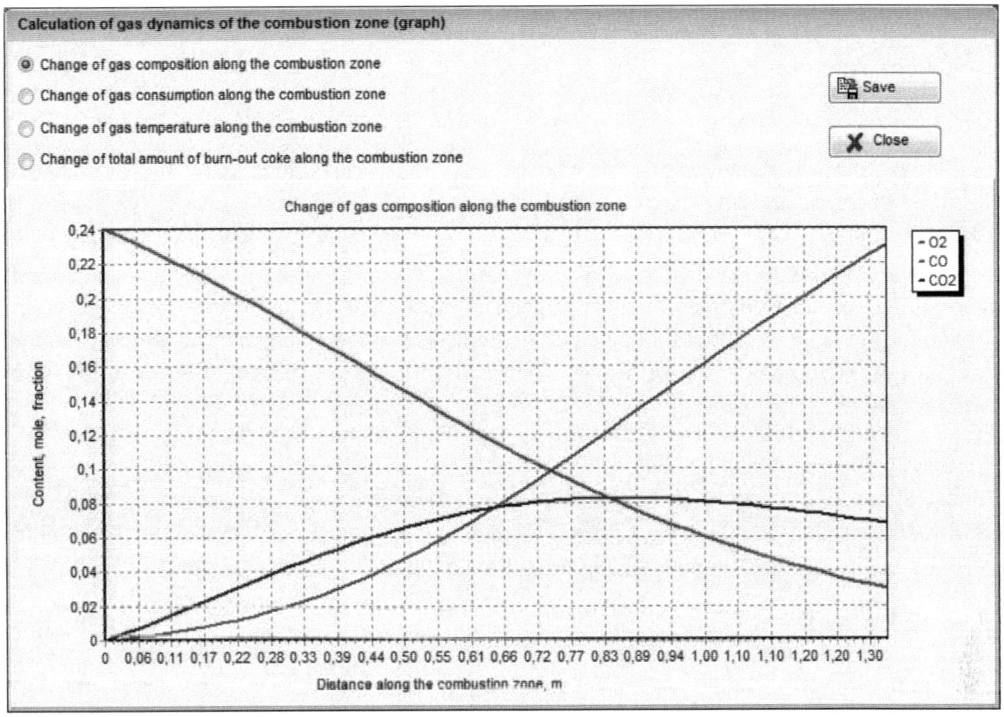

Figure 8. Change in the tuyere gas composition along the combustion zone.

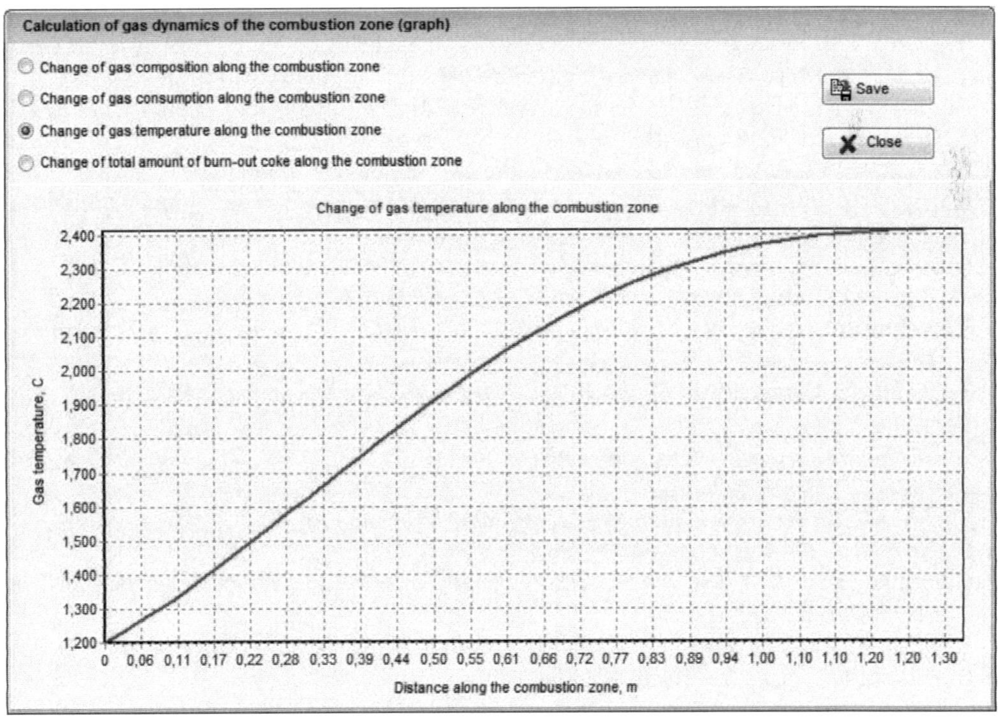

Figure 9. Changes in the tuyere gas temperature along the length of the blast chamber.

Calculations show that the large-scale vortices are intensively damped by the non-deformable layer, and the decrease intensity of vorticity is determined mainly by the gas Reynolds number. With an increase of number Re the area dimensions, where it is necessary to take into account the vortex nature of the gas flow in the layer, increases. If to accept $\tilde{\omega}$ = 0.05 as a boundary of vortex area, then the layer volume, occupied by the vortex flows, is 41%. It is clear that in these cases the vortex nature of the gas flow in the layer cannot be ignored. Note that the zone of two-dimensional gas flow takes in this case 34% of the layer volume, i.e. flow vorticity covers the lower part of one-dimensional flow (~ 12.3% of its volume).

5. Conclusions

Thus, the mathematical model can be used for any metallurgical unit. Of course, in each case, the equations system of the model, and the algorithm of its solution will have its own characteristics, but the methodology of constructing the model remains unchanged.

6. Acknowledgements

The work was supported by Act 211 Government of the Russian Federation, contract № 02.A03.21.0006.

7. References

[1] Matros Yu Sh 1982 *Non-stationary Processes in Catalytic Reactors* (Novosibirsk: Nauka) p 228
[2] Kitaev B I, Yaroshenko Yu G, Sukhanov E L, Ovchinnikov Yu N, and Shvydkiy V S 1978 *Heating Engineering of BF Process* (M.: Metallurgiya) p 248
[3] Gordon Ya M, Bokovikov B A, Shvydkiy V S and Yaroshenko Yu G 1989 *Thermal Operation of Shaft Furnaces and Units With a Dense Layer* (M.: Metallurgiya) p120
[4] Gordon Ya M, Maksimov E V and Shvydkiy V S 1989 *Mechanics of Materials and Gases Motion in the Shaft Furnaces* (Alma-Ata: Nauka) p 144
[5] Lykov A V 1972 *Heat- and Mass-transfer (Reference Book)* (M.: Energiya) chap 5 pp 338–467
[6] Slattery J S 1978 *The theory of Momentum Transfer, Energy and Mass in Continuous Media* (M.: Energiya) p 448
[7] Smirnov V I 1955 *Course of Higher Mathematics* (M.: Nauka) vol 1 p 480
[8] Myshkis A D 1969 *Lectures on Higher Mathematics* (M.: Nauka) p 640
[9] Shvydkiy V S, Ladygichev M G, Shavrin V S 2001 *Mathematical Methods of Thermal Physics: Textbook for Universities* (M.: Mashinostroyeniye) p 232
[10] Serrin J 1963 *Mathematical Foundations of Classical Fluid Mechanics* (M.: IL) p 256
[11] Shvydkiy Yu G, Yaroshenko Ya M, Gordon V S, Shavrin V S and Noskov A S 2003 *Fluid Mechanics: Learners' Guide for Universities* (M.: IKTs "Akademkniga") ed V S Shvydkiy p 464
[12] Shlikhtin G *Tethory of the Boundary Layer* 1974 (M.: Nauka) p 586
[13] Prager S 1961 *Phys. of Fluids* **4** 1477–85
[14] Goldshtik M A 1984 *Transfer Processes in the Grain Layer* (Novosibirsk: Inst. of Thermal Physics SB RAS) p 164
[15] Bogdandi L F and Engel G D 1971 *Iron Ore Reduction* (M.: Metallurgiya) p 520
[16] Brauer H 1960 *Chem. Ing. Tech.* **32** 585 – 590
[17] Eshar R, Bredegeft R, Mavrodis M et al 1971 *Ferrous Metals* **12** 121–40
[18] Patankar S 1984 *Numerical Methods for Solving Problems of Heat Transfer and Fluid Dynamics* (M.: Energoatomizdat) p 152

[19] Shi D 1991 *Numerical Methods in Heat Transfer Problems* (M.: Mir) p 544
[20] Frank-KamenitskyDA 1987 *Diffusion and Heat Transfer in Chemical Kinetics*(M.: Nauka) p 492
[21] Keyns V M *1972 Convective Heat and Mass Transfer*(M.: Energiya) p 448
[22] Kheygman L and Yang D 1986 *Applied Iterative Methods*(M.: Mir) p 448
[23] Samarsky A A and Nikolaev E S 1978*Methods for Solving Grid Equations* (M.: Nauka) p 512
[24] Gosmen A D, Pan V M, Ranchel A K, Spolding D B and Volfshteyn M 1972*Numerical Methods for the Study of Viscous Fluid* (M.: Mir) p 328
[25] Spirin N A, Ovchinnikov Yu N, Shvydkiy V S and Yaroshenko Yu G1995 *Heat transfer and efficiency of blast furnace* (Ekaterinburg: UrSTU) ed Yu G Yaroshenko p 243
[26] Spirin N A, Shvydkiy V S, Lobanov V I and Lavrov V V 1999 *Introduction to the systematic analysis of thermal process industry: Learner' Guide for Universities* (Ekaterinburg: UrSTU – UrPU) p 205

IOP Conf. Series: Materials Science and Engineering **150** (2016) 012014 doi:10.1088/1757-899X/150/1/012014

Mathematical modeling of non-stationary heat and mass transfer in disperse systems

L A Ermakova, S Y Krasnoperov and S N Kalashnikov

Siberian State Industrial University, 42 Kirova Street, Novokuznetsk, 654007, Russia

E-mail: ermakova@sibsiu.ru

Abstract. The work describes mathematical model of non-stationary heat and mass transfer processes in dispersed environment, taking into account the phase transition; presents the results of numeric modelling for conditions of direct reduction in high-temperature reducing atmosphere, corresponding to the direct reduction in the jet-emulsion unit according to the principles of self-organization. The method was developed for calculation of heat and mass transfer of the aggregate of iron material particles in accordance with the given distribution law.

1. Introduction

Today, the most urgent task in the development of metallurgy technology is to obtain the metal by direct reduction directly from fine-disperse materials, including wastes from metallurgical production (scale, sludge, flue dust). These technologies have a number of important advantages: solution of environmental problems through recycling of wastes from metallurgical production, a significant acceleration of reduction due to the large reaction surface of fine materials, reduction of costs and energy consumption by eliminating the stage of pre-processing and pelletizing materials. One of the variants of such processes is the process for producing metal from dispersed materials in the jet-emulsion unit having a reaction chamber and the refining sump [1, 2].

During development of technologies for production metals in the units of a jet-emulsion type, an important task is to determine the feeding mode of charging the main reactors of the unit with iron ore materials, oxygen, fuel and reductants. The solution of this problem requires the creation of appropriate models and techniques allowing the heat and mass transfer processes in disperse systems to be calculated. In this paper we carry out the mathematical modeling of heat and mass transfer of particles with regard to phase transition. On its bases we developed the method of calculation of heating, melting and reduction of the aggregate of particles in the iron ore materials under the given boundary conditions.

2. Calculation of numerical characteristics of raw materials

When calculating the numerical characteristics of dispersed materials it was assumed that the particles have a spherical shape, and the initial experimental data for quantitative description of density of particles distribution according to their diameters in the interval (d_{min}, d_{max}) are the weight fractions, obtained by sieving processing of charge materials on the row of standard sieves. For description we used such characteristics as the density of particles distribution by their quantity $f_n(d)$, surface $f_s(d)$ and the mass $f_m(d)$, as well as the respective distribution functions $F_n(d)$, $F_s(d)$, $F_m(d)$.By $\tilde{f}_n(d)$, $\tilde{f}_s(d)$ $\tilde{f}_m(d)$ we denote the corresponding normalized density distribution, and by $\tilde{F}_n(d)$, $\tilde{F}_s(d)$, $\tilde{F}_m(d)$ – normalized distribution function. Assessment for $\tilde{f}_m(d)$ was performed in the class of regular distributions in the interval (d_{min}, d_{max}) in the form of a lognormal law, taking into account the minimum and maximum diameters of particles d_{min} и d_{max} [3-4].

$$\tilde{f}_m(d) = \frac{d_{max} - d_{min}}{(d_{max} - d)(d - d_{min})\sigma\sqrt{2\pi}} e^{-\frac{\left(\ln\left(\frac{d_{max}-d}{d-d_{min}}\right) - \ln a\right)^2}{2\sigma^2}} \quad (1)$$

Specific type $\tilde{f}_m(d)$ is determined by the values of the parameters a^* and σ^* that minimize the functional.

$$\Phi(a,\sigma) = \min_{a,\sigma} \sum_{i=1}^{N_f} \left(\int_{d_{i-1}}^{d_i} \tilde{f}_m(\xi)d\xi - \frac{M_i}{M} \right)^2 \quad (2)$$

where N_f – the number of fractions; d_i–the maximum diameter of particles in the i-th fraction; M_i mass of i-th fraction.

As a result of the processing of experimental data we obtained and adopted in the simulation the following values $a^*$and σ^* of parameters of the distribution laws and the corresponding values of the numerical characteristics of the dispersed charge materials shown in Table 1.

Table 1.Numerical characteristics of dispersed materials.

Dispersed material	d_{min}, mm	d_{max}, mm	$d(\max f_m)$, mm	$d(\max f_s)$, mm	$d(\max f_N)$, mm	Distribution options a	σ	m_Σ, kg	S_Σ, m^2	$N_\Sigma \cdot 10^{-6}$
Iron ore concentrate	0.025	1.0	0.60	0.49	0.20	0.80	0.79	1.0	5.2	11.9
Sludge	0.025	1.0	0.13	0.08	0.05	3.85	1.00	1.0	16.2	537.4
Scale	0.025	3.0	0.94	0.61	0.23	1.76	0.78	1.0	1.2	1.2
Manganese concentrate	0.03	1.0	0.26	0.21	0.14	2.55	0.64	1.0	16.9	122.3
Coxlittle	0.025	5.0	0.10	0.02	0.02	4.93	1.42	1.0	17.2	444.7
Cox	5.0	50.0	1.24	1.11	0.95	0.65	3.88	1.0	5.5	1.2
Lime	0.05	3.0	0.07	0.04	0.03	4.04	1.64	1.0	18.6	1079.2

3. The mathematical model of heat and mass transfer process of particles with phase transition

The study of the processes in the jet-emulsion unit, when fine materials are used as a charge, showed that heat and mass transfer processes occur in the turbulent environment and have a very high intensity [1]. This allows to pose the problem of mathematical modeling of heat and mass transfer processes based on the solution of internal tasks, taking into account the external conditions of processes [5-6].

Therefore, the process of unsteady heat and mass transfer was described by the equation of heat conduction and molecular diffusion for particles of spherical shape with the boundary conditions of the 3 kind.

The obtained system of equations and the boundary conditions for the dimensionless relative temperature and concentration have the form [7]:

$$\begin{cases} \dfrac{\partial \tilde{C}}{\partial Fo_{dif}} = \dfrac{1}{\tilde{r}^2}\dfrac{\partial}{\partial \tilde{r}}\left(\tilde{r}^2\dfrac{\partial \tilde{C}}{\partial \tilde{r}}\right) \\ \dfrac{\partial \Theta}{\partial Fo_{dif}} = \dfrac{1}{Lu\,\tilde{r}^2}\dfrac{\partial}{\partial \tilde{r}}\left(\tilde{r}^2\dfrac{\partial \Theta}{\partial \tilde{r}}\right) \\ \tilde{C}\Big|_{Fo_{dif}=0} = 0 \quad \Theta\Big|_{Fo_{dif}=0} = 0 \\ \dfrac{\partial \tilde{C}}{\partial \tilde{r}}\Big|_{\tilde{r}=0} = 0 \quad \dfrac{\partial \Theta}{\partial \tilde{r}}\Big|_{\tilde{r}=0} = 0 \\ \dfrac{\partial \tilde{C}}{\partial \tilde{r}}\Big|_{\tilde{r}=1} = Bi_{dif}\left(1-\tilde{C}\right) \quad \dfrac{\partial \Theta}{\partial \tilde{r}}\Big|_{\tilde{r}=1} = Bi\left(1-\Theta\right) \end{cases} \tag{3}$$

where $\tilde{C} = \dfrac{C - C_0}{C^* - C_0}$ – dimensionless relative concentration; C_0 – initial concentration of the considered element in the particle; C^* – concentration of this element in the environment; $\Theta = \dfrac{t - t_0}{t_{en} - t_0}$ – dimensionless relative temperature; t_0 – initial temperature of the particle; t_{en} – ambient temperature; R – radius of the particle; $\tilde{r} = \dfrac{r}{R}$ – dimensionless radius; $Lu = \dfrac{D}{a}$ – Lewis number; $Fo_{dif} = \dfrac{D\tau}{R^2}$ – diffusion Fourier number; $Bi_{dif} = \dfrac{\beta R}{D}$ – diffusion Bionumber; $Bi = \dfrac{\alpha R}{\lambda}$ – Bionumber; D – diffusion coefficient, m^2/c; β – coefficient of mass transfer, m/c; a – thermal diffusivity of particles, m^2/c; λ – coefficient of thermal conductivity of particles, W/(m·K); α – the heat transfer coefficient, W/(m^2·K); τ – time, sec.

The process of phase transition for spherical particles with a diameter of $0.025\cdot10^{-3}\div3\cdot10^{-3}$m, where the criterion Bio is changed in the range of 0,1÷10 is described by the ratio of the form [8]:

$$Fo - Fo_0 = \frac{Ko}{Bi}(\delta - \delta_0) + \frac{1}{n}\left(\frac{1}{n+1} + \frac{Ko}{2}\right)\left(\delta^2 - \delta_0^2\right) \tag{4}$$

where $Fo = \dfrac{a\tau}{R^2}$; $Fo_0 = \dfrac{a\tau_0}{R^2}$ – Fourier number; τ_0 – time of particles heating up to the melting temperature, c; τ – heating time, c; $Ko = L_{mel}/\,c\cdot(t_{sur} - t_{mel})$ – number of Kossovich; t_{sur} – surface temperature, ^{0}C; t_{mel} - temperature of melting, ^{0}C; L_{mel} – latent heat of phase transition, kJ/kg; $\delta = \xi/R$; $\delta_0 = \xi_0/R$; ξ_0 – thickness of the molten crust during τ_0, m; $\xi = \xi(\tau)$ – law of displacement of the flat front of the melt; n – indicator of the parabolic law of temperature distribution.

On the basis of equation (4) we calculated the time of particles melting, which was taken into account in the calculation of heat transfer process.

The boundary value problem was solved by separation of variables using a rapidly converging series. The peculiarity of the solution method consists in the construction of a calculation scheme, which allows the results to be obtained, that characterize the relationship between the determining parameters of the problem in any given interval of variables, presented in the criterion and dimensional physical forms. This technique is described in detail in [6].

4. The results of mathematical modelling

In the numerical realization of the method the change in the system parameters during the phase transition was taken into account as follows. At each iteration, first, by the composition we determined the melting temperature, then counted the current temperature, according to which determined the diffusion coefficient taking into account the phase state of the particle. Then, on the next iteration the new thermophysical material parameters with regard to the obtained composition and phase state were determined.

In the study accepted: the range of sizes variation for materials 0.025÷3 mm; the temperature of the environment in accordance with the conditions of liquid-phase reduction1400÷1600 ^{0}C; thermophysical parameters and diffusion coefficients, which are given in [8-12].

Figure 1 and Figure 2 show the results of heat exchange process at a temperature of gas medium t=1500 ^{0}C for a process time 0.3÷0.6 s.

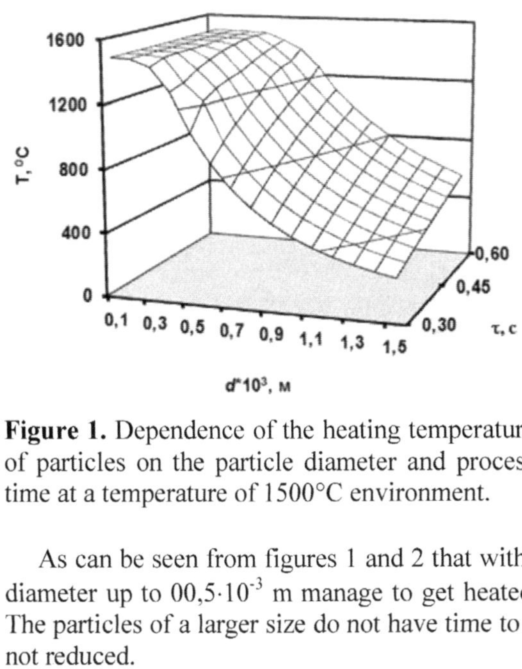

Figure 1. Dependence of the heating temperature of particles on the particle diameter and process time at a temperature of 1500°C environment.

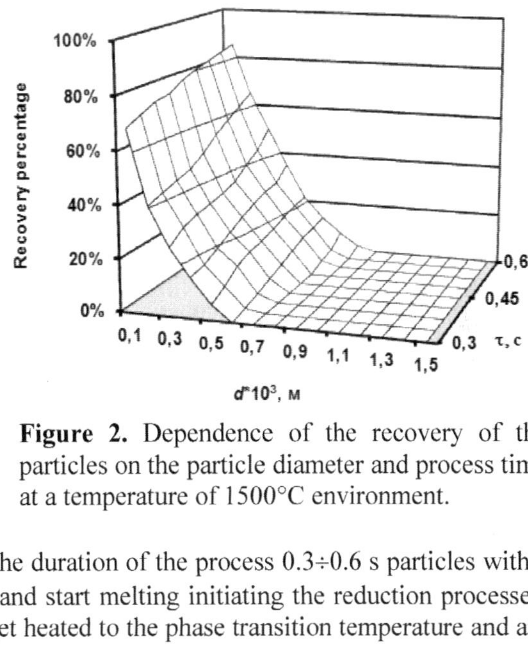

Figure 2. Dependence of the recovery of the particles on the particle diameter and process time at a temperature of 1500°C environment.

As can be seen from figures 1 and 2 that with the duration of the process 0.3÷0.6 s particles with a diameter up to $00{,}5 \cdot 10^{-3}$ m manage to get heated and start melting initiating the reduction processes. The particles of a larger size do not have time to get heated to the phase transition temperature and are not reduced.

Thus, the developed model of heat and mass transfer allows the temperature and composition of particles or droplets at any point of time to be determined which are in the high-temperature reducing environment. This model served as a basis for the methods of calculation of heat and mass transfer of an aggregate of dispersed particles.

5. The calculation method of heat and mass transfer of the aggregate of dispersed particles

To calculate the heat exchange process of the aggregate of dispersed particles of the given distribution law on the basis of the equations (3)-(4) we used the following method: set of initial conditions (process time, ambient temperature, division of the size interval into a number of ranges, so that all particles within one rage would have the same size); calculation of heat and mass transfer processes for particles of the selected range, calculation of the fraction and composition of phases according to the results of heat and mass transfer; determination of costs of heat required for heat and mass transfer of all particles of the current range; definition of the integral values of masses, phase composition and costs for heating, melting and reduction; analysis of material and heat balances.

Figure 3 shows the simulation results of heat and mass transfer processes for 1 kg of the iron-containing material (σ = 1.01 and a = 4.92) at the environment temperature 1550^{0}C.The dependence of fractions is shown here: solid heated charge material (CH_s^h), in which each range of particle sizes has its own temperature; melted charge material (CH_m), the droplets of which have a temperature above the melting temperature; charge material is melted and recovered (CH_m^r).

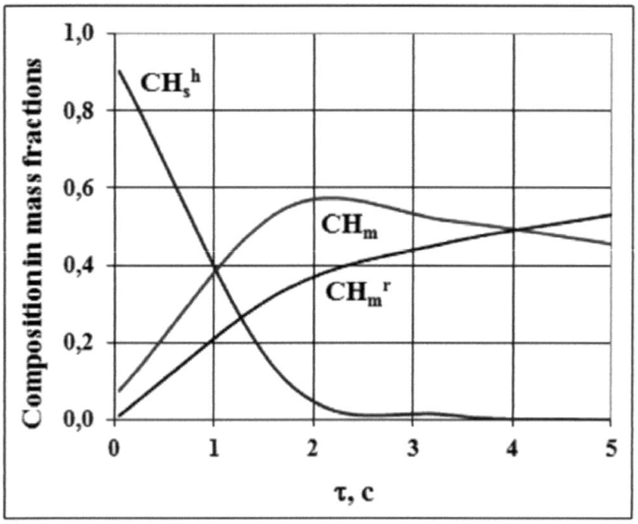

Figure 3. The results of simulation of heat and mass transfer processes of particles aggregate in the iron-containing materials at a ambient temperature 15500C.

As it can be seen from Figure 3 the whole process can be divided into two periods. The first period lasts 2÷3 seconds, and this time is sufficient for complete melting of the dispersed material and reduction of small particles with sizes up to $1 \cdot 10^{-3}$ m, which is 40% by weight of the total mixture. The duration of this period is comparable to the residence time of the materials in the reaction chamber of the jet-emulsion unit. In the second

period the reduction of already molten particles of larger size takes place, it requires several minutes, which might be provided by the design and modes of phases interaction in the emulsion reactor of a column type.

6. Conclusions

A mathematical model of heat and mass transfer of particles with phase transition in the form of a joint system of equations of heat conduction and molecular diffusion with boundary conditions of the 3 kind was offered. The method of modeling heat and mass transfer processes of particles aggregate of iron ore material of the given distribution law was developed. The proposed method allows the heat exchange process of the aggregate of the dispersed iron-bearing particles of the given distribution law to be calculate, and can be used for calculation of processes occurring in the reactors of jet-emulsion units.

7. References

[1] Tsymbal V P, Kozhemyachenko V I et al 2015 *Steel in Translation* **45** (8) 550–554
[2] Ermakova L A and Mochalov S P 2004 *Izvestiya Vysshikh Uchebnykh Zavedenij. Chernaya*

Metallurgiya **10** 48–51

[3] Nichiporenko O S, Naida Y I and Medvedsky A B 1980 *Atomized Metal Powders* (Kiev: Naukova Dumka) p 239

[4] Dityatky Y F, Klyachko L A et al 1977 *Spraying Liquids* (M.: Mashinostroenie) p 263

[5] Kalashnikov S N, Ermakova L A and Mochalov S P 2001 *Izvestiya Vysshikh Uchebnykh Zavedenij. Chernaya Metallurgiya* **6** 67–71

[6] Kalashnikov S N, Mochalov S P, Ermakova L A and Krasnoperov S Y 1999 *Steel in Translation* **29** (10) 19–23

[7] Telegin A S, Shvydky V S and Yaroshenko Yu G 1995 *Heat and Mass Transfer* (M.: Metallurgy) p 440

[8] Mastryukov B S 1996 *Thermal Physics of Metallurgical Processes* (M.: MISiS) p 271

[9] Krivandina V A 1986 *Metallurgical Heat Engineering* (M.: Metallurgiya) vol 1, 2

[10] *Thermophysical Properties of Fuels and Charge Materials of Ferrous Metallurgy. Reference Book* 1982 (M.: Metallurgy) p148

[11] Kafarov V V and Glebov M B 1991 *Mathematical Modeling of Basic Processes of Chemical Productions* (M.: Higher school) p 400

[12] Kamkina L V, Yakovlev Yu N et al 1995 *Izvestiya Vysshikh Uchebnykh Zavedenij. Chernaya Metallurgiya* **2** 8–10

Software for simulation of utilization schemes of secondary energy from the exhaust gases of metallurgical units

A A Olennikov[1] and V P Tsymbal[2]

[1] Institute of Architecture and Construction, Siberian State Industrial University, 42 Kirova Street, Novokuznetsk, 654007, Russia

[2] Institute of Information Technologies and Automation Systems, Siberian State Industrial University, 42 Kirova Street, Novokuznetsk, 654007, Russia

E-mail: oaa@cirkul-m.ru

Abstract. The work is devoted to the program complex intended for designing schemes of secondary energy utilization from metallurgical units. The structure of the software system is based on three levels of complex systems assembled from subsystems. The mathematical models of a complex process of heat transfer and gas dynamics occurring in the energy utilization units and gas cleaning devices. We describe the user interaction with the software package, and show the calculation results in the form of plots.

1. Introduction

In the metallurgical process the cycle efficiency rate is only about 20%. Part of the energy is consumed for the process flow in the main production unit. About 20-30% of the energy goes to heat losses through the unit walls, skull cooling and 40-50% is lost in the flue gases [1-3].

Given that every day fuel extraction and processing is becoming more expensive, and the world reserves are being depleted, energy saving issues are placed on one of the first places. Energy-saving technologies should find a much wider application. Therefore, the most rational option is to create energy-metallurgical complex based on two approaches to energy use. This is the physical and chemical thermal energy regeneration.

The first approach is based on the utilization of thermal energy in the power plants: waste-heat boilers, skull cooling systems, fuel heating installations. Depending on the power of metallurgical unit, thermal energy obtained in the first two units, an average of 2 MW to 15 MW, can be used for heating the production halls. Connection of the recovery boiler with the steam turbine will provide from 50 to 420 kW of electricity. This approach will reduce the load on the boiler or CHP plant, and the energy of exhaust gases utilization will be about 70-85%.

The second approach is based on the use of exhaust gas from the engine fuel, followed by obtaining chemical potential. This process consists of the following steps: first, the exhaust gas enters the reformer, then passes desulfurization, catalytic synthesis is performed, where the synthesis gas is produced. The resulting gas with a volumetric ratio of CO / H_2 close to 1.1 can be used as power fuel and as a raw material for production of synthetic fuel. This utilization option allows all of physical and chemical energy of the exhaust gas to be used with the efficiency up to 98% of energy use. This technology will create a completely smoke-free process.

It is not an easy task to perform the design of any variant of exhaust gas energy utilization. The purposes of the use of utilized secondary energy resources (SER) should be clear understood, and intensive calculations need to be performed. A promising method of this problem solution is the modeling of heat transfer, gas dynamics on a computer that can significantly reduce labor intensity and avoid many mistakes in the design and evaluation.

2. Methods

During the development of models used in the software package we considered the approach to the modeling of complex systems. It is based on the elementwise-physical formulation of the problem, analysis and synthesis of the system in its individual components. This approach is called a direct analogy method [4].

Adequacy of complex models build that way is based on the additional concept about the similarity of complex systems, formulated by V.A. Venikov in the development of the first similarity theory [4].

In the present work we deal with at least three levels of complex systems assembled from subsystems.

First level: processes in the individual structural elements of heat-assemblies (bundles of pipes and ducts, etc.). Here the complex processes of heat exchange between heat-perceiving elements and environments, the gas friction processes on the walls, etc., are considered as basic systems, which are described by the fundamental laws of physics, the validity of these laws is considered to be proved, and the final result depends on the used combinations and connections.

Second level: heat-perceiving units as a whole. They consist of the structural elements described above. And therefore, a complex system of this kind (second) level depends on the combination and method of connecting the individual structural elements (first level systems). Examples of complex systems of the second level are: waste heat boilers, gas turbines, fluidized bed units, etc. This level includes the complex heat transfer model in the recovery boilers [5-8]:

For line bundles

$$\alpha_K = 0.2C_sC_z\frac{\lambda}{d}\left(\frac{wd}{\nu}\right)^{0.65}\mathrm{Pr}^{0.33} + 5.67\cdot10^{-8}\frac{\alpha_c+1}{2}aT^3\frac{1-\left(\frac{T_c}{T}\right)^{3.6}}{1-\frac{T_c}{T}}, \quad (1)$$

where C_s – correction for the geometric arrangement of the beam, determined by the relative pitch pipes; C_z– corrected for the number of rows of tubes in the course of gas, it is determined depending on the average number of packets in individual rows calculated beam; λ – coefficient of thermal conductivity of the material of the tube bundle; d – diameter of the tube bundle; w – gas flow rate; ν – the kinematic viscosity of the flowing medium; Pr – Prandtl number; α_c – emissivity factor of the contaminated walls of receptive surfaces; α – emissivity of gas flow depending on the temperature; T, T_c – gas temperature and the outer surface of the wall, taking into account contamination.

For staggered

$$\alpha_s = 0.36C_sC_z\frac{\lambda}{d}\left(\frac{wd}{\nu}\right)^{0.6}\mathrm{Pr}^{0.33} + 5.67\cdot10^{-8}\frac{\alpha_c+1}{2}aT^3\frac{1-\left(\frac{T_c}{T}\right)^{3.6}}{1-\frac{T_c}{T}}. \quad (2)$$

Model for determination of the frictional pressure loss in view of the gas transporting pressurized

$$\Delta P = P_1 - P_2 = P_1 - \left[P_1^2 - \left(\lambda \frac{l}{d_{equ}} \cdot \frac{\rho w^2}{2} \cdot \frac{2PT_w}{T} \right) \right]^{0.5}, \quad (3)$$

where P_1, P_2 – respectively gas pressure at the beginning and end of the pipeline section; l – length of the duct under consideration; d_{equ} – equivalent channel diameter for the exhaust gas passage; ρ – density of the transported gas; T_w, T – respectively the average within the gas duct wall portion calculated temperature and the exhaust gas.

The third level consists of a combination of the units described above, their set is determined by the criterion of efficiency of heat use, which is defined after calculations of thermal installations included in the gas path and identification of their particular coefficients of thermal energy use as follows [7]:

$$\sum \eta = \eta_{mu} + \eta_{sc} \cdot \eta_{sc}^{eff} + \eta_{u1} \cdot \eta_{u1}^{eff} + \eta_{u2} \cdot \eta_{u2}^{eff} + \ldots + \eta_{uN} \cdot \eta_{uN}^{eff}, \quad (4)$$

where η_{mu}, η_{sc}, η_{u1}, η_{u2}, η_{uN} – respectively, the useful heat factor of a metallurgical unit, skull cooling and successively connected energy utilization units and gas cleaning systems; η_{sc}^{eff}, η_{u1}^{eff}, η_{u2}^{eff}, η_{uN}^{eff} – respectively, efficiencies of skull cooling system, heat-exchangers and gas cleaning systems.

Thus, using successively the indicated additional provision about the adequacy of the derived complex systems from the subsystems of the lower level can be considered (primarily qualitative) proven. Specific parameters setting, especially for units of the second level, naturally appear, but only after the implementation of design solutions.

2. Software package

On the basis of the created models the software SKV_SAPR is being actively developed at the present time using the object-oriented approach [5, 9], as well as multi-threaded applications technology.

The task of the software includes: layout of the exhaust gas energy utilization schemes; structural calculation of thermal power units used in the scheme; calculation of the amount of heat energy received by the utilization units; evaluation of pressure losses along the gas path, including energy utilization units; calculation of physical parameters of the exhaust gases in each section of the gas path; search for the optimal variant of thermal power plants and the whole scheme with the participation of experts; issuance of technical documentation and guidance on the design.

Figure 1 shows the structure of the software system. At the initial stage of the simulation, the specifications and output parameters of the exhaust gas of metallurgical unit are set: composition, flow rate, pressure, temperature. In accordance with the name of the project the working files and paths are generated.

Next, the arrangement of the energy utilization scheme is performed with possible access to the database in order to select the proposed options, calculated previously, or to construct a new one. Then the thermodynamic parameters of exhaust gas on the basis of known thermodynamic laws are calculated and the entering of the necessary data for thermal power plants is described. It is possible to access to the extensible databases that are developed in Microsoft Access environment [10], and contain, regulatory, engineering, thermo physical and energy documentation of standard models offered by the manufacturer, or designed independently. Upon completion of data entry the calculations are made and the results on the current unit in the course of gases movement are displayed.

On completion of work on the last energy utilization unit in the scheme, the calculation is performed, the results of which determine the final design parameters settings, the possible absorption of heat by them, the pressure loss in each section of the scheme, determination of the gas thermodynamic parameters and its calorific value. If the project is expected to choose an effective option from the set of energy utilization schemes, then the analysis on the cross-cutting factor of thermal energy use is performed. The results are evaluated by experts, if necessary, the necessary changes to the system configuration and recalculates are introduced.

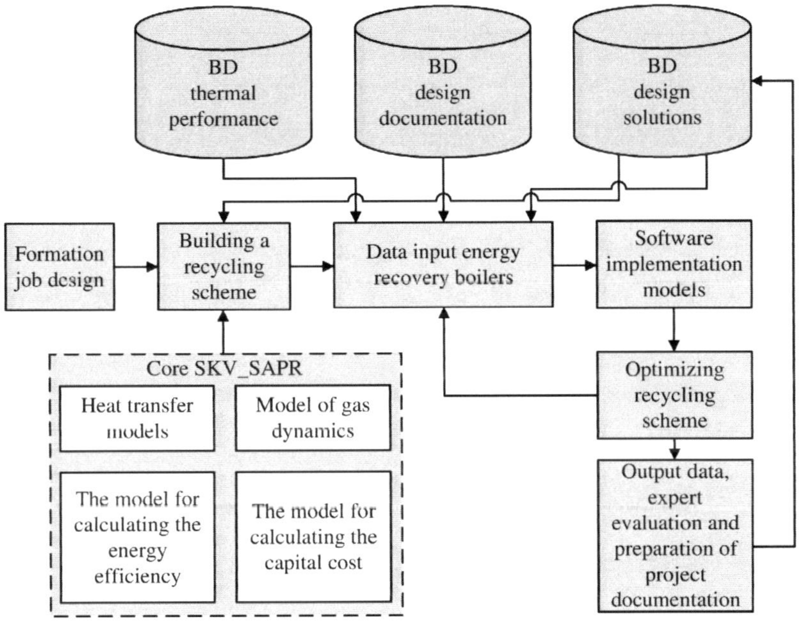

Figure 1. Complex structure SKV_SAPR.

3. Results

Below the fragment of calculation by software is given. Figure 2 and 3 considered provide three possible energy utilization schemes, graphic dependencies of pressure losses and utilized thermal energy are presented (at various design parameters of the heat recovery systems, gas purification, resistance nodes and channels).

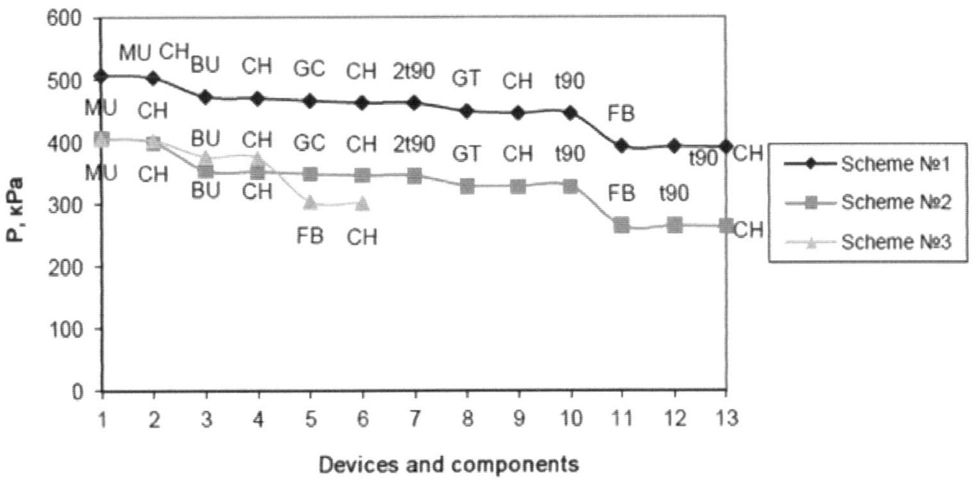

Figure 2. Pressure drop in energy utilization plants, gas cleaning systems, channels and nodes of resistance (MU – metallurgical unit; CH – channel; BU – waste heat boiler; GC – a group of cyclones; t90 – turn to 90^0; GT – gas turbine; FB – fluidized bed dryer).

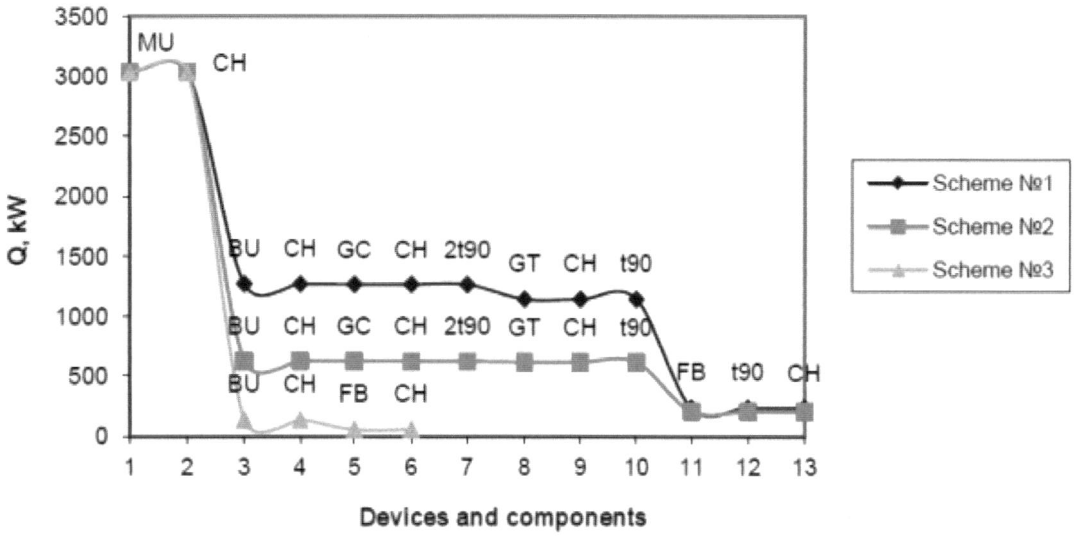

Figure 3. Heat energy usefully used by utilization units.

4. Conclusions

Thus, the use of this software package allows energy utilization schemes to be built, as well as perform calculations, and choose the best option for the cross-cutting factor of heat usage and explore the influence of the main regime parameters.

The proposed model and software system were tested at JSC "Siberian SantekhProekt", Novokuznetsk. The findings were compared with the third-party software products and have a positive feedback from designers working in the thermotechnical field.

5. References

[1] Kurunov I F and Savchuk N A 2002 *State and Prospects of Direct Ore Reduction Process* (M.: Chermetinformatsiya) p 198
[2] Tsymbal V P, Mochalov S P and Shakirov K M 2015 *Metallurgist* **59** (3) 119–125
[3] Ermakova L A and Mochalov S P 2004 *Izvestiya Vysshikh Uchebnykh Zavedenij. Chernaya Metallurgiya* **10** 48–51
[4] Tetelbaum I M and Tetelbaum Y I 1979 *Models of Direct Analogy* (M.: Science) p 834
[5] Voinov A P, Zaytsev V A et al 1989 *Boilers – Utilizators and Power Technological Units* (M.: Energoizdat) 1989 p 272
[6] Olennikov A A 2010 *Mathematical Modeling and Complex Programs Solving Problems of Secondary Utilization of Exhaust Gases in the Energy Metallurgical Units of Jet-emulsion Type* (Novokuznetsk) PhD thes. p 133
[7] Olennikov A A and Tsymbal V P 2008 *Steel in Translation* **38** (6) 439–443
[8] Olennikov A A and Tsymbal V P 2008 I*zvestiya Vysshikh Uchebnykh Zavedenij. Chernaya Metallurgiya* **6** 43–51
[9] Baknell D 2006 *Fundamental Algorithms and Data Structures in Delphi* (M.: Diasoft) p 556
[10] Efremova L I and Glukhova T V 2008 *Database Design DBMS Environment MS Access: Tutorial* (Saransk: Publishing house CBMO) p 121

Thermodynamic modeling of restoring items converter vanadium slag

M A Golodova[1], I D Rogihina[2], O I Nohrina[2] and I A Rybenko[1]

[1] Institute of Fundamental Education, Siberian State Industrial University, 42 Kirova Street, Novokuznetsk, 654007, Russia
[2] Institute of Metallurgy and Material Science, Siberian State Industrial University, 42 Kirova Street, Novokuznetsk, 654007, Russia
[3] Institute of IT and Automation Systems, Siberian State Industrial University, 42 Kirova Street, Novokuznetsk, 654007, Russia

E-mail: golodova_ma@mail.ru

Abstract. Calculations of parameters for equilibrium processes of iron, manganese, vanadium and titanium reduction from converter vanadium slag by carbon from small-sized coke and silica from ferrosilicon were performed by the method of thermodynamic modeling using the program complex "Terra". Calculations were made for each type of reductant separately and in combination. Dependences of process parameters on reducing agents consumption were drawn. The analysis of the results was carried out.

1. Introduction

One of the most effective and demanded alloying elements for production of steels with high performance properties is vanadium, and the perspective technology of steel alloying is processing by converter vanadium slag using carbon and silicon as reducing agents [1-3]. However, industrial converter vanadium slags alongside with vanadium oxides (from 10 to 30% V_2O_5) contain iron, titanium, manganese, silicon and calcium oxides (10 ÷ 30 % SiO_2, 1 ÷ 5 % CaO, 1 ÷ 5 % MgO, 4 ÷ 12 % MnO, 25 ÷ 35 % Fe_2O_3, 2 ÷ 12 % TiO_2, 1 ÷ 3% Al_2O_3). Study of the recovery of elements in such complex multi-component oxide system is possible only with the use of modern methods and software products for thermodynamic modeling [4-7].

2. Research and results

Solving the problem of processes research of elements recovery from converter vanadium slag we used one of the methods for thermodynamic modeling based on the calculation of equilibrium parameters in the model of thermodynamic systems [8-10]. The advantage of these methods consists in the feasibility of multivariate calculations of equilibrium of multicomponent systems during varying in the wide range of control actions. Calculation of thermodynamic equilibrium can reveal the possibility of obtaining substances, the allocation of which is essential in addressing the task of evaluating limit end state, and determine the scope of the valid parameter values. When performing thermodynamic modeling the software product – program complex "Terra" was used, developed at Moscow State Technical University, which allows the equilibrium composition of multi-component heterogeneous thermodynamic system for high temperature conditions to be found on the basis of the principle of maximum entropy [5].

While performing study of elements recovery from converter vanadium slag (16.0% V_2O_5, 5.0% TiO_2, 10% MnO, 25.0% Fe_{tot}) we made calculations of equilibrium compositions of heterogeneous mixtures of slag and reducing agents at a temperature of 1873 K. The boundary conditions were defined according to the calculations of vanadium extraction ratio η_V for small-sized coke consumption in the range from 0 to 0.5 kg/kg of slag and ferrosilicon FeSi75 from 0 to 0.5 kg/kg of slag [11-12]. Vanadium extraction ratio η_V is close to 1 at a specific consumption of small-sized coke more than 0.2 kg/kg of slag and ferrosilicon over 0.25 kg/kg of slag, these consumptions were chosen as boundary values.

Analysis of the calculations results presented in Figures 1 and 2 showed that the recovery of iron from iron oxides ends at a specific consumption of small-sized coke 0.1 kg/kg of slag, and restoration of silicon from ferrosilicon – 0.14 kg/kg of slag. Vanadium recovery begins after full iron recovery. The content of manganese oxides and titanium as the consumption of reducing agents increases varies slightly, indicating the limit of their recovery degree at a given temperature. During vanadium and titanium recovery by small-sized coke smaller values of residue content of their oxides in the recovery products in comparison with the ferrosilicon recovery (0.25 % and 1.3 % for vanadium, 0.08 % and 0.60 % for titanium respectively).

According to the estimated data presented in Figures 3-6, during carbon silicothermal recovery (by carbon from small-sized coke and silicon from ferrosilicon) the content of iron oxides in the recovery products of converter vanadium slag at a specific consumption of small-sized coke more than 0.1 kg does not depend on the consumption of FeSi75. Content of vanadium oxides decreases as the reducing agents (at any proportion) consumption increases and reaches 0.04 % at a specific consumption of small-sized coke more than 0.14 kg/kg of slag and ferrosilicon more than 0.2 kg/kg of slag. Content of titanium oxides in the slag less than 2 % is achieved at the specific consumption of FeSi75 more than 0.1 kg/kg of slag at any consumption rate of small-sized coke. The minimum number of titanium oxides is approximately 0.01 % at maximum consumptions of both reducing agents. Content of manganese oxide in the recovery products of converter vanadium slag is reduced via reduction by small-sized coke only or by ferrosilicon up to 1%.

During joint reduction of manganese from its monoxide by carbon from small-sized coke and silicon from ferrosilicon in the specified range of reducing agents consumptions we can achieve the content of oxide in the slag 0.5 % (Figure 6). It should be noted that the content of vanadium oxides at such reducing agents consumptions is less than 0.05 %. Manganese recovery is limited under these conditions.

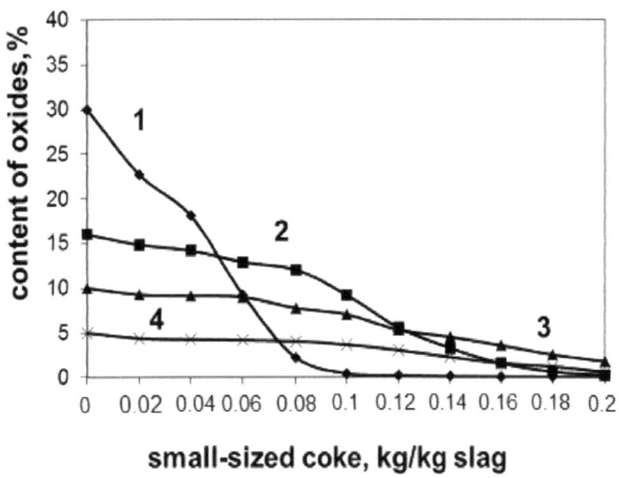

Figure 1. Dependence of the content of iron, vanadium, manganese and titanium oxides in slag on quantity of small-sized coke at 1873K (1 – iron oxides, 2 – vanadium oxides, 3 – manganese oxide, 4 – titanium oxides).

IOP Conf. Series: Materials Science and Engineering **150** (2016) 012016 doi:10.1088/1757-899X/150/1/012016

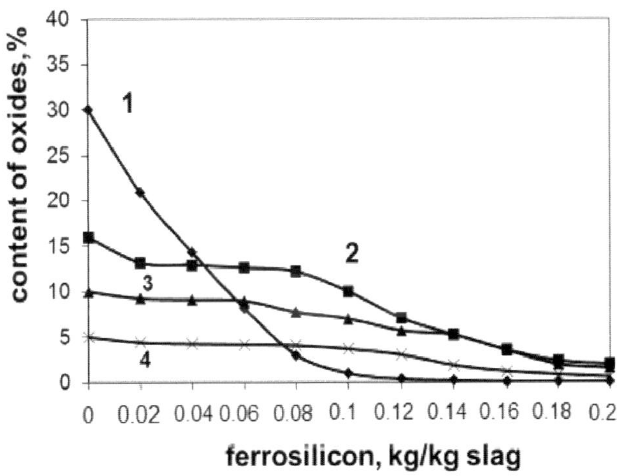

Figure 2. Dependence of the content of iron, vanadium, manganese and titanium oxides in slag on quantity of ferrosilicon at 1873K (1 – iron oxides, 2 – vanadium oxides, 3 – manganese oxide, 4 – titanium oxides).

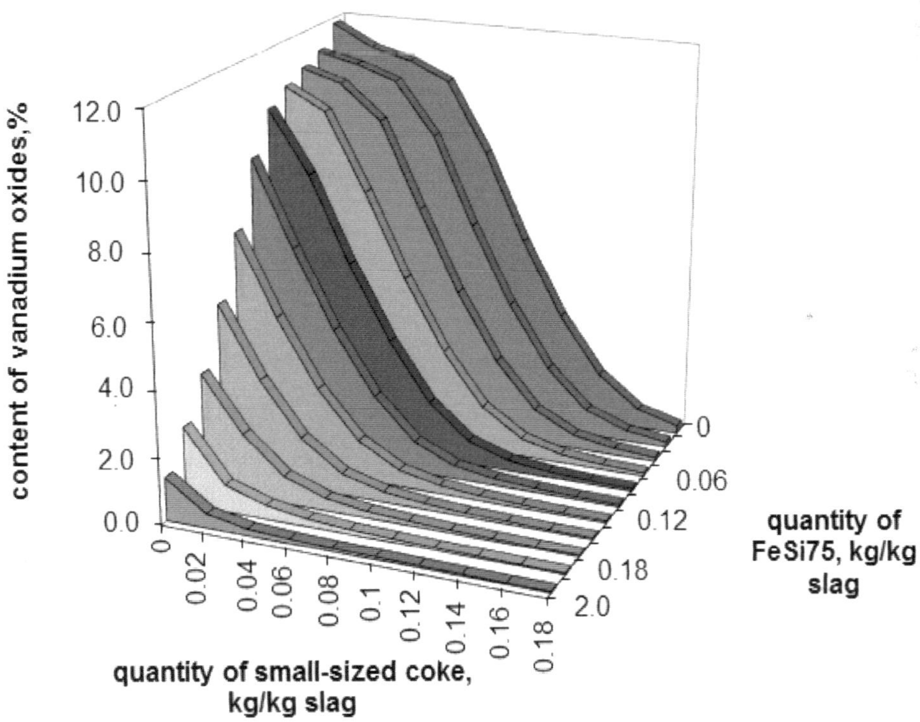

Figure 3. Change in the content of vanadium oxides in the slag during joint reduction by carbon from small-sized coke and silicon from ferrosilicon.

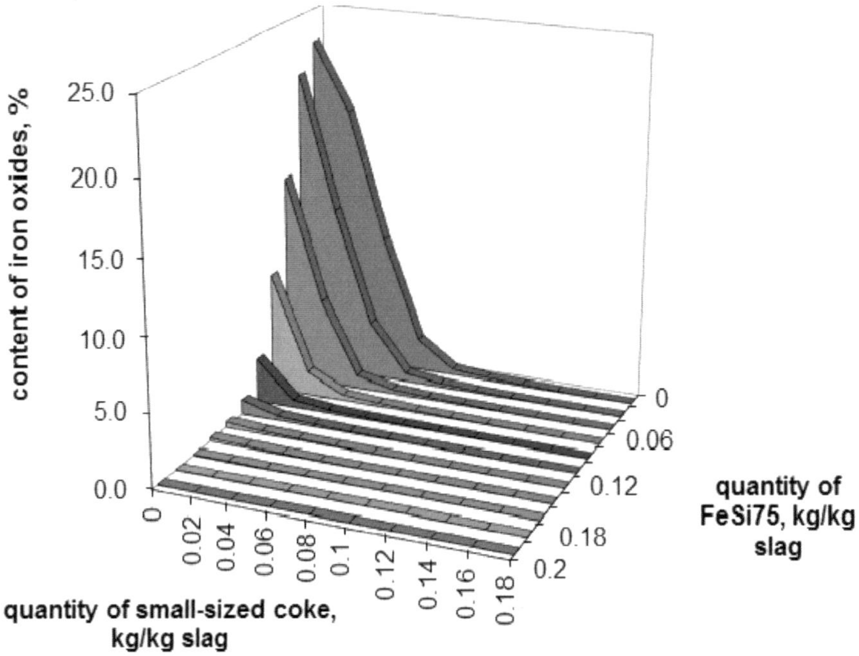

Figure 4. Change in the content of iron oxides in the slag during joint reduction by carbon from small-sized coke and silicon from ferrosilicon.

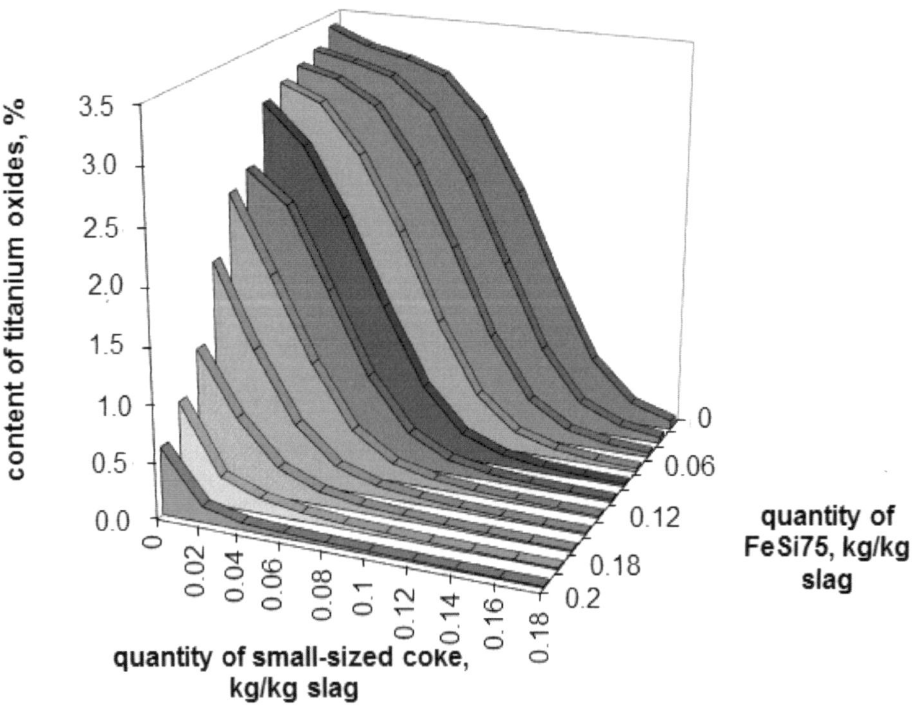

Figure 5. Change in the content of titanium oxides in the slag during joint reduction by carbon from small-sized coke and silicon from ferrosilicon.

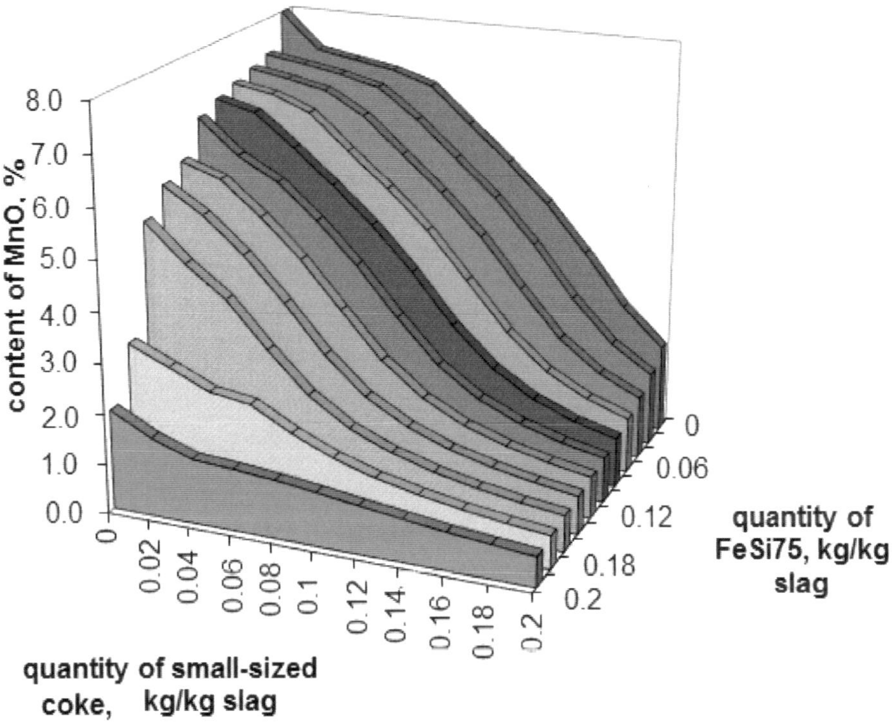

Figure 6. Change in the content of manganese oxide in the slag during joint reduction by carbon from small-sized coke and silicon from ferrosilicon.

3. Conclusions

The results of the calculations of the content of vanadium carbide and metal vanadium in recovery products during joint reduction of converter vanadium slag by carbon from small-sized coke and silicon from ferrosilicon showed that vanadium recovery takes place with the formation of vanadium carbide. At a given quantity of reducing agents (up to 0.2 kg of small-sized coke and 0.25 kg of ferrosilicon) iron and vanadium from converter vanadium slag are almost completely recovered [11-12]. Titanium in recovery products of converter vanadium slag according to estimated data are also in the form of titanium carbide. When the quantity of reducing agents consumption is less 0.20 kg per 1 kg of converter vanadium slag full recovery of titanium and manganese from their oxides does not occur even at maximum consumptions of both reducing agents [11-12].

The results of thermodynamic calculations proves that at given specific consumptions of reducing agents the carbon silicothermal recovery of vanadium and iron from converter vanadium slag by carbon and silicon proceeds quite successfully alongside with the limited recovery of titanium and manganese.

4. References

[1] Langeborg R, Siwecki T et al 1999 Scand. *J. Metallurg* **28** (5) 186–241
[2] Zajac S 2007 *Proc. of Int. Sc. and Tech. Seminar* (Ekaterinburg: UrO RAN) pp. 80–95
[3] Deryabin A A, Kozyrev N A et al 1998 *Stal'* **2** 19–21
[4] Vatolin N A et al *Thermodynamic Modeling of High Temperature Inorganic Systems* (M.: Metallurgy) p 175
[5] Trusov B G 2005 *Proc. of III Int. Symp. on Combustion and Plasmochemistry* (Almaty, Kazakhstan) pp 52–57
[6] Sinyarev G B et al 1982 *Computer Application for Thermodynamic Calculations of*

Metallurgical Processes (Moscow: Nauka) p 132
[7] Belov G et al 2013 *Thermodynamic Modeling of Chemically Reacting Systems* (M.: Bauman MGTU) p 96
[8] Klimov V Yu et al 2006 *Izvestiya vuzov. Ferrous metallurgy* 455–64
[9] Rybenko I A et al 2015 *Simulation and Optimization of Stationary Regimes of Metallurgical Processes* (Novokuznetsk: SibGIU) p170
[10] Rybenko I A et al 2010 *Steel in Translation* **40** (4) 310–313
[11] Golodova M A et al 2011 *Izvestiya vuzov. Ferrous metallurgy* **4** 3–5
[12] Golodova M A et al 2015 *Technical Sciences – From Theory to Practice* **48–49** 70–77

Simulation of heterogeneous flow and the problem of application of the Navier-Stokes equations

V P Tsymbal[1], P A Sechenov[1], A A Olennikov[2] and A G Padalko[1]

[1] Institute of Information Technologies and Automation Systems, Siberian State Industrial University, 42 Kirova Street, Novokuznetsk, 654007, Russia
[2] Institute of Architecture and Construction, Siberian State Industrial University, 42 Kirova Street, Novokuznetsk, 654007, Russia

E-mail: tsymbal33@mail.ru

Abstract. The article describes a dissipative structure – gravity separator in the jet-emulsion reactor. The simulation model is based on "first principles" and the Monte Carlo method of statistical tests. Dispersed particles of the charge and reaction products, their possible transformations and variants of interactions are treated as the first level. The core of this model is the process flow of the condensed particles in the vertical gas flow. Taking into account the experience of simulation and the presented considerations the concept of turbulent viscosity for heterogeneous flow is defined, the task and the direction of further investigations are defined.

1. Introduction

As a result of multi-year work the team of scientists, experts and designers has been developed and brought to the level of a large-scale pilot plant at West Siberian Metal Works (Zapsibmetkombinat) [1, 2] a new metallurgical process and JER unit (self-organizing jet emulsion reactor). Control over the chemical composition of the metal in the unit JER is achieved by a regime of dynamic organization of dissipative structures [3], playing the role of "chambers", which were planned to be designed in the units, when the first work on the creation of continuous steelmaking processes was stared [4, 5].

It was an attempt to use some of the ideas of self-organization theory (synergy), such as the principle of subsidiarity and the least coercion, a large deviation from thermodynamic equilibrium, and others. To implement these principles could be used a number of interesting physical effects, such as dispersion of charge colliding jets of gas, the creation of self-organizing of the reactor-oscillator by using the effect of critical flow of two-phase flow and feedback to change the gas content of the reaction products, lower supply of combustible mixture of oscillator reactor column reactor, the organization of the internal pneumatic working mixture.

The most important issues the implementation of the process are the mathematical models. The band model of the process was discussed earlier [1, 3]. Here we shall focus on one of the most important of dissipative structures – gravity separator, which is formed in a vertical column reactor at a certain flow regimes of two-phase mixture flow, prepared in the reactor-oscillator.

2. Simulation model of gravity separator and the particles flow task in the turbulent viscosity

2.1. Simulation model of gravity separator

Preparing the mathematical description of this important dissipative structure we managed to realize a simulation model based on the use of the Monte Carlo method [6, 7] and "first principles" In this case, as the first level are considered the disperse particles of the charge and reaction products, as well as all possible transformations and their interactions. The core of this model is the process flow of the condensed particles of vertical gas flow.

First, let us consider the mechanism of soaring and interaction of dispersed particles of the charge and reaction products flow in the vertical carrier gas.

Figure 1 shows the forces acting on the particle: the power flow pressure directed from bottom to top, gravity and buoyancy force. Depending on the ratio of these forces the particle can raise, fall or remain motionless.

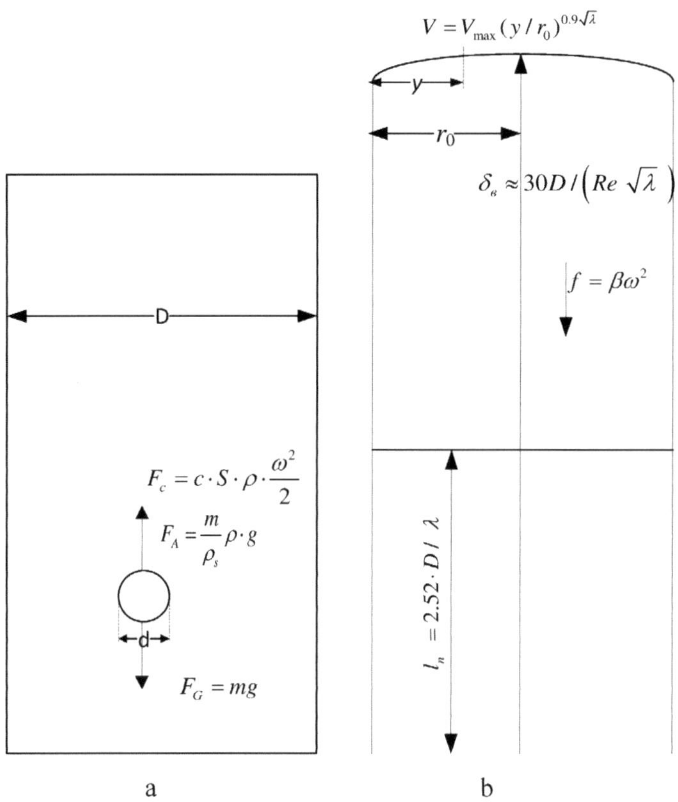

Figure 1. Forces acting on a particle in the gravitational separator a), velocity profile for turbulent flow b).

In the context of the dynamic equilibrium d'Alembert principle for moving particles leads to the equation [8]:

$$F_G - F_C - F_A = m\frac{d\omega_s}{dt} \quad (1)$$

where ω_s – the motion speed the spherical particles. Gravity is expressed as follows:

$$F_G = mg \quad (2)$$

Flow pressure force:

$$F_c = c \cdot S \cdot \rho \cdot \frac{\omega^2}{2} \quad (3)$$

where S – sectional area of the heat streamlined along amidships (projection area of the body in a plane perpendicular to the vectors of the incoming flow rate); ρ – density of the mixture; ω – mixture flow velocity; c – drag coefficient, which depends on the shape of the streamlined body and the Reynolds number. For spherical particles in the turbulent regime, the ratio $c = 0.44$.

Lifting or buoyancy force:

$$F_A = \frac{m}{\rho_s} \rho \cdot g \quad (4)$$

From equations (1) – (4) we get:

$$\frac{d\omega_s}{dt} = g\left(1 - \frac{\rho}{\rho_s}\right) - \frac{c \cdot \rho \cdot \omega^2}{2 \cdot m} \cdot S \quad (5)$$

The parabolic velocity distribution in turbulent flows in round pipes takes place not at the beginning of the pipe, but at some distance from the inlet section of l_n according to the formula [9]:

$$l_n = 2.52 \cdot D / \lambda \quad (6)$$

where D – diameter of the channel, λ – Darcy coefficient.

For the turbulent motion the coefficient of Darcy is determined by formula G. Blasius:

$$\lambda = 0.3164 / Re^{0.25} \quad (7)$$

The rate curve is calculated as Altshul [10]:

$$V = V_{max} (y / r_0)^{0.9\sqrt{\lambda}} \quad (8)$$

where y – the distance from the point in which speed to the tube wall is calculated; r_0 – column reactor radius.

Normal law of distribution for grain size composition of the ore, coal and other particles is generated using the method of Neumann.

Let us turn to the consideration of particle collisions. In the simulation model, two types of stress are distinguished: elastic (for unmelted particles) and inelastic (for molten particles). For inelastic collision, the rate of the combined particle:

$$\omega = \frac{m_1 \overline{\omega}_1 + m_2 \overline{\omega}_2}{m_1 + m_2} \quad (9)$$

For elastic shock from the law of conservation of momentum and energy we obtain:

$$\omega_1 = \bar{\omega}_1 \frac{(m_1 - m_2)}{(m_1 + m_2)} + \bar{\omega}_2 \frac{2m_2}{(m_1 + m_2)}$$

$$\omega_2 = \bar{\omega}_1 \frac{2m_1}{(m_1 + m_2)} + \bar{\omega}_2 \frac{(m_2 - m_1)}{(m_1 + m_2)} \tag{10}$$

The processes occurring in the upper part of the reactor [1, 3] determine the character of changes in the composition of condensed substances and are mostly reducing.

There are following main reactions:

$$(FeO) + \{CO\} = \{CO_2\} + [Fe] \tag{1}$$

$$(FeO) + C_s = \{CO\} + [Fe] \tag{2}$$

$$\{CO_2\} + C_s = \{CO\} \tag{3}$$

Based on these reactions, the following mechanisms of interaction particles were selected. Into the lower part of the column reactor with the gas flow the iron ores of predetermined granulation composition are fed according to the normal distribution law. The composition of the iron ore and gas entering the reactor is taken in accordance with the thermodynamic and balance calculations [1]. Also solid carbon particles can flow into the reactor. At the collision of two types of particles: iron ore - iron ore, carbon - carbon and iron ore – carbon the impacts are considered elastic. As soon as the iron ore particles are melted by more than 50% at a collision iron ore - carbon impacts are considered inelastic. At a collision of two particles, such as slag-slag and iron-iron the impacts are inelastic. After the specific particle of iron ore melts, two new particles of iron and slag are formed: iron and slag, the average composition and density of which are adopted in accordance with the calculations of balance and thermodynamics, and then in accordance with the normal distribution law.

Figure 2 shows a fragment of the work created by the simulation model as well as a graphical representation of the simulation results.

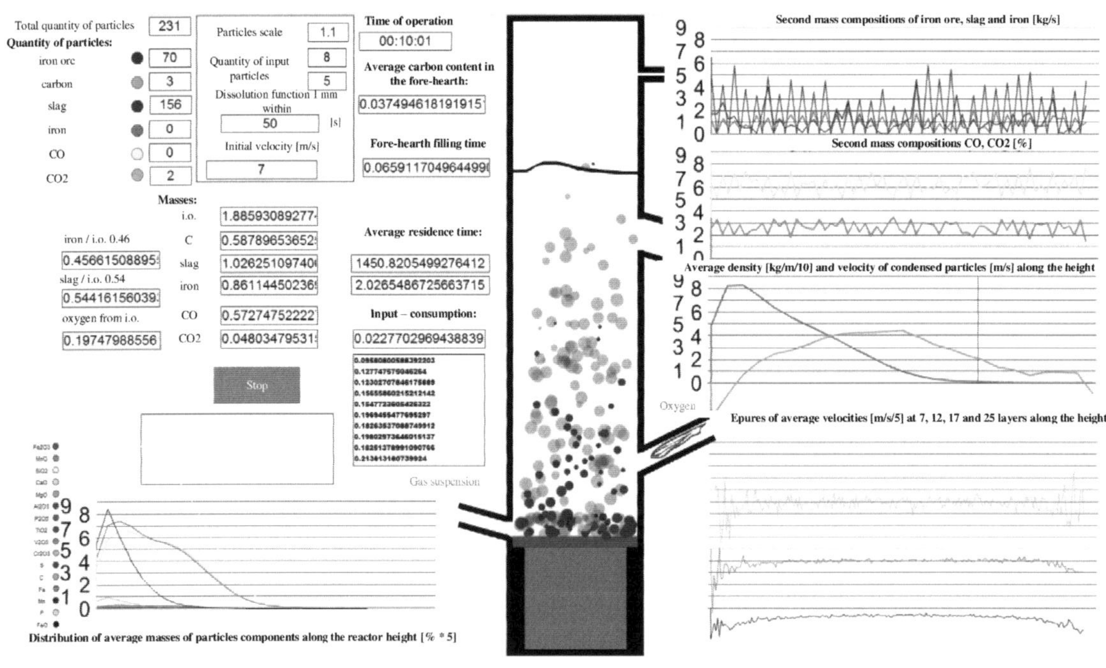

Figure 2. Fragment of implemented simulation model of a gravity separator.

Alongside with displaying the results of movement and interactions of particles the information is displayed: total number of particles in the reactor, amount of iron ore, carbon, slag, iron, CO, CO_2; average masses of iron ore, slag, and iron per minute; graphs of mass of the corresponding particle. Also the quantity of feeded particles of iron ore and carbon can be changed.

Testing showed efficiency of this algorithm and satisfactory agreement of final simulation results with the law of conservation of matter. The developed simulation model, which is to a certain extent the object substitute, allows the mechanism of processes studying in the gravity separator to be visualized with different particle size distribution, the average residence time of the particles to be estimated, the average consumption of iron and slag particles at different predetermined amount of carbon particles to be determined.

2.2. The task of the particle flow and turbulent viscosity

Naturally, the problem of flow even of a homogeneous turbulent flow is quite complex, in our case, there is a heterogeneous flow around the particles. It is obvious that a closed analytical solution of this problem is hardly possible, especially if to take into account the fact, as shown by O.A. Ladyzhenskaya [11], that in the case of turbulent motion the unique solution of the Navier-Stokes equations for the general form of the initial and boundary conditions does not exist. In this regard, one approach, as illustrated above by creating a specific gravity separator simulation model, to this problem is decomposed into two tasks:

- the task of a specific particle flow-over by a flow at a sufficiently short interval of length, depending on the level of turbulence;
- statistical modeling of the nature of particle interaction by Monte Carlo method.

Therefore, it is very important to understand what such turbulent viscosity and viscous friction coefficient in the Navier-Stokes equations are.

J. Stokes received the solution of the problem of a spherical body flow-over with by viscous fluid, found the distribution of velocities field around the ball and the force acting on it (Stokes' law):

$$F = 6\pi\eta R\upsilon_0 \tag{11}$$

where R – particle radius; υ_0 – ball movement relative to fluid velocity.

During the body flow-over alongside with the drag the components of viscous forces might evolve, orthogonal to the velocity vector of the body motion relative to the fluid (lift force of Zhukovsky).

Normal viscosity – a macroscopic manifestation of molecular motion and mixing. Therefore, equation (11) for a single particle, as well as in the given below equation (12) for the pressures field $\rho(\vec{r},t)$ and velocities field $\vec{U}(\vec{r},t)$

$$\rho\left[\frac{\partial\vec{U}(\vec{r},t)}{\partial t} + (\vec{U}(\vec{r},t),\vec{\nabla})\vec{U}\right] = -\vec{\nabla}\rho - \eta\Delta\vec{U}(\vec{r},t) \tag{12}$$

are valid only if the speed does not exceed the critical value determined by the Reynolds number, i.e. until the real flows of current lines do not become "tortuous".

L. Prandtl in 1925 proposed a formula for the turbulent viscosity:

$$\eta_T = l^2\left|\frac{\partial\bar{U}}{\partial y}\right| \tag{13}$$

where l – mixing length is determined from experience;

$\bar{U}$ – average flow velocity, and T. von Karman based on the similarity of the velocity fluctuations at different points assumption fluid concluded that the shear stresses (viscous force) depend only on the average velocity and its derivatives and added to the expression of Prantl with the mixing ratio of length

$$l \cong k\left|\frac{\partial \bar{U}}{\partial y} \Big/ \frac{\partial^2 \bar{U}}{\partial y^2}\right| \tag{14}$$

where k – dimensionless coefficient.

In these expressions two important concepts are formulated: mixing length (the correlation length of the field of turbulent fluctuations) and velocity fluctuations similarity property.

Turbulent viscosity can be studied by measuring the increasing force of the flown-over body with flow, which is conveniently characterized by one parameter – the Reynolds number.

The simplest approach is measuring the force acting on the flown-over cylinder. Instead of force [11] it is decided to fix the drag coefficient C_d, representing the ratio of the force acting on the cylinder to the value $\rho\upsilon_0^2 Dl$ (D – diameter, l – length of the cylinder), which characterizes the pressure force exerted by the flow onto the orthogonal to it cylinder section:

$$C_d \approx \frac{F_\eta}{\rho\upsilon_0^2 Dl} \tag{15}$$

Since the viscous force acting on the laminar flow in a streamlined cylinder is proportional to the viscosity and the average speed, as in the Stokes' law $F = 6\pi\eta R\upsilon_0$, then taking into account (15) in laminar flow entrainment factor must decrease with the increase υ_0: $C_d \approx 1/\upsilon_0$.

3. Conclusions

With regard to the described above experience of modelling and based on the presented considerations which allowed the concept of turbulent viscosity to be clarified, including for heterogeneous flow, the task setting and directions for further investigations identified.

1. Improvement of the developed simulation model of the vertical flow movement of a two-phase medium in relation to the column reactor of unit SER:
   - verification of the compliance of the statistical results of this model with conservation laws;
   - obtain specific statistical regularities: the rate curve over the radius and height, the mixture density, the residence time of the charge particles and other reaction products.

Figure 3 shows a fragment of the experiment targeting determination of the turbulent viscosity on the installation of physical modeling.

2. Creation of a similar simulation model for the system – air, water, solid particles to be implemented in the automation unit of physical modeling and the matching of the results to the Reynolds number:
   - study the viscosity by lowering the speed of spherical particles, as well as the coefficient of hobbies measurement (force);
   - study on the diffusion rate (dissolution rate) of spherical particles at different Reynolds numbers, verification of the relationship between viscosity and diffusion coefficient;
   - measuring the sizes and velocities of fluctuations at different Reynolds numbers to justify the application of (13) and (14).

3. Transfer of the results to the real unit model.

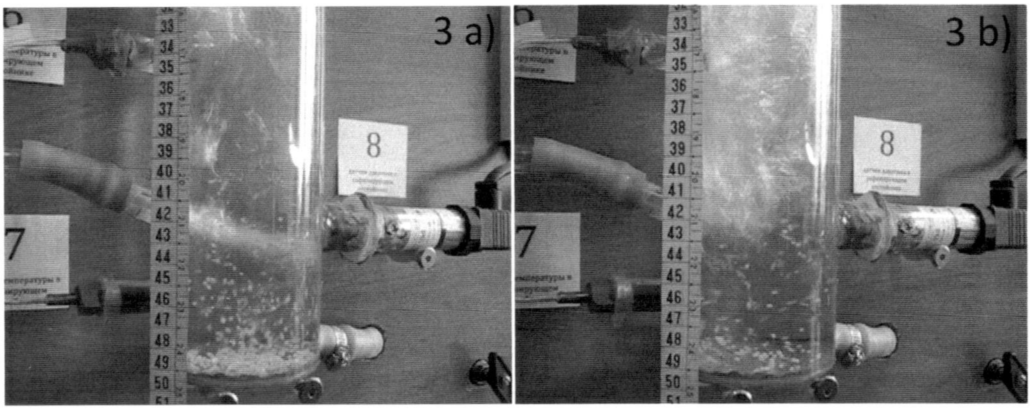

Figure 3. Start the gas suspension feeding (gas and water) into the column reactor (water and particles) as well (a); turbulence of the gas suspension and solid particles (b).

Getting generalized laws to solve problems of this type.

To solve this problem we plan to use the developed at the Department of ITM the automated installation of low-temperature physical modeling, which gives an opportunity to verify the adequacy of the specially created a similar simulation model for the system – air, water, solid particles, and also link them to the Reynolds number for possible transfer to the actual process.

4. References

[1] Tsymbal V P et al 2014 *JER Process – Metallurgical Jet Emulsion Reactor* (M.: Metallurgizdat) p 488

[2] Tsymbal V P et al 2015 *Springer Journals* **59** 119–125

[3] Tsymbal V P et al 2009 *Steel in Translation* **39** (4) 310–312

[4] Baptizmanskij V I et al 1975 *Proc. Meeting on Continuous Smelting Processes of Ferrous and Nonferrous Metals* (M.: Publishing Science) pp 24–28.

[5] Bigeev A M 1986 *Continuous Steelmaking Processes* (M.: Metallurgy) p 136

[6] Kroese D P et al 2014 *Why the Monte Carlo method is so important today* (WIREs Comput Stat 6) pp 386–392

[7] Kroese D P et al 2011 *Handbook of Monte Carlo Methods* (Wiley) p 743

[8] Lavrenko A T *Analysis of the Direct Coke Oven Gas Separator abstr.* Master's thesis http://www.uran.donetsk.ua

[9] Kramarenko V V et al 2009 *Hydraulics Part 1: Methodological Materials for the Course "Hydraulics"* (Tomsk: TPU) p 124

[10] Al'tshul' A D et al 1977 *Examples of Calculations on Hydraulics. Textbook for High Schools* (M.: Strynzdat) p 248

[11] Bakay A S et al 1980 *Multi-faced Turbulence* (M.: Knowledge) p 48

Background and principles of self-organizing jet-emulsion metallurgical unit

V P Tsymbal[1], V I Kozhemyachenko[1], I A Rybenko[1], S P Mochalov[1], A G Padalko[1], S N Kalashnikov[1], S Yu Krasnoperov[1], L A Ermakova[1] and A A Olennikov[2]

[1] Institute of Information Technologies and Automation Systems, Siberian State Industrial University, 42 Kirova Street, Novokuznetsk, 654007, Russia
[2] Institute of Architecture and Construction, Siberian State Industrial University, 42 Kirova Street, Novokuznetsk, 654007, Russia

E-mail: tsymbal33@mail.ru

Abstract. The basic principles laid in the creation of a new jet-emulsion process and metallurgical unit were considered. Development of self-organizing oscillator reactor, bottom feed of the prepared in it combustible mixture to the column reactor and a large deviation from the thermodynamic equilibrium allow dissipative structure to be created, and thus control the ratio of reducing and oxidizing processes, the carbon content in the metal.

1. Introduction

The authors on the basis of their multi-year research set a goal of describing in a compact and consistent manner the basic principles which were laid in the creation of a new jet-emulsion metallurgical process and a unit.

The idea, later realized in the self-organizing jet-emulsion process, appeared in 1983-85, when our research team began to study and comprehend the just appeared in the Russian translation of basic works on the theory of self-organization by I. Prigogine [1] and G. Haken [2], and then the examples [3] of practical application synergy concept and principles.

There was an idea to try to create a new process and machine, in which to some extent these principles could be implemented. Beside, the analysis of development and status of traditional metallurgy showed that the tendency to increase the single capacity of units in combination with the accepted in the iron production principle of charge materials pelletizing led to the clunkiness and energy inefficiency of the multistage technological scheme.

The idea of charge pelletizing is in contradiction with the principle of global energy saving (entropy) as a huge reaction surface of ore materials, crushed at the stage of enrichment, is reduced by one thousand times at a high cost of material and energy resources in agglomeration of coke production. Heating and melting of agglomerated materials require a significant residence duration and, consequently, a large amount of aggregates. The processes occur in the vicinity of the state of thermodynamic equilibrium.

Eliminating this irrationality and saving the large initial reaction surface of the charge aggregates was the first task, which was to be implemented in combination with the principles of self-organization, the most important of which is a large deviation from thermodynamic equilibrium,

Gauss's principle of least constraint (Le Chatelier-Brown) and the circular congruence. These principles, as it turned out later, we managed to implement constructively with the right technological parameters.

In more detail the sequence of the development and implementation of the idea of creating a new process we described in our monograph [4]. Below we explain the basic principles of the functioning of the process and the unit.

2. Basic principles

To implement the task of creation of a greater reaction surface we used the principle of charge dispersion by the counter gas jets followed by the organization of the forced movement of the formed gas slurry under pressure. As it turned out later with the scientists from the Institute of Thermal Physics SB RAS (Novosibirsk), the flow rate of the forced two-phase medium in the channel is in a substantially non-linear dependence on the gas content [5], as shown in Figure 1.

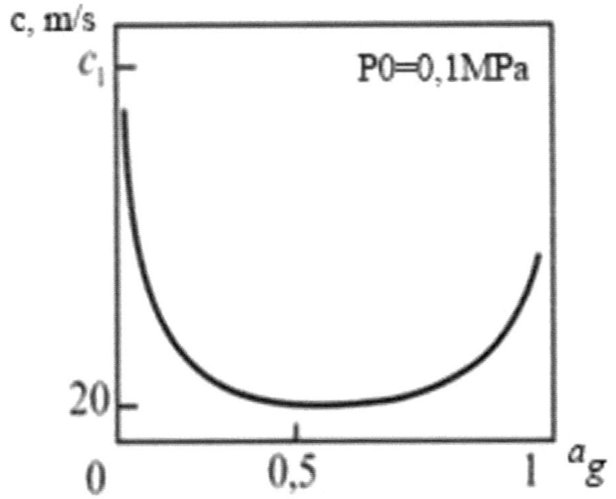

Figure 1. The dependence of the speed of sound in the water-air mixture on the gas content.

This phenomenon was used to create a gas-dynamic output channel lock in the developed based on this idea the reactor-oscillator (Figure 2), which implements an internal feedback due to the critical flow of two-phase medium [6] and the effect of pressure (on the principle of Le Chatelier-Brown) on chemical reactions with gassing. This leads to a change in the reactor volumetric gas content, $\alpha_g = \frac{v_g}{v_g + v_{kp}}$, where v_g – the amount of gas, v_{kp} – volume of the condensed phase, and then to a change in the outflow of a two-phase medium in the connecting chanel, which in turn leads to a change in pressure in the spherical reactor. The process "runs" on the right-hand member of the nonlinear dependence of the sound velocity *C* from the gas content (Figure 1), the outflow velocity of the two-phase medium from the connecting channel 3 (Figure 2) can be changed by several times.

The result is a possibility of creating a stationary oscillatory mode at any given pressure level. Figure 3 shows the nature of the agreed processes of vibrational pressure P and volumetric gas content αg in the reactor-physical model of the oscillator, and Figure 4 shows an example of recording in real time the pressure in the piston pump fuel supply of oil plasticized charge in the spherical reactor of the pilot unit. It can be seen here that in the real high-temperature reactor, with an inner diameter of 2 meters, due to the above-described two-phase gas-dynamic effect of locking jets the pressure of two atmospheres was created. Due to changes in the consumption of substance flows at the input and output, as well as changes in the diameter of the connecting channel 3 (Figure 2), a fixed level of pressure support in this reactor can be set.

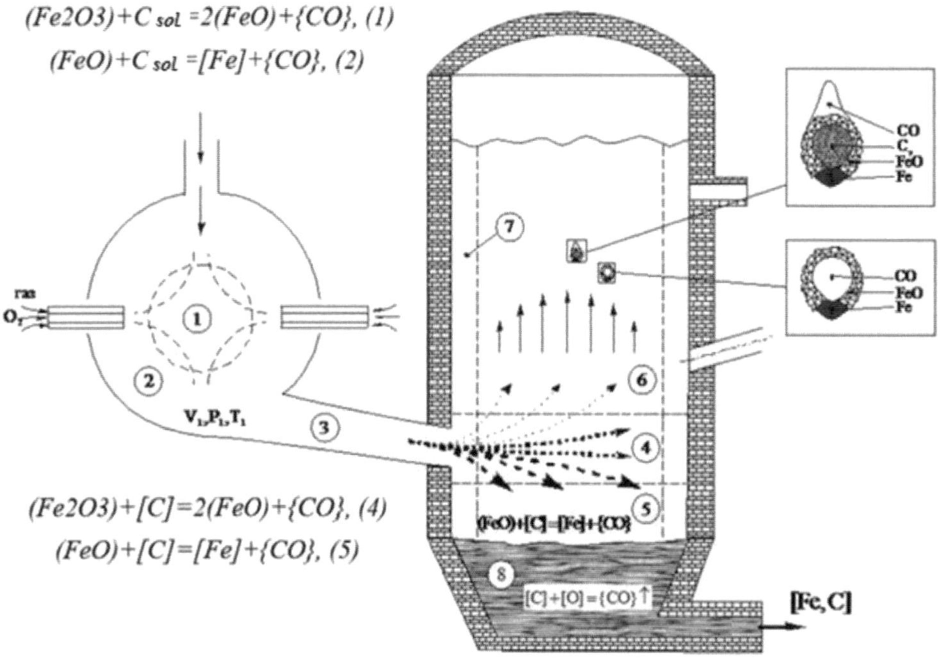

Figure 2. Zone model of the jet-emulsion unit.

Thus, based on the reactor-oscillator a stimulus flow (a kind of carb-compressor) was created, by which (in combination with the transition process in the area of the gas suspension and emulsion) it was possible to organize the internal transport of the working mixture through all elements of the unit, including heat-utilizing installations (waste heat boiler, gas turbine, fluidized bed, smoke reformer in the synthesis gas). This creates conditions for using deep fuel energy source (70-80%). High pressure and isolation from the atmosphere of the process allowed the energy of the expansion of the working mixture to be fully used (8.31 J/mol), i.e. "cork the imp inside the bottle", which helped to make the reaction mixture work in the right mode and perform different process operations, thanks to that the unit has a certain degree of flexibility.

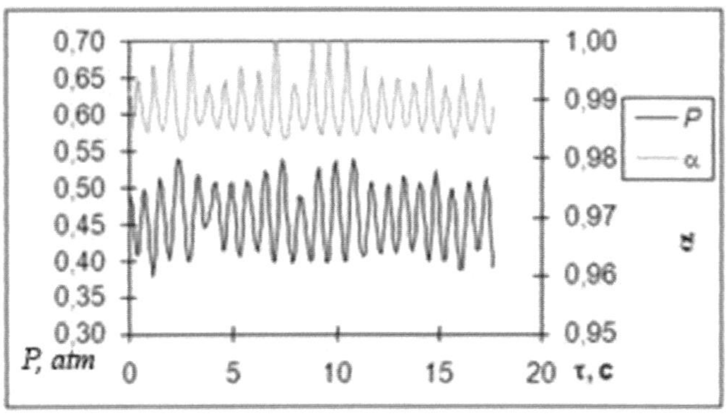

Figure 3. Oscillating processes in low-temperature physical model of the reactor-oscillator.

Another important decision was the lower supply of the prepared in the reactor-oscillator of the working mixture to the column reactor, on the slag-metal boundary. At the same time, the kinetic energy of the ejected from the connecting channel 3 jets is spent on the intensive bubbling gas slag emulsion together with embedded therein particles of ore materials, turning into the potential energy of pressure, forms a "pillow", a kind of a failing grid, which maintains a high post (about 2-3 m) of gas-slag-metal emulsion and gas-suspension (zones 6 and 7 in Figure 2). In zone 5a there are mainly regenerative processes occurring on the very developed contacts surfaces, on the iron oxide films, on CO bubbles and on the surface of oxide slag – solid carbon, according to the reactions:

$$(FeO)+\{CO\}=[Fe]+\{CO_2\};$$
$$(FeO)+C_s=[Fe]+\{CO\};$$
$$C_s+\{CO_2\}=2\{CO\}.$$

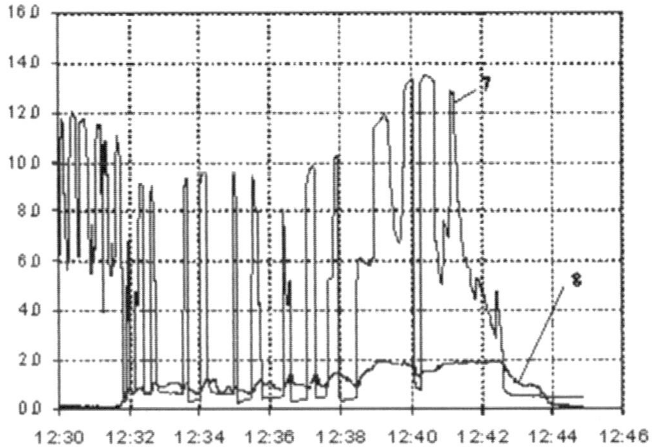

Figure 4. Graphs of pressure changes during the experimental smelting at the pilot unit (7 – pressure piston pump supplying the plasticized charge; 8 – pressure in the spherical reactor-oscillator, atm.

In the initial stages of continuous steelmaking process hearth and converter types [6, 7], an important and essential advantage of this process, together with the lack of intermediate energy and raw material losses, considered the existence of separate chambers (zones) for the implementation of the necessary technological operations (decarbonization, desulfurization, dephosphorization, etc.). However, at that time, due to the flow of the process is close to equilibrium and, consequently, low rates of chemical reactions, these cameras were produced quite cumbersome, and the unit as a whole capital-intensive.

Due to entering of the bottom feed mixture into the zone of gas suspensions and emulsion, and keeping the large deviation from the thermodynamic equilibrium, we manage to create (instead of physically realizable chambers) definite zones in the form dynamic dissipative structures that exist only under the condition of a certain kinetic mode. The desulfurization is achieved due to a large reaction surface on the gas bubbles and the dephosphorization problem can be solved by controlling the particle size distribution in the raw materials fed to the metal-slag boundary. On the problem of carbon content control we shall focus separately.

One of the important dissipative structures discussed in more detail in [8, 9], is a gravitational separator of metal, slag and gas, which is formed in the upper half of the column reactor (zone 6 and 7 in Figure 2). Due to the influence of the gravitational component and close to the parabolic distribution of emulsion speed vector in the cross section of the column reactor there is a separation of particles according to their density. Because of this phenomenon, the reduced iron particles having a density in two to three times greater than the particles of iron oxide, "slide" on the flow periphery and form the near-wall layer with negative (reverse) velocity (zone 7). We visually observed this phenomenon on a physical model.

Thus, taking into account counter gravitational and aerodynamic forces in the upper half of the column reactor (zone 6) there is a separation of metal and slag, and also uneven (in density, chemical composition and gas content) parameter distribution along the unit height takes place. Distributed content of iron oxides in height, which plays an important role in control over the ratio of oxidation and reduction processes, is also due to the nature of non-equilibrium thermodynamic processes, including the intensity of the flow of matter and energy coming from below through the connecting channel 3 from the reactor-oscillator 2.

3. Control over the metal chemical composition

Now, after getting acquainted with the basic principles of the functioning of the process and the role of dissipative structures, we can concentrate on the control over chemical composition of the metal, primarily carbon. To do this, let us turn again to the zone model (Figure 2).

Stressing that chemical reactions occur in non-equilibrium conditions with removal of the reaction products, we shall note that the reduction reactions of iron oxide by solid carbon

$$(Fe_2O_3) + C_{sol} = 2(FeO) + \{CO\}; \quad (1)$$
$$(FeO) + C_{sol} = [Fe] + \{CO\}; \quad (2)$$
$$C_{sol} = [C], \quad (3)$$

occur in zone 6 (the upper part of the column reactor), and the reactions with the carbon dissolved in metal

$$(Fe_2O_3) + [C] = 2(FeO) + \{CO\}; \quad (4)$$
$$(FeO) + [C] = [Fe] + \{CO\}, \quad (5)$$

take place at the boundary of zones 5 and 8 (turbulent layer on the boundary metal-slag) and have a velocity an order of magnitude greater rate [10], than reactions (1, 2). The zones 5 and 8 are separated by dynamic “pillow” (zone 4) from zone 6. The proportion of iron oxides processed at the boundary slag-metal (zones 5 and 8) can be increased by rising the fractions of larger ore materials and additional induction heating of the forehearth, and in zone 6 to raise the reduction potential due to supplying coke fines or coal to the column reactor to the upper level of gas-slag emulsion. Thus, possibility for creation the above-described dynamic dissipative structures under certain gasdynamic modes allows the speeds ratio of redox processes to be controlled separately and, consequently, control pver the carbon content in the produced metal.

Most of the experimental melts in the pilot unit of the new process at Zapsibmetkombinat gave the carbon content in the range 0.2-1.7%, and after some melts the carbon content was 0.04-0.10%, but this was due either to long taphole opening, or a significant slag overoxidation due to excess of oxygen. It should be noted that the pilot unit could not realize continuous but continuous periodical mode, also in the actual technology, due to the low inertia of the process, the implementation of such mode (with periodic release and leaving the “swamp” in the forehearth) is possible.

The following table 1 shows two examples of the direct production of metal with carbon contents 0.44 and 1.72%, corresponding to these values of concentration of iron oxides in the slag, and the main control values: mass ratio of the charge to mass ratio of reducing agents.

Table 1. Examples of direct production of metal.

Experiment No.	C	FeO	Fe_2O_3	Mass ratio charge / reductant	Notes
20	0.44	37.84	8.36	2.89	Slag on the level of the connecting channel
20		34.04	5.45	Slag from the wall of the column reactor at a height of 2 meters from the connecting channel	
29	1.72	6.56	2.29	2.31	The slag on the metal release

Also the concentrations of iron oxides are given in two slag samples taken after the experiment No. 20 (on the occasion of the machine stopping for reconstruction) at the upper edge of the forehearth and at the height of 2 m on the inner wall of the column reactor, indicating that the iron oxide content is distributed along the reactor height.

4. Comparison with the known processes of direct reduction

In the last 20-25 years, a lot of attention is paid to the development of direct reduction processes (see the review in [4]). More than 100 direct reduction plants have been built or are in the process of construction. Among the solid-phase direct reduction processes "Midrex" is the most advanced but, in fact, this process is hardly straightforward, as the same global energy illogic related to pelletization is repeated as in the conventional technology. Moreover, if the iron in the blast furnace is obtained in the liquid form, in this process the solid intermediate is produced. We should not forget that the arc energy used for the subsequent pellets or briquettes melting comes mainly from thermal power plants, where the thermal efficiency of about 37%, that is, only one tonne of coal reaches the arc, while the blast furnace coke consumption is less than 400 kg/t. In the solid phase reduction the economy is obtained only due to the lack of coke production, but only rich ores can be processed by this process.

In this sense, the liquid-phase reduction processes are being developed: Romelt [11], Hismelt, Ausmelt, DIAS, etc. [4], and have certain advantages. In these processes, which do not require sintering, dusts may be processed, including poor ores and metal-containing waste (sludge, scale). As a reducing agent energy coal is used; the constructive decisions on the use of exhaust gas energy are developed.

However, by none of the known processes the direct reduction of steel cannot be obtained. Cast iron with a carbon content of 2 to 4% is produced. This is due to the fact that the processes are close to thermodynamic equilibrium, and the metal and slag contact with each other for a long time in. Independent control over the metal composition and slag as well as the period of their residence in these units is practically impossible.

Among the combined processes of direct reduction Corex process should be mentioned, which was industrially implemented and has achieved some progress. But it also turned out to be rather clunky, it also has failed to avoid the stage of pelletizing. For the pre-reduction of oxidized pellets in this unit a shaft furnace with volume of 1000 cubic meters is used installed over the liquid-phase reduction unit. All together this building is about 100 m high.

It is interesting to draw an analogy with the considered in this article unit SER, which, due to the refusal of the agglomeration stage, the role of the preparatory unit plays a spherical reactor-oscillator with volume only a few cubic meters. It is also the compressor, which prepares and pushes the mixture through all devices of the unit.

5. Design and technological features of SER unit

After such comparison in the conclusion of the paper we shall consider the main features of the design and technological solutions in SER unit.

The basis of the technological scheme of the unit (Figure 5) is as follows: charge supply system 1-5; reactor-oscillator 6; a connecting channel with a gas-dynamic self-locking 7; refining sump 8; forehearth simultaneously acting as a first stage of wet gas purification 9; and skull cooling system 10, a channel for gas flow 11; slag tank 13 with a granulator 14; the heat utilization system in the fluidized bed or the flue gas reformation in the synthesis gas 17 and gas cleaning system 18.

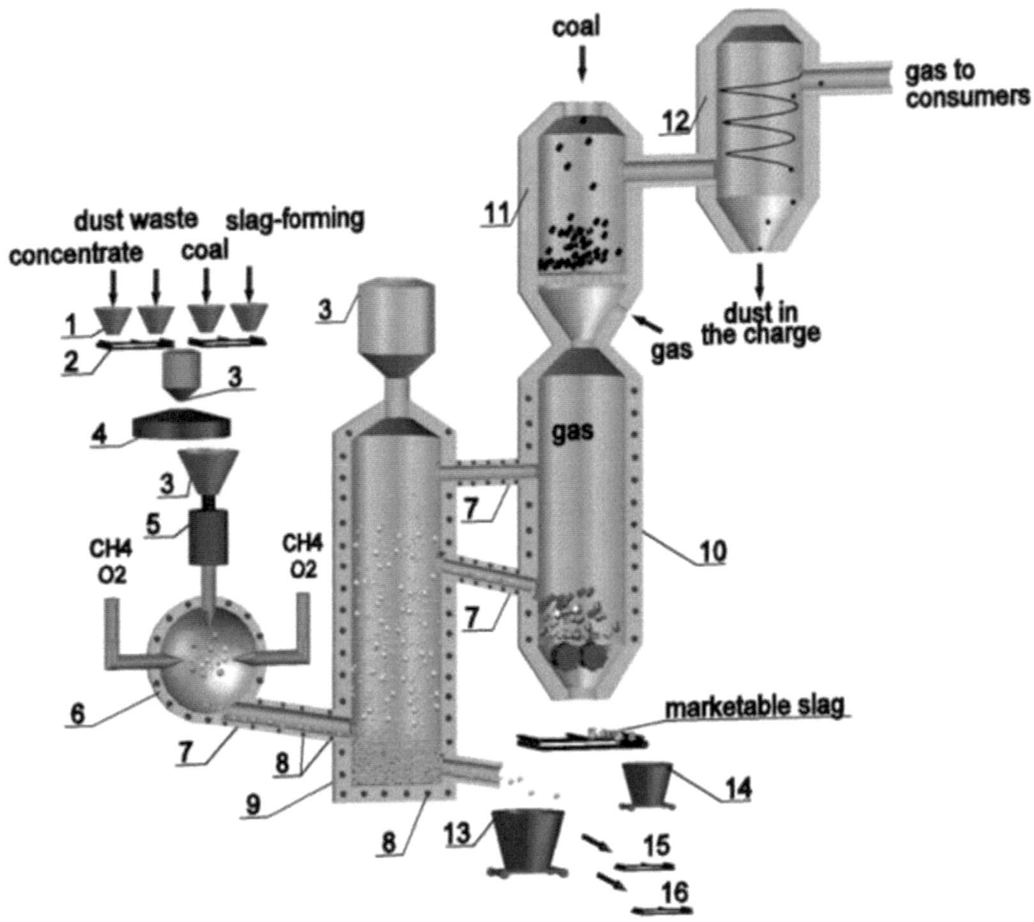

Figure 5. Technological mini-module scheme based on SER process.

Use as a refining sump of the vertical column reactor with bottom feeding of the reaction gas suspension in combination with a significant deviation of processes from thermodynamic equilibrium is the major factor that determines the possibility of separating the iron metal flow flowing along the periphery of the column reactor and settling in the forehearth 9 (Figure 5), and iron depleted slag flow withdrawn via an inclined channel 11 into the slag receiver 13.

Dustlike charge consisting of a mixture of metal oxides and solid reductants is effectively dispersed in the reaction chamber 6. This creates a suspension of gas volume fraction of the order of 0.99. As an effect of dependence of the speed of the two-phase medium flow on the gas content (Figure 1), in the connecting channel 7 aerodynamically lockable gate is formed (jet braking), and in combination with the feedback in gas content (by changing the conditions of chemical reactions, as well as the supply conditions in the batch and oxygen feeding into the reactor) it is possible to create a stationary oscillatory mode (self-organizing reactor-oscillator). The created in reactor-oscillators 6 a high pressure potential and the total process isolation from the atmosphere, allows products through reactions to be pushed through all energy utilization devices, located after the main technological device, without the usage of high flow boosters.

Figure 6 shows a general view of the experimental unit SER in the converter shop of the West Siberian Metallurgical Combine.

Figure 6. General view of unit after regular reconstruction.

In the foreground there is a column reactor, at the top of the "stack" is a system of metering and feeding the batch into the reactor-oscillator located at the bottom in the closed chamber.

Based on the considered unit a mini-mill of a new structure in the from of operating independently from each other and in parallel processing units can be designed, instead of succession of large units in the smelting cycle. The implementation of the principle of continuity eliminates the need for heavy removable equipment (casting ladles, heavy-duty bridge cranes, heavy foundations, etc.). The units can be placed in a relatively light buildings, enabling fast return on investments. The modules may operate independently from each other and be stopped in a few minutes. The proposed development opens the possibility for creating a mini-metallurgy of "full cycle", without use of scrap, shortages and price of which is steadily rising. At the same time the problem of obtaining primary metal without non-oxidized impurities (copper and nickel) is solved, as well as using the powdered metal waste (scale, sludge, small chips, etc.). Creation of own mini industry is also very important for the machine-building plants, many of which have metallurgical industry in an outdated state.

6. Conclusions

- The basic principles that were laid in the creation of a new jet-emulsion process and metallurgical unit are considered.
- It is shown that due to the lower supply of the prepared in the reactor-oscillator working mixture (gas suspension) into the a column reactor, as well as a large deviation from thermodynamic equilibrium, there are dissipative structures, allowing the ratio of reducing and oxidizing processes to be controlled, and, consequently, control over carbon content in metal.
- The features of the structure and functioning of the SER unit are considered, allowing the described principles to be implemented.
- Thanks to the capacities of direct the processing in this unit of dusty ores and wastes (without agglomeration), the framework for designing of compact mini-mills with a full production cycle, from ore to steel, is developed.

7. References

[1] Nikolis G and Prigoshin I 1979 *Self-organization in Nonequilibrium Systems* (M.: Mir) p 512
[2] Khaken G 1980 Synergy (M.: Mir) p 406
[3] Kurdyumov S P and Malinetskii G G 1983 *New in Life, Science, Technology, Series "Mathematics, Cybernetics"* (M.: Knowledge) **6** p 48
[4] Tsymbal V P et al 2014 *The Process of SER – Metallurgical Jet Emulsion Reactor* (M.: Metallurgy) p 488
[5] Nakoryakov V E et al 1990 *Wave Dynamics of Gas and Liquid-vapor Environments* (M.: Energoatomisdat) p 248
[6] *Proc. Meet. On Continuous Process of Smelting Ferrous and Nonferrous Metal* (1975) (M.: Nauka) pp 14–28, 55–64, 160–167.
[7] Bigeev A M 1986 *Continuous Steelmaking Processes* (M.: Metallurgy) p 136
[8] Tsymbal V P et al 2014 *Proc. XIII Int. Cong. Steelmakers* (M.: Polevskoy) p 472–477
[9] Tsymbal V P et al 2015 *Metallurgist* **59** (3) 119–125
[10] Telegin I A et al 1993 *Ferrous metallurgy* **6** 10–4
[11] Roments V A 2005 *Process Romelt* (M.: MISiS) p 400

Study of dioxins formation in the basic oxygen furnace during organic waste disposal

S N Kuznetsov, E P Volynkina and E V Protopopov

Institute of Metallurgy and Material Science, Siberian State Industrial University, 42 Kirova Street, Novokuznetsk, 654007, Russia

E-mail:kaf.erc@sibsiu.ru

Abstract. Incineration is one of the most common methods used for organic waste utilization. However, there is a danger of the secondary generation of such supertoxicants as dioxins and furans. The results of the investigations of experimental and comparative converter meltings with the use of paper and plastic wastes under conditions of JSC "EVRAZ ZSMK" show the absence of influence of wastes on the concentration and isomeric profile of dioxins and furans in the converter gases.

1. Introduction

High temperature incineration is one of the most widely accepted methods used in organic wastes disposal [1-5]. As noted in [1], the incineration is applicable to virtually all types of organic wastes, since it transforms the main elements of most organic compounds (carbon, hydrogen, oxygen) into environmentally safe CO_2 and H_2O, and in the presence of chlorine – into chlorine hydride recycled in the wet cleaning system. However, to ensure the full decomposition of hazardous organic compounds it is necessary to observe a number of conditions:

- incineration temperature not less than 1200 °C;
- residence time of incineration products in the incineration chamber is not less than 2 s (rule
- excess of oxygen in the combustion zone;
- rapid cooling of converter gases to a temperature below 200 °C, which makes secondary generation of dioxins impossible.

Failure to comply with these requirements may result in the formation of polycyclic aromatic hydrocarbons, many of which are carcinogens, such as benzopyrene and high toxic polychlorinated dibenzodioxins (PCDDs) and dibenzofurans (PCDFs).

Over the past decade incineration technology in the world has come a long way from an open incineration and simple incinerators, which heavily polluted the environment, to the modern efficient incinerators equipped with a complex system of flue gases purification and limiting emissions of pollutants into the environment.

Nowadays, in Russia several types of incinerators are produced. They are offered for incineration of different kinds of solid and liquid organic wastes, including chemical, medical wastes, biological, oil sludge, as well as organochlorine wastes that belong to the list of POPs, such as PCB oils, pesticides, herbicides. Incinerator is a chamber lined with bricks and equipped with a diesel / natural gas burner. In the incinerators usually a two-step technique of waste disposal is used, that is incineration of wastes at a temperature 850-900 °C, post-incineration of gases at a temperature 1100-1200 °C followed by mechanical and chemical cleaning.

Rotary cement furnaces are widely used abroad for disposal of hazardous organic wastes. In 2003 in the "Report of the first session of the expert group on the best available techniques and best environmental practices" of the UNEP UN [6] cement furnaces are recommended for decomposition of PCDDs and PCDFs. In cement furnaces it is technologically possible to meet the described above requirements and ensure the complete decomposition of organic compounds in wastes, including organochlorine compounds.

A well-known international document containing data on PCDDs and PCDFs emissions in various technological processes is "Standardized Toolkit for Identification and Quantification of Dioxin and Furan Releases" developed by UNEP in the framework of UN environment program [7]. It states that "PCDDs and PCDFs can be supplied for incineration alongside with the disposed wastes, be generated in the process of incineration or after its completion during the flue gas cooling".

In this regard, [6] states the following primary methods and measures for optimization of the production process to reduce PCDDs / PCDFs formation:

- continuous supply of fuel and wastes with the registration and accounting of the content of heavy metals, chlorine, sulfur;
- maintenance of control over the materials loaded into the furnace;
- pre-treatment of wastes in order to provide a more uniform load, which contributes to efficient incineration under more stable conditions (drying, fining, mixing, grinding);
- loading of wastes into the main chamber or the secondary one in the furnaces with preliminary heating (temperature > 900 °C);
- wastes cannot be loaded alongside with a raw material mixture, if they contain organic substances;
- stabilization of technological parameters (constancy of fuel properties, constant dosage, an excess of oxygen, CO control);
- wastes loading at the stages of furnace start-up and shut-down is not permitted;
- rapid cooling of flue gases to temperatures below 200 °C.

The possibility of achieving PCDDs content in the flue gases at a level less than 0.1 ng/m3, permitted by European regulations, due to primary measures at the existing plants has been proved. If all these methods do not result in the value decrease to the level less than 0.1 ng/m3, in [6] as secondary measures it is recommended to improve the purification methods of flue gases from dust, for example, to use activated carbon filters or selective catalytic reduction.

However, the cement production technology in rotary furnaces cannot guarantee the complete decomposition of organic wastes. In cement furnaces there are several temperature zones continuously transforming into each other: heating zone (500-600 °C), calcination (900-1200 °C) zone, sintering area (1450 °C), cooling zone (1000-1200 °C). Only one of these zones ensures the prevention of PCDDs and PCDFs formation, but only by maintaining the time conditions. However, it cannot be guaranteed, because the flue gases pass successively through all temperature zones, and the time spent in each zone is unlikely to be sustained precisely because of their continuity. According to the EU standards the geometry of the furnace hot zone should provide residence of gases in the zone at a temperature not lower than 850 °C for at least 2 seconds (rule of 2 seconds) with an oxygen concentration not less than 6 %. This is a very stringent requirement which is not easy to meet. The increase in emissions is observed at the beginning of furnace operation, so the EU standards allow the beginning of pyrolysis or waste incineration only after the furnace is heated to 850 °C [6].

High-temperature steel units provide a more safe incineration of combustible waste components than incinerators or cement furnaces. It is conditioned by the fact that the incinerators and cement furnaces cannot compete with high-temperature metallurgical aggregates such as blast furnace or basic oxygen furnace (BOF) in conditions of toxic wastes disposal, that have temperature up to 2500-2700 °C, active oxidizing or reducing atmosphere, intensive heat- and mass-transfer, availability of high active basic slags, providing binding of non-combustible residue left from the waste incineration and its vitrification. In such circumstances dioxins formation is practically impossible. Besides, the powerful metallurgical units are equipped with a full set of equipment for the capture and cleaning of the released gases, including the wet gas cleaning and a closed water supply system. The UNEP "Standardised Toolkit for Identification and Quantification of Dioxin and Furan Releases" [7] gives the following values of PCDDs / PCDFs emissions in various technological processes (Table 1).

Table 1. Emissions of PCDDs / PCDFs from a variety of sources, μg TEQ / tonne of product.

Type of a source	Air	Products	Wastes	
			Ash	Slag
Solid waste incineration in the waste incineration plant with a complex system of gas purification	0.5	n/t**	15	1.5
Medicine waste incineration in the incinerators with a complex system of gas purification	1	n/t	150	-
Cement furnace	0.15-5.0	n/a***	0.003-1.0	-
Blast furnace	0.01	n/t	-	n/t
BOF	0.06	n/t	-	1.5

* TEQ – International toxic equivalent
** does not transform
*** data is not available

These data indicate that high-temperature metallurgical aggregates (blast furnace, basic oxygen furnace) provide minimal emissions of dioxins and furans – by one-two orders of magnitude below of any known most modern units, currently used in the world for incineration of hazardous organic waste. At the same time, incineration of organic waste in metallurgical units provides additional heat generation and save traditionally used energy supplies (coke, coal, fuel oil, natural gas, etc.). Technologically in the blast furnace neutralization of liquid and solid organic waste with sizes less than 100 mm is possible, and in the BOF – solid organic waste with size from 5-10 mm.

2. Experimental research and results

Let us consider the technological possibility of neutralization of hazardous organic wastes in the BOFs of "ZSMK".

Waste incineration in the converter proceeds in the following temperature conditions:

- period of scrap heating – 1250-1300 °C;
- period of hot-metal charging – over 1380 °C;
- blowing period – from 1440 to over 2000 °C.

Wastes incineration in the converter occurs in the atmosphere of commercially pure oxygen with a purity not lower than 99.5 %, supplied with high intensity and providing a high rate of heat- and mass-transfer. It allows an effective complete combustion of all organic components of the fuel and raw material and destruction of the lattice structure of complex organic compounds, including dioxins. Wastes reside in the converter during the whole period of smelting, complete incineration takes place during scrap heating and subsequent blowing. From the very first moment of waste input into the converter they are in close contact with fluxes, the main of which is lime. During blowing the lime ensures formation of the basic Ca-containing slag. The presence of significant amounts of lime in the converter accelerates oxidation and decomposition reactions of complex organic compounds and destruction of dioxins. Residues from waste incineration transforms into the molten slag, which absorbs all the products of combustion with subsequent vitrification during cooling and is not harmful to the environment.

Converters are equipped with gas exhaust ducts serving for extraction, cooling, purification and post-incineration of converter gases. Gases, exiting the converter, go through the boiler-cooler, where the temperature drops to 850 °C, and then are directed to the first stage of purification – irrigated gas duct. In the irrigated gas duct the further cooling of gases to 250-300 °C takes place, as well as coarse cleaning and wetting until their complete saturation. Saturated gases then pass the I and II stages of rectangular Venturi tubes, which is their primary treatment, and, after the drops separator drained enter the post-incineration device. Wet cleaning of converter gases ensures a high degree of dust collection – 99.5 %. The process water in the wet gas cleaning system is characterized by high pH value at the level of 11-12 units, that indicates a high content of lime. High alkalinity of the process water will contribute to the high capture efficiency of chlorine-containing gases (chlorine hydride, gaseous chlorine) from the converter gases. Thus, in the existing gas treatment system of converters there are favorable conditions for collecting chlorine-containing impurities from the flue gases at the moment of their formation, for example, when products from PVC get into the wastes or disinfection of wastes by chlorine-containing disinfectants. Table 2 provides the comparative characteristics of the conditions of dioxins formation in various technological processes used for the incineration of hazardous organic wastes – incinerator, rotary cement furnace, as well as in the high temperature metallurgical unit (BOF).

Table 2. Comparative characteristics of the conditions of dioxins formation and technological parameters of units used for the incineration of hazardous organic wastes.

Technological parameters	Conditions of dioxins formation	Technological units		
		incinerator	rotary cement furnace	BOF
Method of waste charging	Charging of large amounts	Charging of large amounts	Wastes pre-treatment, that provides even charging of small amounts	Charging of small amounts with consumption from 0.1 to 2 kg/tonne of steel, that along with the high rate of oxygen blast ensures good heat- and mass-transfer during their incineration
Incineration temperature, °C	Less than 900°C	850-900°C	Temperature zones from 500°C to 1450°C	High temperature incineration mode at temperatures 1700-2700 °C
O_2 content, %	Lack of oxidizer	21	21	Excess oxygen, oxygen blast – 99.5 % O_2
Purification of converter gases	Dust collectors, functioning at high temperatures (more than 200°C)	Dry purification, in some cases wet one in the scrubber with dust removal efficiency 93 %	Purification in electrofilter	Three-stage wet purification, providing high degree of dust removal (99.5 %) and rapid gases cooling to 40-60 °C
Post-incineration of converter gases	Absence	Post-incineration in the flare with temperature 1500°C	Absence	Post-incineration in the flare with temperature 1500°C
Formation of ash/slag		Ash	Ash	High-active liquid slag, containing more than 50% of CaO, creates good conditions for sulfur and halogens absorption. During solidification-vitrification.

The given data show that the smelting conditions in the steel converter process comply with the conditions that ensure non-dioxins waste disposal:

- high-temperature incineration mode at temperatures above 1250 °C;
- intense heat- and mass-transfer in the converter;
- active oxidizing atmosphere with an excess of commercially pure oxygen;
- high-active liquid slag, absorbing sulfur and halogen containing compounds, during solidification – vitrified slag;
- rapid cooling of the flue gases in a wet gas purification system to 40-60 °C;
- post-incineration of gases in the flare with a temperature 1500 °C.

Three-stage wet purification of converter gases, providing a high degree of dust removal (99.5 %) and their rapid cooling to 40-60 °C, as well as post-incineration of converter gases exclude the possibility of dioxins re-synthesis.

Thus, high-temperature metallurgical units, in particular oxygen converters, are the safest among currently known units used for incineration of hazardous organic wastes including chlorine-containing ones.

Under conditions of "EVRAZ ZSMK" the experimental-industrial smelts were conducted using specially prepared pilot lots of solid municipal wastes (SMW): compacted waste paper, cardboard and polyethylene; crushed electronic equipment enclosures, consisting of polystyrene, with a particle size less than 20 mm.

The experimental meltings were performed in the Oxygen Converter Plant – 2 on the 350 tonne converter No. 4. Blast of the experimental and comparative melts was carried out according to the standard operating procedures. SMW consumption ranged from 1.0 to 8.8 kg/tonne of steel. It was established that the basic technological parameters of experimental and comparative meltings slightly differ. Meltings with SMW were carried out practically without use of carbon-containing heat-transfer materials, the steel temperature of the turn-down exceeded the steel temperature of comparative meltings. As a result of the calculations of material and heat balance of the converter melting it was found that by introduction of combustible components of SMW into the converter with SMW consumption rate 7.2 kg/tonne of steel additionally 8338.1 kJ of heat energy is introduced into the converter, or 4.21 % of the total heat generation.

The technology of experimental meltings presupposes supply of the compressed bales of paper and bags of polyethylene with the crushed polystyrene into the furnace alongside with scrap and fluxes (lime). After charging the burden it is heated by wastes incineration in the jet of technically pure oxygen for 5-6 minutes. Then, the liquid pig iron was poured into the converter, and then the melting was carried out by standard method of melt blowing by oxygen for 17-18 min. and addition of fluxes during smelting.

The results of industrial tests, described in detail in [8-10], showed that the basic technological parameters of experimental and comparative meltings differed very slightly. A higher steel temperature of the turn-down during experimental meltings was noted – 1706 and 1655 °C, while during comparative meltings it reached accordingly 1619-1661 and 1629 °C, that indicates a higher level of temperatures in the pool during melting due to the high heat of waste combustion. The average dust level of the converter gases after purification was 117 mg/m3.

During the experimental and comparative meltings the dust analysis at the outlet of gas emissions flow for dioxins was made. The analysis of the collected dust was carried out by an accredited laboratory of Bashkir Republican Scientific Research Ecological Center in Ufa using the USEPA methodology – 1613, adapted to the Russian conditions.

The methodology consisted in the dioxins extraction by solvents from the objects in the environment, dioxins extraction from the matrix, their cleaning from the impurities present on the sorbents. The quantitative analysis was performed by the method of isotope dilution using chromatography-mass spectrometer of high resolution. All 17 isomers of dioxins and furans marked by the isotope of carbon C^{13} were used as internal standards. We studied 5 isomers of tetra-chlorinated dioxins – 1,3,6,8 – TCDD, 1,2,3,4 – TCDD, 1,2,3,7 – TCDD, 2,3,7,8 – TCDD, 1,2,3,9 – TCDD.

The results of quantitative isomer-specific analysis of polychlorinated dibenzo-p-dioxins and dibenzofurans are shown in Table 3.

Table 3. Quantitative isomer-specific analysis of polychlorinated dibenzo-p-dioxins and dibenzofurans in the collected dust of the experimental and comparative meltings.

PCDDs/Fs	SampleNo. 1 (experimental melting)		SampleNo. 2 (comparative melting)	
	Concentration, pg/g of dust	TEQ-WHO, pg/g of dust	Concentration, pg/g of dust	TEQ-WHO, pg/g of dust
2378-Tetra-CDD	5.36	5.36	5.78	5.78
12378-Penta-CDD	4.90	4.90	6.99	6.99
123478-Hexa-CDD	2.89	0.29	1.78	0.18
123678-Hexa-CDD	4.49	0.45	3.54	0.35
123789-Hexa-CDD	0.86	0.09	0.06	0.01
123678-Hepta-CDD	8.95	0.09	11.18	0.11
Octa-CDD	18.71	0.00	18.47	0.00
2378- Tetra-CDF	898.51	89.85	651.86	65.19
12378-Penta-CDF	262.34	13.12	247.41	12.37
23478-Penta-CDF	269.53	134.77	342.77	171.39
123478-Hexa-CDF	86.53	8.65	96.35	9.64
123678-Hexa-CDF	99.76	9.98	123.17	12.32
123789-Hexa-CDF	6.61	0.66	8.46	0.85
234678-Hexa-CDF	43.69	4.37	64.36	6.44
1234678-Hepta-CDF	54.50	0.55	50.58	0.51
1234789-Hepta-CDF	15.44	015	13.71	0.14
Octa-CDF	18.80	0.00	16.35	0.00
Total amount of PCDDs, pg/g of dust	**46.16**	**11.18**	**47.8**	**13.42**
Total amount of PCDF, pg/g of dust	**1755.71**	**262.09**	**1615.02**	**278.82**
Total amount of PCDD/F, pg/g of dust	**1801.87**	**273.27**	**1662.82**	**292.24**

For each sample the concentrations of isomers in the unit of measurements pg (1·10-12 g) per 1 g of dry sample weight are given, as well as the toxicity coefficients of each isomer in terms of toxicity 2,3,7,8 – TCDD in accordance with factors of equivalent toxicity (TEF).

As a result of the studies no significant differences in the concentrations or in the isomeric profile for both samples were found. Value of the cumulative rate of sample toxicity TEQ (amount of the reduced toxicity of all isomers) also differed slightly and amounted to 273.27 pg/g TEQ for the experimental melting and 294.24 pg/g for the comparative melting.

The experimental data on the content of PCDDs / Fs in the ash samples can be used to assess the pollution of the converter gases. If to assume that the analyzed samples are the settled dust emitted from the pipe with the flue gases into the atmosphere, then the amount of dioxins emitted, connected with dust particles, will depend on the concentration of suspended particles in the converter gases.

So, if the concentration of suspended particles in the converter gases is 117 mg/m3, according to the study the total content of dioxins and furans in the flue gases of experimental melting was 31.94 pg/m3, comparative melting – 34.40 pg/m3. It should be noted that the European standard for flue gases from stationary sources is 0.1 ng/m3 (1 ng=1000 pg=1·10-9 g) or 100 pg/m3.

3. Conclusions

Thus, the analysis of the conditions of high-temperature processes in the metallurgical units and the research results show that high-temperature metallurgical units are by far the safest for disposal of hazardous organic wastes, including those containing such hazardous to human health supertoxicants as dioxins and furans, as well as carcinogenic hydrocarbons that are formed in large amounts even during incineration of some kinds of wastes at incineration plants, a fortiori, during un-organized burning at landfills.

4. References

[1] Dawson G and Merser B 1996 *Neutralization of Toxic Wastes* (M.: Stroyizdat) p 288

[2] Kraynov I P 2002 *Ecol. Tech. & Res.* **4** 45–57

[3] Orris P, Chary L K, Perry K and Astury J 2000 *Persistent Organic Pollutants (POPs) and Human Health* (Washington WEPHA) p 38

[4] Birman Yu A and Vurdovoy N G 2002 *Environmental Engineering* (M.: Association of Construction Universities) p 269

[5] Vlasov O A, Mechev V V and Mechev P V 2015 *MSW* **11** 19-21

[6] *Report of the Co-chairs of the Expert Group on Best Available Techniques and Best Environmental Practices* http://mognovse.ru/loo-organizaciya-obedinennih-nacij-ep-uneppopsegb-33.html

[7] *Standardized Toolkit for Identification and Evaluation of Emissions of Dioxins and Furans. UNEP* http://www.chem.unep.ch/pops/pdf/toolkit/toolkitru.pdf

[8] Volynkina E P, Klimovskaya I A and Protopopov E V et al 2008 *Proc. 2nd Int. Conf. on Waste Management – Basis for the Restoration of the Ecological Balance in Kuzbass* (Novokuznetsk: SibSIU) pp 241–251

[9] Kuznetsov S N, Volynkina E P and Protopopov E V 2015 *Bulliten of Universities. Ferrous Metallurgy* **5** 352–358

[10] Kuznetsov S N, Volynkina E P, Protopopov E V and Zorya V N 2014 *Metallurgical Technologies for Disposal of Technogenic Deposits, Industrial and Domestic Wastes* (Novosibirsk: SB of the RAS) p 294

Study of the possibility of iron extraction from waste coal by magnetizing roasting

E P Volynkina[1], V N Zorya[2] and V V Korovushkin[3]

[1] Institute of Metallurgy and Material Science, Siberian State Industrial University, 42 Kirova Street, Novokuznetsk, 654007, Russia
[2] LLC "TeckhResurs", 37 Kirova Street, Novokuznetsk, 654007, Russia
[3] National University of Science and Technology "MISiS", 4 LeninskyProspekt, 119991, Moscow, Russia

E-mail: kaf.erc@sibsiu.ru

Abstract. The efficiency of magnetizing roasting of waste coal in converting iron, contained in the mineral part, from weak-magnetic and non-magnetic minerals (siderite, pyrite, marcasite) into high-magnetic ones (magnetite, maghemite) is shown. Products of magnetizing roasting are used as raw material for the subsequent magnetic separation.

1. Introduction

Wastes from mining and processing of mineral recourses are the most widespread types of solid wastes. Thus, in Russia total generation of wastes in 2014 was 5.15 billion tonnes including mining wastes – 4.7 billion tonnes, or 91.2 % [1]. In Kemerovo region in 2014 2.64 billion tonnes of wastes were generated, among them are mainly overburden and host rock from open pit coal mining which is 99.33% or 2.62 billion tonnes of the total amount [2]. In recent years the coal processing has rapidly increased. In 2010 in Kuzbass at the processing plants 127.1 million tonnes of coal were processed and 49.8 million tonnes of coal concentrate were produced [3], which resulted in 77.3 million tonnes of waste coal from the mineral rock processing.

Over a long period of operation coal processing enterprises located in various regions of the country accumulated more than 20 billion tonnes of waste coal. During designing new centers of coal mining a mandatory construction of modern processing plants is required. In general, in Russia it is planned to increase the level of coal enrichment from 40 % to 60%. Annually as a result of operation of coal processing enterprises about 240 million tonnes of waste coal will be generated. [4]

Currently, the main method of coal mining wastes disposal is their use for stowing and waste coal – for production of construction materials: bricks, porous aggregates (agglоporite, lightweight expanded clay aggregate), crushed aggregate [5, 6]. New directions for utilization of coal rocks are still insufficiently developed. The waste dumps have accumulated huge reserves of some metals comparable to the volumes of natural mineral deposits. Therefore, mining wastes can be considered as an alternative way for replenishing natural resources. One of the promising areas of processing coal rocks is the iron extraction from them.

It is known that waste coal in Kuznetsk Basin are characterized by relatively stable mineral composition, dominated by silicon oxides, aluminum and iron. The content of iron oxide in the mineral part of coal averages 5-15 %, reaching in some cases up to 25 % [7].

2. Experimental research and results

Complex studies of coal tailings in Kuzbass, stored in the tailings reservoir of JSC "EVRAZ ZSMK", conducted by the authors [8-10] showed that they contain iron mostly as a component of siderite, as well as pyrite and marcasite. Thus, iron containing minerals are attached to carbonaceous particles. In [10] it is noted that for iron extraction from waste coal it is advisable to use the method of magnetic separation with preliminary magnetizing roasting in the weak oxidizing medium, in the process of which the burn-out of carbonaceous particles, destruction of attachments, decomposition of carbonates and iron sulfides with formation of magnetite will take place.

Samples of waste coal with sizes 0-1 and 1-3 mm were subjected to roasting in the laboratory muffle furnace at 600 °C under conditions of air access restriction for 1.5 hour until constant mass was achieved. The results of chemical and X-ray phase analysis of the initial and roasted samples given in Tables 1 and 2 prove the full siderite decomposition with formation of magnetite in both samples. In the small fraction of size 0-1 mm a small amount of hematite was detected, that indicates its increased oxidation. Thus, the thermal treatment of samples in the laboratory conditions can be characterized as magnetizing roasting.

Table 1. Content of iron and iron-containing phases in the waste coal of various sizes.

Particles size, mm	Fe_{tot}	Fe_{met}(fraction from Fe_{tot}, % rel.)	Fe^{2+} (fraction from Fe_{tot}, % rel.)	Fe^{3+}(fraction from Fe_{tot}, % rel.)
Initial 0-40	14.0	0.98 (7.0)	12.92 (92.3)	0.10 (0.7)
0-1	13.5	0.23 (1.7)	13.5 (78)	3.91 (20.3)
1-3	14.6	1.37 (9.4)	13.71 (73.2)	3.62 (17.4)

Table 2. Results of X-ray phase analysis of waste coal.

Name of the sample	Particles size, mm	Dominant mineral	Other minerals	Impurities
initial	0-40	Hydromuscovite $KAl_2(AlSi_3O_{10})(OH)_2 \cdot nH2O$, siderite $FeCO_3$, quartz SiO_2	Hydromica, montmorillonite, X-ray amorphous materials (AM)	Calcite, feldspar, kaolin or chlorite, dolomite, graphite (carbon)
initial	0-1	Siderite $FeCO_3$, quartz	AM, hydromica, montmorillonite	Calcite, kaolinite or chlorite, dolomite
initial	1-3	Siderite $FeCO_3$, quartz	AM, hydromica, montmorillonite	Calcite, kaolinite or chlorite, dolomite
after roasting	0-1	Magnetite Fe_3O_4	Hematite α-Fe_2O_3, silica, AM, hydromuscovite	Calcite, feldspar and other impurities
after roasting	1-3	Magnetite Fe_3O_4	Quartz, AM, hydromuscovite	Calcite, feldspar and other impurities

Mossbauer spectra, their parameters and interpretation of the initial and roasted samples with sizes 1-3 mm (PK34-1 and PK-34-3) and 0-1 mm (PK34-2 and PK34-4) are provided in Figure 1 and in Table 3.

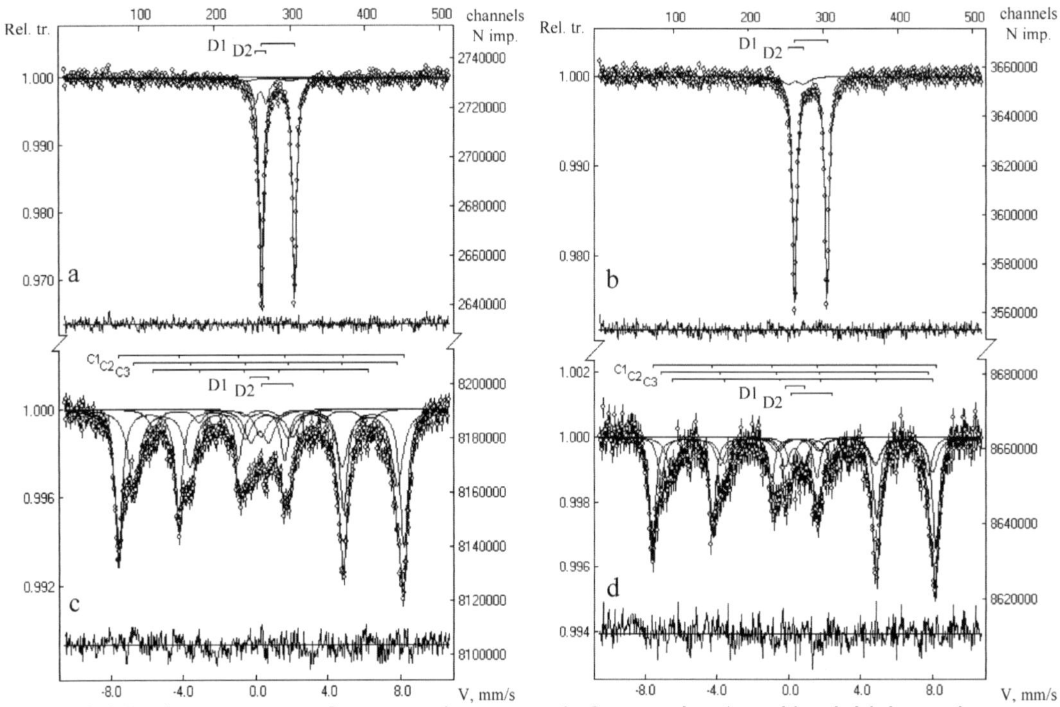

Figure 1. Mossbauer spectra of waste coal before and after roasting (a and b – initial samples of sizes 0-1 mm and 1-3; c and d – roasted samples with sizes 0-1 and 1-3 mm).

Table 3. Mossbauer spectra parameters of waste coal before and after the magnetizing roasting and their interpretation.

Name of the sample	Spectrum component	Isomer shift δ, mm/s	Quadrupole splitting Δ, mm/s	Magnetic fields on nuclei Fe^{57} H, kOe	Component area S, %.	Interpretation
Initial 1-3 mm	D1(Fe^{2+})	1.23	1.78	0	87.5	Siderite
	D2(Fe^{2+})	0.29	0.61	0	12.5	Pyrite
Initial 0-1 mm	D1(Fe^{2+})	1.23	1.78	0	93.5	Siderite
	D2(Fe^{3+})	0.26	0.60	0	6.5	Pyrite
Roasted 1-3 mm	C1(Fe^{3+})	0.32	0.00	489	43.4	Magnetite-maghemite
	C2(Fe^{2+}+Fe^{3+})	0.58	-0.04	452	36.0	
	C3(Fe^{3})	0.32	-0.03	369	7.6	
	D1(Fe^{3+})	0.22	1.00	0	6.8	Hydromica
	D2(Fe^{2+})	1.16	1.73	0	6.2	Siderite
Roasted 0-1 mm	C1(Fe^{3+})	0.31	-0.01	489	43.6	Magznetite-maghemite
	C2(Fe^{3})	0.42	-0.16	462	20.7	
	C3(Fe^{2+}+Fe^{3+})	0.69	-0.12	451	22.2	
	D1(Fe^{3+})	0.34	1.07	0	6.8	Hydromica
	D2(Fe^{2+})	1.27	2.28	0	6.7	Siderite

The obtained results showed high efficiency of magnetizing roasting in converting iron from non-magnetic (pyrite) and weak-magnetic (siderite) forms into high-magnetic ones (magnetite and maghemite γ - Fe_2O_3). In the roasted products non-stoichiometry of magnetite is revealed, Fe^{3+} ions of tetrahedral position form two non-equivalent positions, which indicates the possible isomorphic substitutions leading to their formation. Doublet D1 of Fe^{3+} ions has a tetrahedral coordination, perhaps it is related to hydromicas. Doublet D2 is related to siderite.

Mossbauer spectra of the roasted samples of various sizes, as well as of the initial ones, differed slightly, mainly, in cationic distribution in magnetite. In the roasted sample with size 1-3 mm in comparison to the sample with size 0-1 mm there is no additional non-equivalent tetrahedral position of Fe3+ ions, it appears in octahedral coordination. The total fraction of iron in the composition of high-magnetic phases (magnetite and maghemite) depended little on the particle size of the initial samples and was 87 % rel. for the sample with size 1-3 mm, and 86.5 % for sample with size 0-1 mm. As it follows from Figure 1, as a result of roasting the bulk of iron transformed from divalent into trivalent state, part of the iron ions is represented by the ions Fe3+ and Fe2+ of octahedral position, which are connected by electron exchange. The roasted sample of fine fraction 0-1 mm was characterized by a higher content of ferric iron, which indicates a greater degree of its oxidation. However, the oxidized ferric iron was present in the composition of high magnetic phases (magnetite and maghemite).

The initial and roasted samples of waste coal were studied by the method of combined thermal analysis (CTA), which included a differential scanning calorimetry (DSC) and the definition of weight loss (TG). The measurements were performed in the argon atmosphere at a heating rate of 20 deg/min to 1000 °C. DSC and TG curves are given in Figure 2. Comparison of the thermogravimetric curves of the initial and roasted samples (Figure 2 a and b) shows that these samples are completely different in mineral composition.

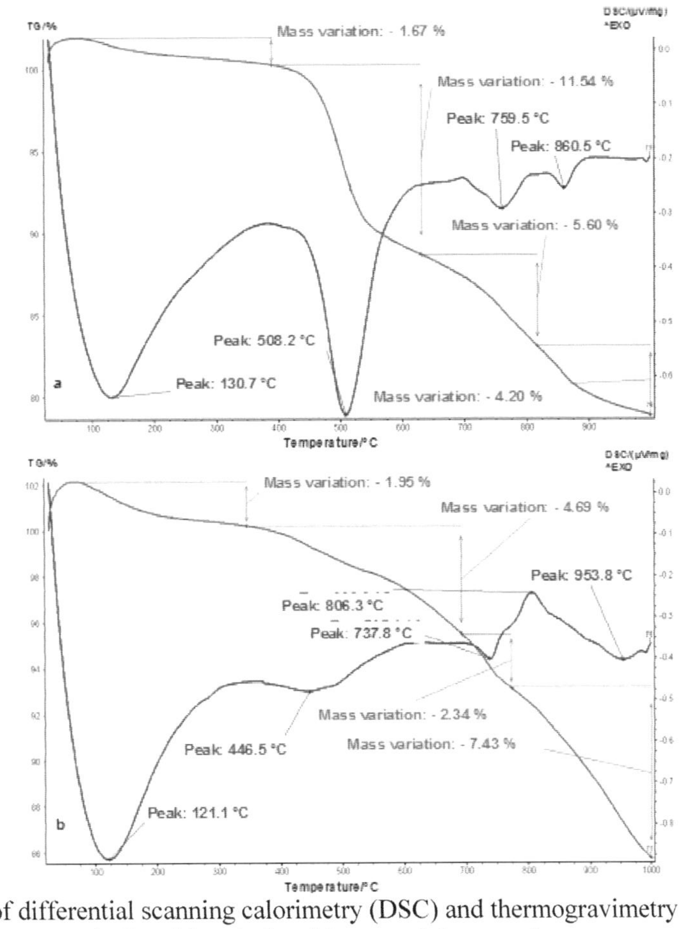

Figure 2. Curves of differential scanning calorimetry (DSC) and thermogravimetry (TG), waste coal before (a) and after (b) magnetizing roasting.

While on DSC curve of the initial sample three distinct endothermic peaks with maxima at temperatures 508.2 °C (siderite decomposition), 759.5 °C and 860.5 °C (reduction of iron to wustite and metal) are observed, on DSC curve of the roasted sample two endothermic effects – flat in the range of 400 – 500 °C are observed, which is probably due to the dissociation of residual siderite and other low temperature impurities, for example, calcite and dolomite, and the peak with a maximum at 737.8 °C indicating the wustite formation.

Thus, unlike the initial sample the roasted sample is characterized by the exothermic peak at 806.3 °C most likely associated with the emission of heat during the oxidation of carbonaceous argillite, and in the range 850 – 975 °C – by the endothermic effect due to iron reduction to α-Fe and γ-Fe. Mossbauer spectra of the sample after heating under conditions of CTA confirmed the presence of these iron phases.

Figure 3 shows the diagram of the calculated phase composition of iron-containing components of the examined samples with the distinct high-magnetic (magnetite, maghemite) and weak-magnetic (siderite) phases.

■ - high-magnetic(magnetite, maghemite); ▨ - low-magnetic (siderite)

Figure 3. Distribution of high-magnetic and low-magnetic iron-containing phases in the waste coal.

As Figure 3 shows, even with during heating under laboratory conditions at a rather low for magnetizing roasting temperature (600 °C) from the waste coal was obtained a product, in which approx. 90 % of iron-containing phases were high-magnetic. It is expected that in industrial conditions complying with the technological regime as a result of magnetizing roasting all iron contained in this material would convert from weak-magnetic (siderite) and non-magnetic (pyrite) minerals into high-magnetic minerals (magnetite, maghemite), which conditions the feasibility of their extraction by magnetic method. It is appropriate to carry out magnetizing roasting in a low-oxidizing atmosphere in order to oxidize carbon of carbonaceous particles, which would provide formation of CO and destruction of their attachments to iron-containing minerals.

3. Conclusions

Thus, the received results confirm the usefulness of the magnetizing roasting of waste coal for conversion of iron, contained in the mineral part, from weak-magnetic siderite into high-magnetic magnetite and maghemite. At the same time the contained in waste coal carbon would provide the roasting process with its own heat, which would help to reduce the costs of the product. The products of magnetizing roasting, enriched by high-magnetic minerals, are the raw material for the subsequent magnetic separation.

4. References

[1] *State report on the state and environment protection in the Russian Federation in 2013*

https://www.mnr.gov.ru/upload/iblock/6c7/gosdokladeco.pdf
[2] *Report on the state and environment protection in Kemerovo region in 2014* http://ecokem.ru/wp-content/uploads/2015/08/NEW_DOKLAD-2014.pdf
[3] Antipenko L A and Ermakov A Yu 2012 *Mining Machinery: Business Directory* **9** 136–140
[4] Efimov V I, Nikulin I B and Rybak V L 2014 *Proceedings of Tula State University. Earth sciences* **1** http://elibrary.ru/item.asp?id=23025053
[5] Afanasiev M M and Tkachev O A 2014 *Economy and Entrepreneurship* **6** 662–665
[6] Stolboushkin A Yu and Berdov G I 2011 *Proceedings of the Higher Educational Institutions. Construction* **1** 46–53
[7] Vdovchenko VS, Martynov M I, Novitsky N V and Yushina G D 1991 *Energy fuel of USSR: Directory* (M.: Energoatomisdat) p 184
[8] Volynkina E P and Zorya V N 2012 *Proceedings of the universities. Ferrous metallurgy* **4** 60–64
[9] Kuznetsov S N, Volynkina E P, Protopopov E V and Zorya V N 2014 *Metallurgical Technology of Processing of Technogenic Deposits, Industrial and Municipal Wastes* (Novosibirsk: SB of the RAS) p 294
[10] Zorya V N, Korovushkin V V, Permyakov A A and Volynkina E P 2015 *Proceedings of the universities. Ferrous metallurgy* **5** 359–366

Gazification of coal dust particles in the blast furnace tuyere apparatus

V S Shvydky, Yu G Yaroshenko, N A Spirin and V V Lavrov

Department of Thermophysics and IT in Metallurgy, Ural Federal University, 19 Mira Street, Yekaterinburg, 620002, Russia

E-mail: n.a.spirin@urfu.ru

Abstract. The mathematical statement of the problem on gasification of coal dust particles in the blast-furnace tuyere apparatus is given, which includes the motion equation of a variable mass particle, heat equation of a particle and the heat-balance equation of the blast flow. The results of calculations are obtained by using mathematical software packages (Mathcad, Maple). Relatively weak effect of the volatiles combustion process on the thermal state of the tuyere zone is shown.

1. Introduction

At the present time, the interest in the use of pulverized coal fuel in the blast furnace process is rising [1-3]. Typically, the injection of pulverized coal into the blast furnace is carried out at a distance of 1 m from the tuyere outlet section [4, 5]. Prior to getting into the tuyere cavity in the coal dust particles (CD) the processes of heat- and mass-transfer take place, that leads to a change in the boundary conditions of mathematical (numerical) model of the blast furnace combustion zone (CZ) [6, 7]. In general, simulation of combustion processes in CZ is rather complicated due to the high turbulence of the recycle gas flows, spatial heterogeneity of CZ walls temperatures, temperature and composition of the recycle gas, constant lowering of coke layer, which leads to entry of coke particles into the CZ zone.

The degree of CD burn-out inside the tuyere (if the coal is counter-injected to the blast flow) and inside CZ determines the temperature of CZ, and the unburned material, which, like the ash (coke residue), collides with the CZ walls, is the potential cause of problems arising because of the low gas permeability.

Accounting for a change in the boundary conditions of mathematical model of blast furnace CZ is connected with the solution of the problem.

2. Calculation of the motion and heat of CD particles

In some section $x = 0$, CD particles enter into blast flow in the amount W, g/m^3 of the blast. We need to determine the change in particles velocity and their temperature, amount of volatile yield, gas (blast) temperature along the length of the tuyere pipe with diameter d_t.

According to the basic provisions of the theory of coals combustion [8] the chemical reactions of carbon with oxidants start when the greater part of the volatiles yields from the particles. CD particles, incoming into the blast flow, interact with the blast, accelerate and get heated. During this process volatiles are released and burned. The mass of CD particles decreases in the process of their movement and heating. The particles are heated by gas, mainly by convection, since carbon monoxide CO_2 in the blast is practically absent, and content of H_2O does not exceed 2.5%. In principle, the heating of particles is possible by radiation from the walls of the tube, for its accounting it is necessary to know the change in the temperature of pipe walls along the blast flow, but it will greatly complicate the task.

The mathematical statement of the problem includes the motion equation of a variable mass particle, heat equation of a particle and the heat-balance equation of the blast flow. Each of these equations becomes more complicated as the described phenomena are accompanied by the release and burning (oxidation) of CD volatiles.

According to the 2nd Newton's law [9] the change in momentum equals to the impulse of effective forces. In the one-dimensional representation the only effective force is the force of aerodynamic resistance. For a variable mass particle the dependence is valid [6, 9]

$$\frac{d(m_p v_p)}{d\tau} = \frac{1}{2} C_D \rho_g |v_g - v_p| (v_g - v_p) \frac{\pi d_p^2}{4}, \qquad (1)$$

where m_p and d_p – weight and size of a particle, kg and mm; v_g and v_p – flow rates of blast and CD particles, m/s; C_D – drag coefficient; ρ_g – blast density, kg/m^3.

For CD particles of typical sizes the drag coefficient can be determined from relations [6, 9, 10-12]

$$C_D = \begin{cases} \frac{24}{\mathrm{Re}_p}\left(1 + 0.15\,\mathrm{Re}_p^{0.687}\right) & \text{if} \quad \mathrm{Re}_p < 1000; \\ 0.44 & \text{if} \quad \mathrm{Re}_p \ge 1000; \end{cases}$$

here Re_p – Reynolds number of a CD particle

$\mathrm{Re}_p = \dfrac{\rho_g |v_g - v_p| d_p}{\mu_g}$, μ_g – dynamic viscosity of blast, Pa·s.

Left side of the equation (1) can be written as $\dfrac{d(m_p v_p)}{d\tau} = m_p \dfrac{dv_p}{d\tau} + v_p \dfrac{dm_p}{d\tau}$.

The coal particle consists of coke residue, weight of which is unchanged at this stage, and volatiles. The initial amount of volatiles V_{l0} kg is determined by the relation [6, 8] $V_{l0} = 0.01\, m_{p0} V_l$, here m_{p0} is the initial mass of a particle; V_l – amount of volatiles in the initial coal, % (by weight). The current mass of a dust particle $m_p(\tau)$ can be represented as follows:

$$m_p(\tau) = m_{p0}\left[1 - 0.01 V_{l0}(\tau)\right] = m_{p0}\left[1 - 0.01 V_l \frac{V_{l0}}{V_l}\right] = m_{p0}\left[1 - 0.01 V_l \delta(\tau)\right],$$

where $V_{l0}(\tau)$ – current amount of volatiles; $\delta(\tau)$ – fraction of the released volatiles. According to the two-stage scheme [8] of volatiles release the equation is valid

$$\delta(\tau) = \frac{V_{l0}(\tau)}{V_l} = C_{01}\left[1 - \exp\left(-\int_0^\tau k_{1v}\, d\tau\right)\right] + C_{02}\left[1 - \exp\left(-\int_0^\tau k_{2v}\, d\tau\right)\right],$$

where C_{01} and C_{02} – empirical coefficients (fractions); k_{1v} and k_{2v} – rate constant rate of reaction of volatile release at the stage of release of capillary-bound volatiles and the stage of destruction of chemical bonds, s^{-1}. Under this scheme the rate of change in the weight of a coal particle is determined by the dependence

$$\frac{dm_p}{d\tau} = -0.01V_l \frac{d\delta(\tau)}{d\tau} = -0.01V_l\varphi(\tau) =$$
$$= -0.01V_l\left[C_{01}k_{1v}\exp\left(-\int_0^\tau k_{1v}d\tau\right) + C_{02}k_{2v}\exp\left(-\int_0^\tau k_{2v}d\tau\right)\right].$$

Here $\varphi(\tau)$ – effective reaction rate of volatile release.

For the use of Runge-Kutta method [13] the equation (1) needs to be written as $dv_p/d\tau = f(...)$, and it is more convenient to replace the time derivative by the coordinate derivative $d\tau = dx/v_p$. The equation (1) in its final form will be

$$\frac{dv_p}{dx} = \frac{0.01V_l\varphi(\tau)}{1-0.01V_l\delta(\tau)} + \frac{\left[1+0.15\left(\frac{\rho_g|v_g-v_p|d_p}{\mu_g}\right)^{0.687}\right](v_g-v_p)}{\tau_p v_p[1-0.01V_l\delta(\tau)]},$$

(2)

where $\tau_p = \frac{\rho_p d_p^2}{18\mu_g}$ – time of particle relaxation.

Heat equation of a CD particle in the case of convective heat-transfer has is [6, 7, 13]

$$c_p m_p \frac{dt_p}{d\tau} = \alpha(t_g - t_p)\pi d_p^2,$$

where c_p and c_g – specific heat capacity of a particle and gas, J/(kg·K); t_p and t_g– particle temperatures and flow of blast, °C; α– heat-transfer coefficient, W/(m^2·K). If we use the expression of Nusselt number [6, 7, 13] $Nu = \alpha d_p/\lambda_g = 2 + 0.634Re^{1/2}Pr^{1/3}$, where λ_g is the coefficient of thermal conductivity, W/(m·K), to determine the heat-transfer coefficient, the following dependence will be obtained

$$\frac{dt_p}{dx} = \frac{6\lambda_g\left[2+0.634\left(\frac{\rho_g|v_g-v_p|d_p}{\mu_g}\right)^{0,5}\left(\frac{\mu_g c_g}{\lambda_g}\right)^{0,33}\right](t_g-t_p)}{c_p v_p d_p^2[1-0.01V_l\delta(\tau)]}. \quad (3)$$

Equations (2) and (3) are sufficient to answer the question concerning the change in the velocity of a CD particle and its temperature. Below in Figure 1 the results of calculations are given for the particular case of W = 2.5 g/m3, V_g = 1.27 m3/s (the initial gas flow rate in the CZ), d_p = 0.15 mm d_t = 160 mm. In Figure 1 , a the change in the velocity of motion of a CD particle is shown as the distance from the place of its entry into the blast increases. The initial velocity of the particle is 5 m/s. The particle under the action of the blast flow rather quickly gains velocity. Relatively quickly rises the temperature of a CD particle (Figure 1, b).

3. Calculation of the processes of volatiles release and blast heat balance

While forming the heat equation it is necessary to take into account all CD particles contained in 1 m^3 of blast. It is required to determine the amount of heat released during the volatile combustion, which depends on the coefficient of Bunte [6, 7, 8]. Bunte coefficient is determined by Bu = 2.37 (H_r–0.125O_r)/C_r, where H_r, O_r and C_r– content (by weight) of carbon, hydrogen and oxygen in the coal working mass. Amount Q_v of heat released during the volatile oxidation is $Q_v = 10^6$ (105 Bu – 33), J/kg of volatiles.

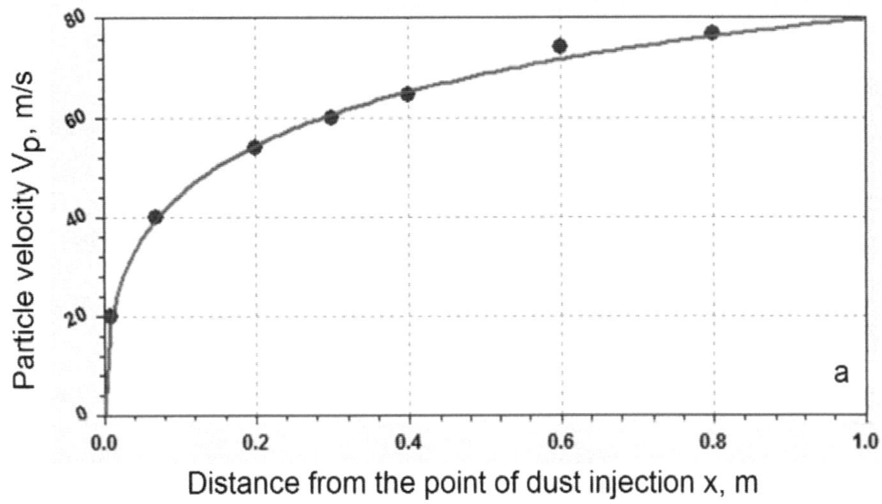

a

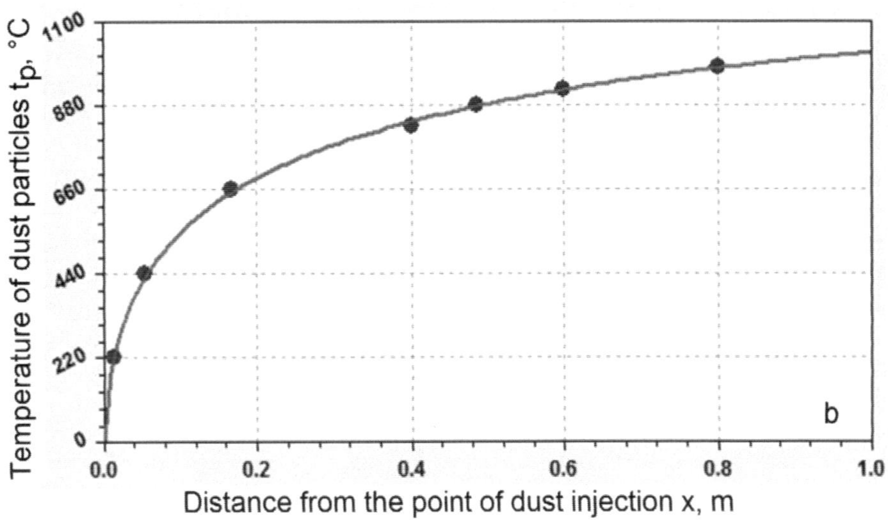

b

Figure 1. Change in the motion velocity of a CD particle (a) and its temperature (b) as the distance from the injection point to the blast flow increases.

1 m^3 of blast contains 0.001W kg of coal particles with 0.01V_r·0.001W kg of volatiles. The total surface area F of CD particles in 1 m^3 of blast is $F = 0.001W \, {}^{6}\!/_{\rho_p d_p}$, m^2/m^3 of blast.

Thus, the following dependence can be written as:

$$\frac{dt_g}{dx} = \frac{\lambda_g F\left[2+0.634\left(\frac{\rho_g\left|v_g - v_p\right|d_p}{\mu_g}\right)^{0,5}\left(\frac{\mu_g c_g}{\lambda_g}\right)^{0,33}\right]\left(t_p - t_g\right)}{C_{vg} d_p v_p} + \frac{0.001W \cdot 0.01V_l Q_v \varphi(x)}{C_{vg} v_p},$$

where C_{vg}– the heat capacity per unit volume of gas, J/(m^3·K);

$$\varphi(x) = C_{01}k_{1v}\exp\left(-\int_0^x \frac{k_{1v}}{v_p}dx\right) + C_{02}k_{2v}\exp\left(-\int_0^x \frac{k_{2v}}{v_p}dx\right), \text{s}^{-1}$$

To "launch" the process of volatiles release it is necessary to overcome an energy barrier. As an example for the calculation the coal with the following composition was chosen: 74.30 % C_r, 4.82 % H_r, 8.01 % O_r; coal ash was 7.1 %, 38.7 % volatiles content. The rate constants for the reactions of volatiles release are determined by the relations [6, 8, 14]

$$k_{1v} = 1.34 \cdot 10^5 \exp\left[\frac{74000}{8.3143(t_p + 273)}\right]; \quad k_{2v} = 1.46 \cdot 10^{13} \exp\left[\frac{251000}{8.3143(t_p + 273)}\right].$$

The values of coefficients C_{01} and C_{02} are 0.33 and 0.67 respectively. The effective rate $\varphi(x)$ of the reaction of volatiles release is maximal at the stage of a sharp increase in the particle temperature (Figure 2).

At a distance of 0.3 m the volatiles practically are not released (Figure 3, a). Temperature of 770 °C in this case can be considered as the temperature of the beginning of volatile release for the coal grade under consideration. High rate of volatiles release should be noted – more than 96 % of volatiles contained in the coal are released at a distance of 1 m from the point of injection of CD particles into the flow.

In Figure 3, b the change in temperature of the gas flow, when it interacts with the CD particles, is shown. In this example, the pulverized coal is introduced cold into the blast, thus, initially the gas temperature slightly decreases (less than 1 °C, which practically cannot be seen in the figure). However, during the volatiles combustion (within 0.35 m) the gas temperature is recovered and then goes up at 7.8 °C.

4. Conclusions

The given in this work results characterizes only the first stage in the development of heat- and mass-transfer process during the pulverized coal injection – volatiles release; calculation is valid for the chosen CD consumption. There is a relatively weak influence of the volatiles combustion on the thermal state of the tuyere zone. Thus, with the increase in CD consumption by 60 % (up to 4 g/m^3) the gas temperature at the end of the particle's trajectory goes up to 1212.4 °C (only 4.6 °C).

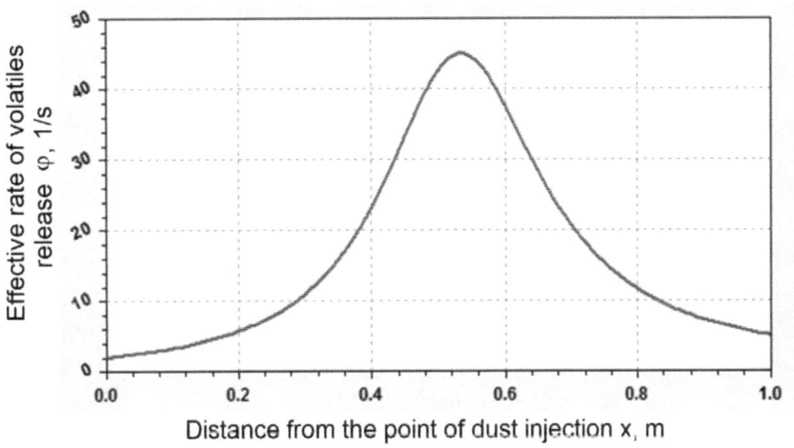

Figure 2. Influence of the path travelled by the particle on the effective reaction rate of volatiles release.

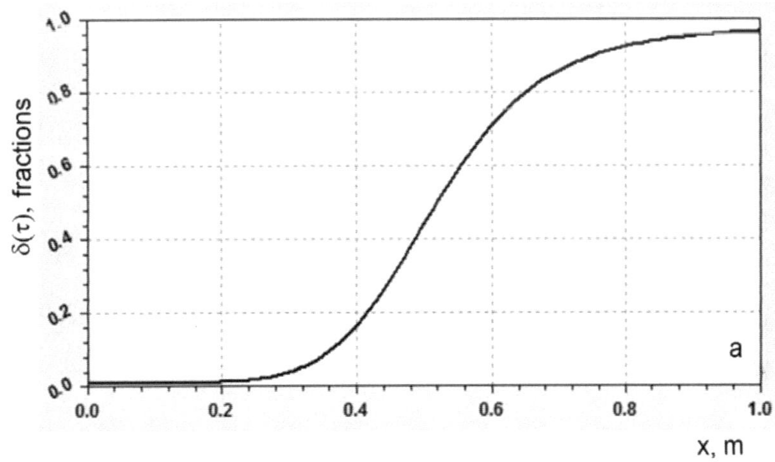

a

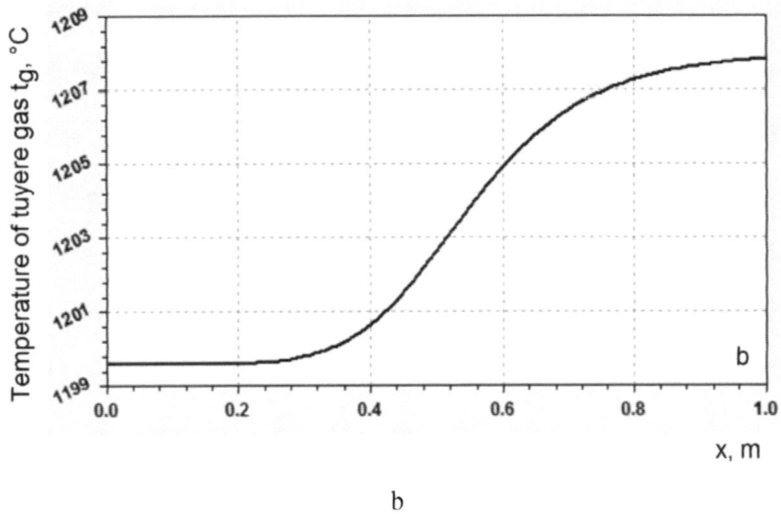

b

Figure 3. Change in the fraction of released volatiles (a) and the temperature of tuyere gases (b) as distance from the enjection point of coal dust particles into the blast flow increases.

5. References

[1] Standish N 1975 *Blast Furnace Aerodynamics* (Wollongong) p 220
[2] Svidkii V S, Gordon Ja M and Bokovikov B A 1981 *Blast Furnace Hearth and Raceway. Newcastle* pp 171–177
[3] Stendish Ni 1972 *Simp. on Blast Furnace Injection* (Australian Institute of Mining and Metallurgy)
[4] Jamaluddin A S, Wall T F and Truelove J S 1986 *Ironmaking and Steelmaking* **132** 91–99
[5] Ramm A N 1980 *Modern Blast Furnace Process* (M.: Metallurgiya) p 302
[6] Spirin N A, Ovchinnikov Yu N, Shvydky V S and Yaroshenko Yu G 1995 *Heat Transfer and Efficiency of Blast Furnace Smelting* (Ekaterinburg: USTU – UPI) p 243
[7] Spirin N A, Shvidkiy V S, Lobanov V I and Lavrov V V 1999 *Introduction to the System Analysis of Thermal Processes in Metallurgy: Learner's Guide for Universities* (Ekaterniburg:USTU) p 205
[8] Pomerantsev V V 1986 *Basics of Practical Combustion Theory: Learner's Guide for Universities* (L.: Energoatomisdat) p 312
[9] Shvydky V S, Yaroshenko Yu G, Gordon Ya M, Shavrin V S and Noskov A S 2003 *Fluid Mechanics: Learner's Guide for Universities* ed VS Shvydky (M.: IKTs "Akademkniga") p 464
[10] Parker A S and Hottel H C 1936 *Ind. Eng. Chem* **28** 1334–40
[11] Heunert G and Willems J 1959 *Stahl und Eisen* **79** 1545–52
[12] Peter V W 1964 *Stahl und Eisen* **84** 979–985
[13] Shvydky V S, Ladygichev M G and Shavrin V S 2005 *Mathematical Methods in Thermal Physics: A Textbook for Univeristies* (M.: Teplotekhnik) p 232
[14] Bogdandi L and Engel G Y 1971 *Reduction of Iron Ores* (M.: Metallurgiya) 1971 p 520

Development of heat-transfer circuits in the blast furnace

N A Spirin, Yu G Yaroshenko and V V Lavrov

Department of Thermophysics and IT in Metallurgy, Ural Federal University, 19 Mira Street, Yekaterinburg, 620002, Russia

E-mail: n.a.spirin@urfu.ru

Abstract. The development of heat-transfer circuits in the blast furnace as the technologies of blast-furnace smelting are improved are considered. It is shown that there are two zones of intense heat-transfer, and in modern conditions, when different kinds of iron ore are smelted, the use of combined blast with high parameters is a prerequisite for the stability of blast-furnace smelting operation and the smelting efficiency.

1. Introduction

Almost all physical and chemical processes occurring in the blast furnace are pyrometallurgical. Hence their development depends on the amount of heat, that is removed from the combustion zone, the distribution of temperature height wise and sections of blast furnaces. This relationship is complicated by the fact that the development of thermal processes that determine the level of temperature, in turn, are influenced by chemical processes changing during the smelting not only the composition and physical state of the charge components, but also their weight. Thus, heat- and mass-transfer are interrelated that determines the approach to the analysis of the blast-furnace smelting operation.

In the course of improvement of the blast-furnace smelting, development of the theory of heat transfer in the blast furnace and deepening of knowledge about the processes, significant clarifications have been made. The main outcome of the evolutionary development of the blast-furnace smelting operation is the reduction of coke consumption and increase of the smelting intensity.

2. Fundamentals of the theory of heat transfer in the blast furnace

The mathematical model, describing the thermal and physical-chemical phenomena of blast furnace smelting, was proposed by prof. B.I. Kitaev in 1944 [1]. The full description of the model is given in the monographs [2-5]. B.I. Kitaev considered the blast furnace as a countercurrent heat exchanger, where heat- and mass-transfer processes occur simultaneously. He found that the temperature distribution in the blast furnace is determined by the heat capacities of charge and gas flows, namely, by their ratio. The heat capacity of a flow of charge $-W_m$ or gas – W_g (kJ/(hr °C) equals to the product of the charge flow rate $-G_m$, (kg/h) or gas – V_g (m^3/h) and the heat capacity of the substance in this flow – c_m (kJ/kg °C) or c_g (kJ/m^3°C), i.e.

$$W_m = G_m c_m \quad \text{or} \quad W_g = V_g c_g \tag{1}$$

To take into account the thermal effects of exothermic or endothermic reactions B.I. Kitaev introduced the concept of "seeming" heat capacity of charge – cs, and proposed a formula for its calculation

$$c_s = c_f + (\pm c_{mas}). \quad (2)$$

The first summand in this expression – c_f is a physical heat capacity, and the second one – ($\pm c_s$) is a conditional heat capacity that takes into account the sinks and sources of heat during the heat- and mass-transfer processes of blast furnace smelting. If the value c_f in (2) is a volumetric heat capacity, than the value

$$\pm c_{mas} = \pm \Sigma q_{mas} / (v_m \times \Delta t_m), \quad (3)$$

where Σq_{mas} – the sum of heat-absorptions in the direct reduction processes, evaporation and decomposition of moisture and other processes requiring heat inputs (sign "+") for their development, and heat emissions in the process of indirect reduction, secondary oxidation and other processes accompanied by heat emission (sign "–"). Its value is determined by the quantity of the reacted substance and the thermal effect of the reaction;

v_m – volume of charge per 1 tonne of pig iron;

Δt_m – temperature range in which the heat absorptions or heat emissions are calculated.

The mass and volumetric "seeming" heat capacity can be used for the calculations of temperature fields in blast furnaces depending on the flow rates and thermo-physical properties of their constituent components. In the analysis of heat-transfer processes in the blast furnace the determining factor is the ratio of heat capacity of the charge flow – W_m to the heat capacity of a gas flow – Wg. Quantity $m = W_m/W_g$ can take values $m<1.0$, $m>1.0$, and $m = 1.0$. Thus, the quantity m determines the difference in the nature of temperature fields along the column of charge.

Exploring the change in heat capacities of charge and gas flows in the blast furnace, Prof. B.I. Kitaev managed for the first time to determine the thermal and physical characteristics of the blast furnace which are as follows:

- heat capacity of a gas flow along the height of the blast furnace remains practically the same, since its decrease during the gas cooling is compensated by the increase in the amount of gas flow components of higher heat capacity, which are formed as a result of reduction of the iron-bearing charge material, evaporation and other possible processes;
- heat capacity of the charge flow has a more complicated nature. Changes of this characteristic are connected to the chemical aspect of the blast furnace process. In the upper parts of the furnace, where the physical processes of charge heating take place, this characteristic depend only on the temperature, and this dependence is smoothed by the action of a heat source during the indirect reduction. According to the estimates of B.I. Kitaev ratio W_m/W_g in this area of the furnace varies between 0.8...0.9. In the lower parts of the furnace, alongside with the physical processes of the charge melting and the melt superheating, direct reduction processes occur, which require less amount of heat. Such heat consumption causes the two- and threefold increase of seeming heat capacity of the charge. This process, as well as other endothermic processes lead to the fact that ratio W_m/W_g for the bottom of the furnace depending on the grade of the smelted pig iron can increase to 2.0...3.0. Change in the heat capacities of gas and charge flows along the height of the blast furnace is shown in Figure 1a. The lines, characterizing the heat capacities of gas and charge flows, indicate that for the top of the furnace $W_m<W_g$, i.e. $m<1.0$, and for the bottom at a considerable height $W_m>W_g$, i.e. $m>1.0$. Maximum of the development of direct reduction processes corresponds to the maximum of the heat consumption and, consequently, to the highest value of heat capacity of the charge flow. With further lowering of charge the development of the direct reduction processes weakens due to the decrease in non-reduced iron. As a result, the heat capacity of

the charge flow begins to decrease. Depending on the grade of pig iron the heat capacity of the charge flow at the level of tuyeres can be very close to the heat capacity of gases flow, and sometimes the condition may arise when $W_m<W_g$, that is indicated in Figure 1a as variations in the heat capacity of the charge flow.

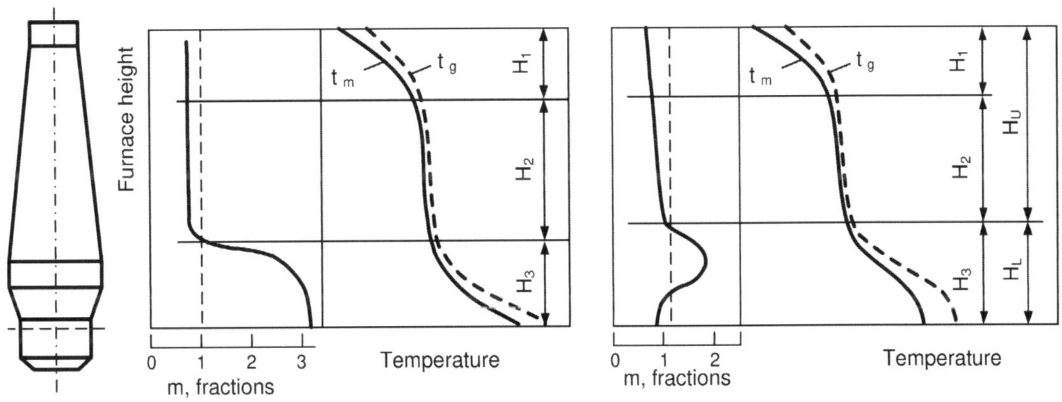

Figure 1. Heat-transfer circuit in the blast furnace: a – heat-transfer circuit of 1944; b – circuit of 1961; t_m, t_g – the temperature of charge and gas; m – ratio of the heat capacities of the charge and gas flows; H_1 – the upper stage of the heat-transfer; H_2 – thermal reserve zone r; H_3 – the lower stage of heat-transfer; H_U – the upper thermal zone; H_L – the lower thermal zone.

Analysis of changes in the heat capacities of charge and gas flows helped to conclude that the heat-transfer in the blast furnace follows the circuits shifted along the height and different in ratios of heat capacities of flows. Laws of heat-transfer in the counter-flow heat exchangers allowed the temperature distribution along the height of the blast furnace to be presented in a generalized form in Figure 1b. As appears from the analysis of temperature fields and flows of gas:

- in the upper part of the blast furnace (the upper stage of heat-transfer –H_1) the heat capacity of the charge flow is less than of the gas flow, i.e. $W_m<W_g$. Under these conditions the gas flow always contains more heat than the charge can absorb. The excess heat is carried away through the furnace mouth, and the heating of the charge to the temperature of gas practically occurs at a depth of 5-8 m below the charge level;
- in the lower parts of the furnace (lower stage of heat-transfer – H_3) heat capacity of the charge flow is more than the heat capacity of the gas flow, namely $W_m>W_g$. With this ratio the charge cannot be heated to the temperature of the combustion zones. After the heat-transfer the charge will have a temperature lower than the initial temperature of gases. Thus, within the lower stage of heat-transfer the temperature differential between the gas and the charge, as the charge being lowered, increases and reaches the maximum in the combustion zones. This area of heat-transfer is characterized by high values of heat-transfer coefficients, significantly developed surface of heat-transfer due to the melting of iron ore and slag formation and, that is very important, by the development of endothermic reduction processes of CO_2 with coke carbon. These conditions lead to the fact that the relatively small height of the column of charge materials is sufficient for the heat energy exchange between the gas and charge. At the exit from the lower stage of heat-transfer – H_3, the temperatures of the gas and charge flows become practically the same;
- in the middle part of the blast furnace (moderate temperature zone – H_2) the heat capacity of the gas and charge flows converge $W_m \approx W_g$. The charge is heated at the exit from the upper stage of heat-transfer almost to the temperature of gas leaving the lower stage of heat-transfer. But due to lack of heat reserve in the gas flow the heat-transfer rate decreases to almost zero values.

The full description of the model is given in monographs [2-6]. Using the analysis results of changes in the heat capacities of charge and gas flows along the height of the furnace, B.I. Kitaev formulated the basic provisions of the theory of heat-transfer in the blast furnace, which are as follows [1, 2]:

- The gas-material heat-transfer rate is carried out in two stages of heat-transfer: in the upper and lower stages. The stages of heat-transfer are separated by the thermal reserve zone, where the temperature gradient along the height reaches a minimum value.
- Thermal performance of each stage of heat-transfer is notable for independence (autonomy) and is characterized by a significant difference in ratios of heat capacities of charge W_m and gas flows W_g. The thermal reserve zone, separating the active heat-transfer zones, is characterized by the ratio of heat capacities of flows $m = W_m/W_g$ close to 1. The upper stage of heat-transfer has $W_m<W_g$, i.e. m<1.0, and the lower stage $W_m>W_g$, i.e. m>1.0.
- Heat-transfer in the blast furnace is completed which is proved by the existence of the thermal reserve zone. In this zone the heat capacities of the gas and charge flows converge so that $W_m \approx W_g$, $m \approx 1.3$

The nature of changes in the gas and charge temperatures, the ratio of their heat-capacities along the height of the furnace are presented in Figure 1a. Thus, along the height of the blast furnace at a regular performance the typical S-shaped temperature distribution is observed.

3. Development of heat-transfer circuit in the blast furnace

The experiments on the operating furnaces in subsequent years showed that the height of the lower stage of heat-transfer is actually greater than the values obtained from the calculations made according to the circuit in Figure 1a. The accumulated data on heat-transfer in the lower stage created the conditions for the further development of heat-transfer circuit. B.I. Kitaev, B.L. Lazarev, Yu.G. Yaroshenko [4-7] expanded the heat-transfer circuit, according to which the upper stage of heat-transfer and the thermal reserve zone remained unchanged, but another character of change in the ratio of heat capacities along the height of the lower stage of heat-transfer was accepted. With approach to the level of blast tuyeres the ratio, like in the old circuit, initially increased, but as the intensity of the direct reduction reaction decreased this ratio became less than one. At the same time the temperature curves of charge and gas in the lower stage of heat-transfer got the typical downward bend of temperature gradient. The revised circuit of heat-transfer along the height of the blast furnace shown in Figure 1b differs from the circuit in Figure 1a due to the division of the lower stage of heat-transfer into two parts: for the upper stage $W_m>W_g$, and the lower stage $W_m<W_g$, that is conditioned by the end of the process of direct reduction of iron oxides to the lower level and, consequently, the overall heat absorption of the charge is substantially reduced.

The research of blast furnaces performance in the next decades shows that the temperature field along the height of blast furnaces undergoes a significant deformation. The essence of this tendency lies in the presence of several local thermal reserve zones in the temperature curves along the furnace height. N.N. Babarykin and F.A. Yushin were the first to note it while considering the temperature field obtained by the method of vertical sounding of MMK blast furnaces. A series of measurements during smelting of different iron ore in some cases showed the presence of two zones of hardly varying temperature along the stack height (Figure 2) [8].

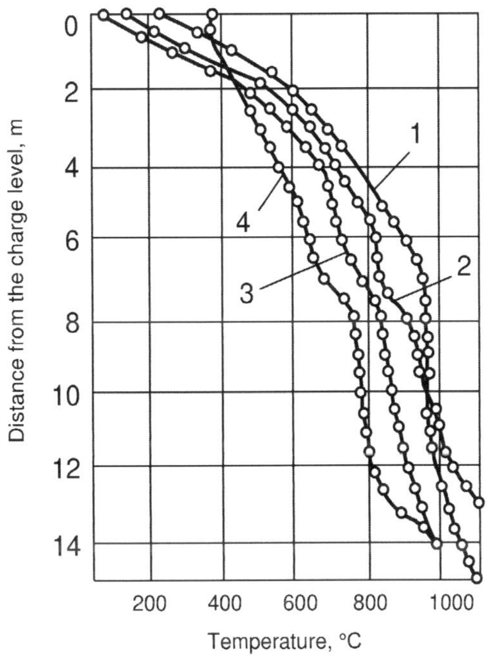

Figure 2. Change of the gas temperature along the blast furnace height during the smelting of lime fluxed pellets (1), dolomite fluxed pellets (2), non-fluxed pellets (3) and sinter (4).

Summarizing the experimental data on thermal and reduction performance of blast furnaces, prof. N.N. Babarykin [8] proposed the heat-transfer circuit given in Figure 3. According to this circuit the observed in the experiments two thermal reserve zones along the stack are formed as a result of chemical reactions with negative thermal effects. Formation of the upper section is conditioned by the effect of endothermic reactions of magnetite reduction with carbon monoxide, and the formation of the lower section is connected with the heat absorption during a substantial development of reaction of coke carbon gasification with carbon dioxide.

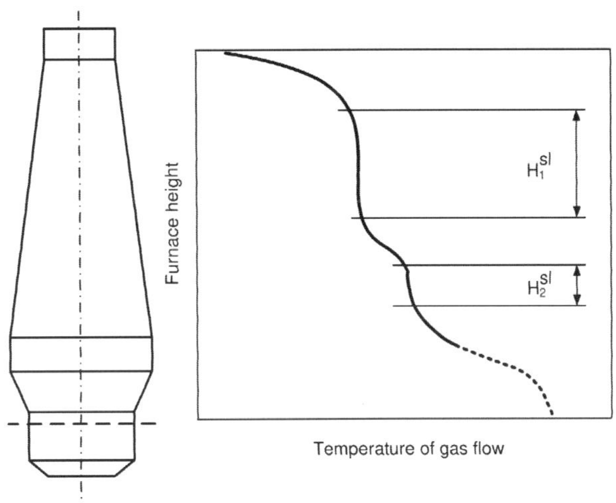

Figure 3. Scheme of temperatures distribution along the furnace height according to N.N. Babarykin: H_1^{sl}, H_2^{sl}– upper and lower sections of slowed heat-transfer.

The published data on mathematical modeling of heat- and mass-transfer in the blast furnaces show that two thermal reserve zones do not always appear during the formation of temperature fields in the furnace shaft [6]. This is understandable, because the simulation results largely depend on the adopted coefficients values of the model settings and, in particular, on the kinetic characteristics of iron ore raw material and heat-transfer coefficients. One of the key parameters in the calculations is the heat-transfer coefficient. Its values in the earlier calculations determined the low temperature difference of charge and gas [1-4]. According to the calculations of A.N. Ramma at the underestimated values of the heat-transfer coefficient this difference reaches 50-150 °C [9].

In this connection the results of simulation of the joint development of heat-transfer and reduction processes along the stack height, performed by B.I. Kitaev and his disciples, are of specific interest. Along with that the assumption was made about the stepwise nature of the iron oxides reduction. In Figure 4 and 5 the simulation results relating to the operating conditions of blast furnace No. 3 with volume 3000 m^3 at West Siberian Metal Plant are presented [10].

The overall picture of heat-transfer and reduction processes shown in Figure 4 is the most complicated case of manifestation of their relationship in the upper part of the charge, when the charge flow is overheated by the opposing gas flow. This area is characterized by the relatively small temperature difference between the gas and the charge. Therefore, this area, as well as the adjacent horizons, differ in the relatively low heat flow from the heating medium to the heated material and form a thermal reserve zone in the upper part of the blast furnace. The temperature level of this zone varies in the range 700-760 °C.

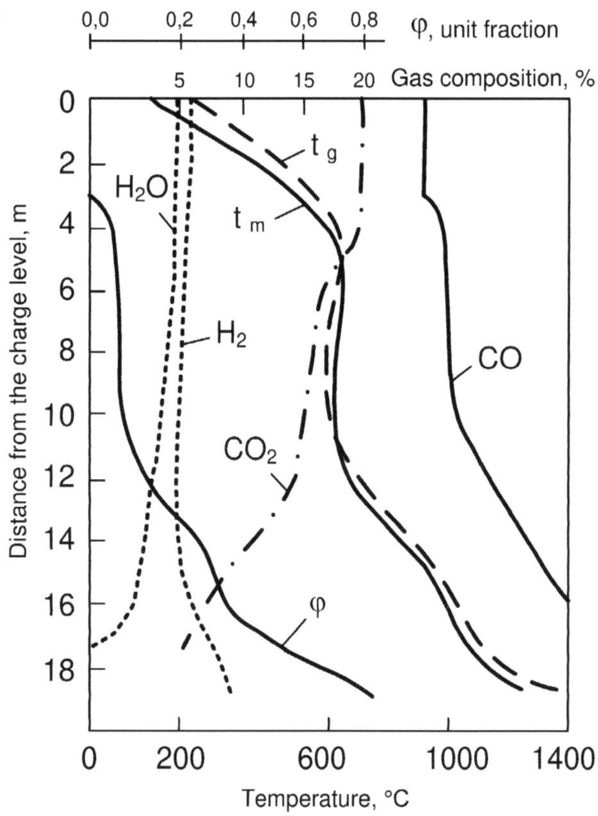

Figure 4. Change in the basic parameters of reduction along the height of the blast furnace shaft in the presence of local overheating of charge relative to gas: φ – the degree of materials reduction; q_g – consumption of the gasified coke carbon.

In Figure 5 the results of mathematical modelling of other operating conditions are presented, when the charge overheating relative to gas does not occur. If there is no overheating of charge compared to gas in the upper layers of the blast furnace shaft, the vertical and low-inclined sections of the temperature curves also form a thermal reserve zone. The reduction processes in this area are resented by underdeveloped reactions of the second stage of reduction. Below the upper thermal reserve zone the heat flow from gas to charge increases, the charge temperature goes up rather rapidly.

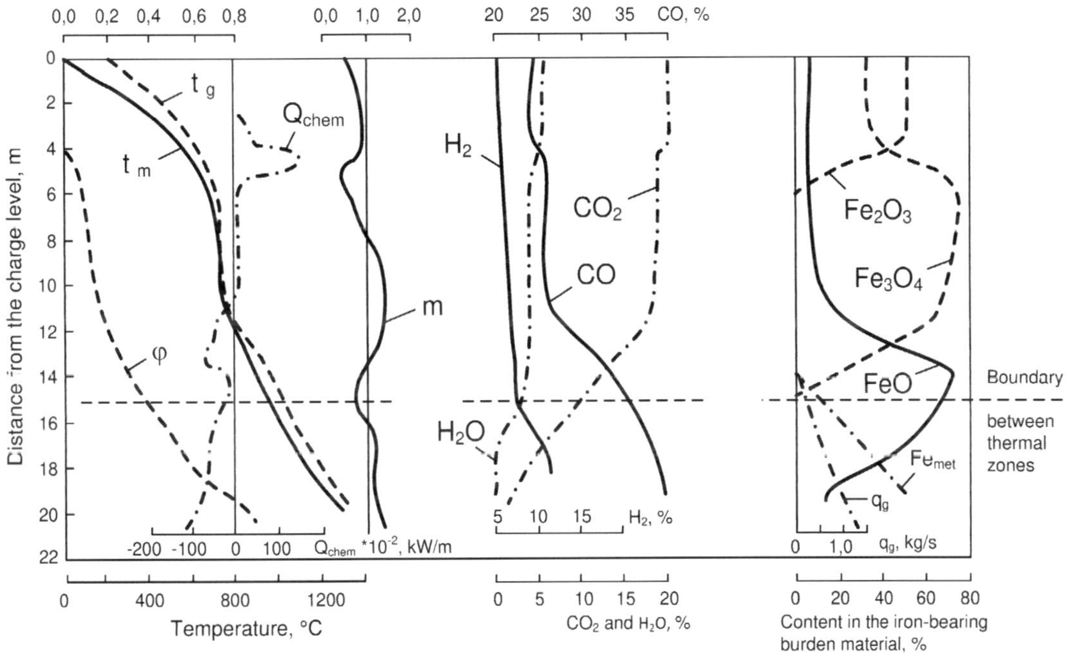

Figure 5. The simulation results of the joint development of heat-transfer and reduction processes in the absence of local overheating of charge relative to gas.

The height of the second (lower) thermal reserve zone depends on the ratio of thermal effects of exo- and endothermic reactions at the final stage of iron oxides reduction. However, a much greater role in the heat consumption in this part of the shaft height begins to play the reaction of coke carbide gasification, which is characterized by high values of thermal effect.

The features of heat-transfer development in the shaft of the blast furnace obtained during the modelling show that the difference between the gas and charge temperatures along the heating medium motion can repeatedly decrease and increase, reflecting a complicated process of mutual influence of heat-transfer and reduction.

4. Modern understanding of heat-transfer circuit in the blast furnace

Summarizing the experimental and theoretical studies of the processes in the furnace shaft and the previously obtained data on the operation of the lower part of the blast furnace, B.I. Kitaev with his followers offered a refined circuit of heat-transfer along the furnace height [5; 10] given in Figure 6, allowing the development of heat-transfer in the blast furnace to be shown for the modern technology of iron pig production.

Specific modes of blast-furnace smelting might differ in horizons of the selected circuits in some areas of heat-transfer and reduction, as well as the particular temperatures in these areas.

Two thermal reserve zones bring up the question concerning the earlier established concept of separation of blast furnace along the height into thermal zones. Therefore, we should note the following point.

The upper thermal reserve zone according to the results of mathematical modeling is characterized by the low heat-transfer rate and low-speed reduction reactions. Thus, the upper part of the thermal reserve zone is the reserve height providing improvement of thermal and reduction operation of furnace.

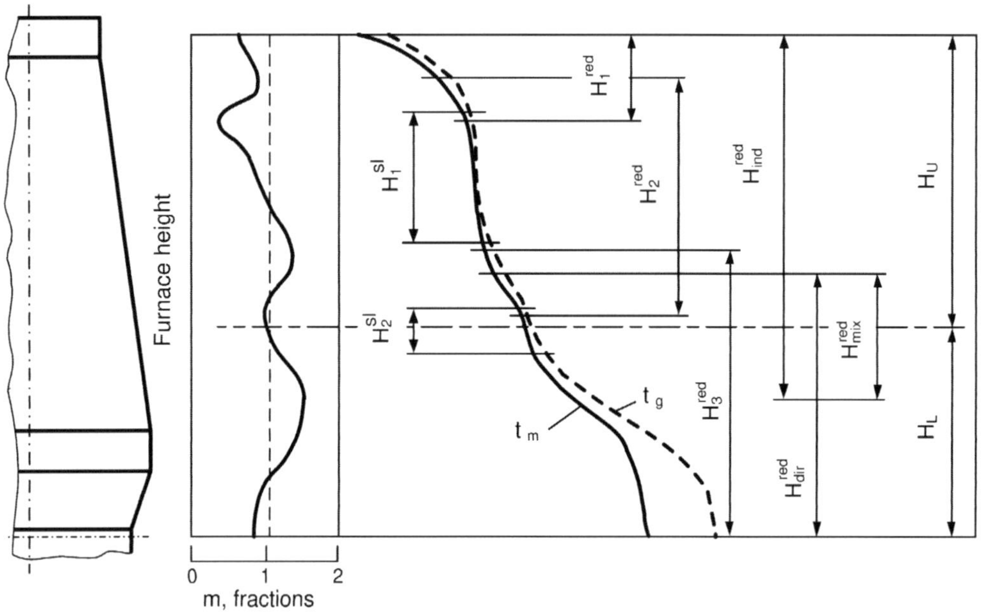

Figure 6. Heat-transfer circuit along the blast furnace height in modern technology of blast furnace smelting: H_1^{sl}, H_2^{sl} – the upper and lower parts of thermal reserve zone; $H_1^{red}, H_2^{red}, H_3^{red}$ – reduction zones of hematite, magnetite and wustite; $H_{ind}^{red}, H_{dir}^{red}, H_{mix}^{red}$ – zones of indirect, direct and mixed reduction; H_U, H_L – the upper and lower heat zones.

The lower part of the thermal reserve zone is characterized by the fact that at low thermal flows from gas to charge the rates of chemical reactions are quite high. Intensification of reduction in this area of the shaft will cause acceleration of coke carbon gasification with carbon dioxide and water vapor, which in turn will lead to the increased coke consumption. On this basis, it can be argued that the lower part of the thermal reserve zone as a reserve for improvement of thermal and reduction processes in the blast furnace should not be used.

In this regard, for the study and evaluation of the thermal state of the blast furnace as a control object it is reasonable to continue division into two thermal zones, the upper and lower one. The interface between them is located in the upper area of the mixed reduction, between the level of beginning of coke carbon gasification and the horizon, below which the iron oxides are reduced directly.

During intensification of the smelting processes due to the use of combined blast there is a change in the heat-transfer processes. On the basis of analytical calculations [17; 18] it is concluded that as the blast is enriched with oxygen and the theoretical combustion temperature goes up, the ratio of heat capacities of gas and charge flows – m in the upper stage of heat transfer increases. When approaching the critical limit of oxygen enrichment the degeneracy of the upper stage of heat-transfer is possible (Figure 7).

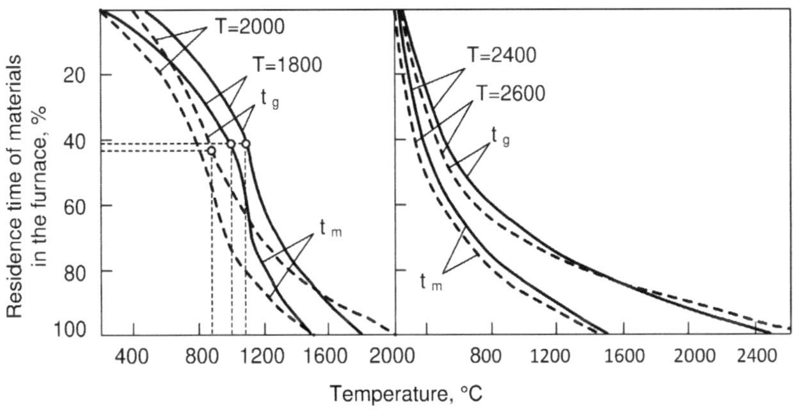

Figure 7. Change in the time and temperature of the materials and gases while the materials remain in the blast furnace at different values of the initial gas temperature at tuyeres (blast oxygen enrichment) according to the data [17].

However, degeneracy of the upper stage of heat-transfer during deep oxygen enrichment is possible only theoretically without taking into consideration the intrinsic properties of the blast furnace smelting operation and effect of feedback. In practice this circuit cannot be realized. In this case, the area of indirect reduction if substantially reduced, the degree of direct reduction inevitably increases, for this purpose additional carbon is required. As a result coke consumption or injected fuel should be increased, that will lead to the recovery of two-stage heat-transfer circuit. This provision is widely confirmed by both experimental and theoretical studies [12-18]. Thus, the concept of "autonomy of the upper and lower heat-transfer zones" should not be understood literally, they interact through the degree of direct reduction.

In Figure 8 (a, b, c) I.G. Tovarovskiy's distribution charts of temperatures and heat capacities of materials and gases [11] are provided for three cases:

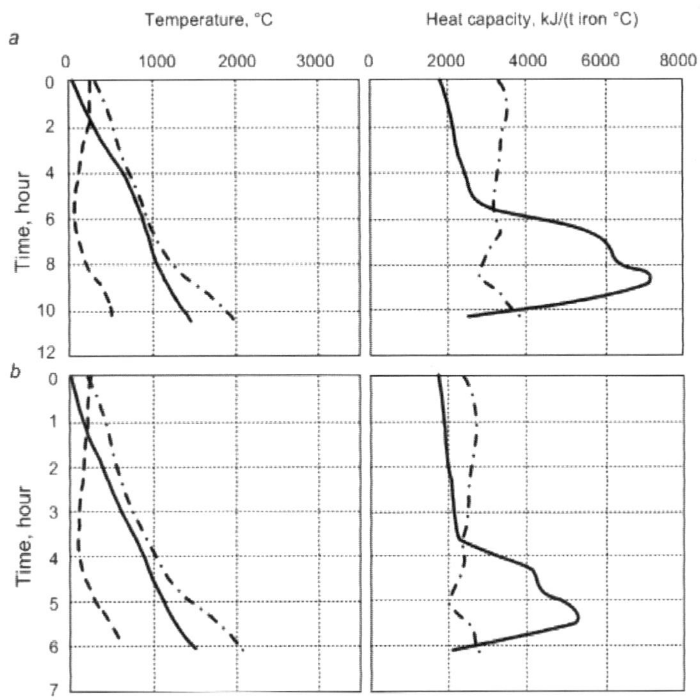

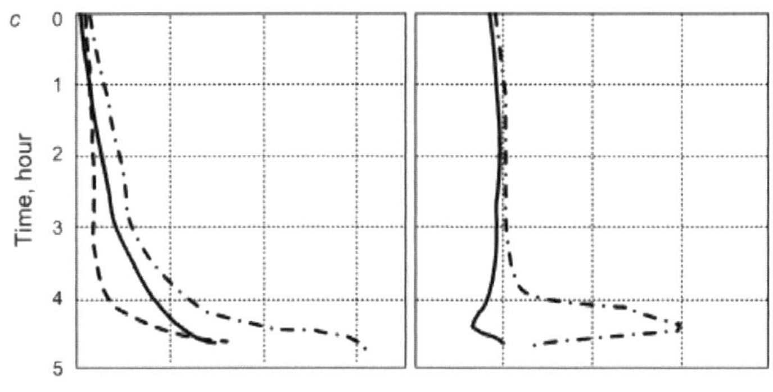

Figure 8. Distribution of temperatures and heat capacities of gas and charge flows while the materials remain in the blast furnace:

· — · — · — · —temperature and heat capacity of the gas flow;

——— temperature and heat capacity of the charge flow;

– – – – temperature difference between the gas and the charge

a – general mode with injection of atmospheric blast at temperature 1100 °C and natural gas (100 m^3/tonne) with output to 1.5 tonne/(m^3d);

b – intensive mode with injection of oxygen-enriched blast up to 40% at temperature 1300 °C and natural gas (180 m^3/tonne) with output 2.7 tonne/(m^3d);

c – the same as b, but without natural gas and output – 3 tonne/(m^3d).

In (a) there is a conventional two-stage heat-transfer, in which the heat capacity of gases flow in the shaft (upper stage) is greater than the heat capacity of materials flow, and during the lowering of charge and heating of materials the temperature difference between the gases and materials decreases to some minimum reaching m=1, after which m>1.0 and the temperature difference increases.

Case (*b*) is characterized by the fact that due to the increase in the quantity of gases the heat-transfer rate in the upper area decreases and the temperature difference between the gas and the charge on the boundary of heat-transfer zones goes up along with the transfer of thermal performance to the bottom of the furnace.

In (c) the limiting conditions for heat-transfer are observed, in which due to the small amount of gas per unit of charge the shaft "is turning cold", the temperature difference between gases and materials goes up and changes little during the lowering of materials, and m≈1 along the shaft height. "Degeneracy" of the upper stage of heat transfer and localization of all processes in the lower stage takes place. This case demonstrates the typical process instability and does not occur in actual practice.

5. Conclusions

Summarizing and analysis of experimental and theoretical studies showed that a two-stage heat-transfer circuit in the modern conditions of the blast furnace operation during smelting of various types of iron ore, use of the combined blast of high parameters, significant reduction of coke consumption and increase of the smelting rate, is a prerequisite for stability of the blast furnace operation and smelting efficacy. Any changes in the technology of modern blast furnace smelting should not change this circuit. The upper and lower stages of heat-transfer are connected with each other by the direct reduction processes, which eliminates the full autonomy of these stages of heat-transfer.

6. Acknowledgements

The work was supported by the Government of the Russian Federation, Act No. 211, contract No. 02.A03.21.0006.

7. References

[1] Kitaev B I 1944 *Steel* 7–8
[2] Kitaev B I *Heat-transfer in the Shaft Furnace* (Sverdlovsk-Moscow: Metallurgizdat) p 152
[3] Kitaev B I, Yaroshenko Y G and Suchkov V D 1967 *Heat Exchange in Shaft Furnaces* (London: PERGAMON PRESS) p 299
[4] Kitaev B I et al 1966 *Heat-transfer in the Blast Furnace* (M.: Metallurgiya) p 355
[5] Kitaev B I et al 1978 *Thermal Engineering of Blast Furnace Smelting Operation* (M.: Metallurgiya) p 248
[6] Dmitriev A N 2011 *Mathematical Modeling of the Blast Furnace Process* (Ekaterinburg: Ural Branch of RAS) p 162
[7] Ovchinnikov Yu N et al 1989 *Non-stationary Processes and Improvement of the Efficiency of Blast Furnace Smelting* (Chelyabinsk: Metallurgiya) p 120
[8] Babarykin N N 1981 *Steel* **3** 5–9
[9] Ramm A N 1980 *Modern Blast Furnace Process* (M.: Metallurgiya) p 304
[10] Spirin N A 1995 *Heat-transfer and Increase of Blast Furnace Smelting Efficacy* (Ekaterinburg: USTU) p 243
[11] Tovarovsky I G 2015 *Study of Processes and Development of Technology of Blast Furnace Smelting* (Dnepropetrovsk: Porogi) p 913
[12] Ando R et al 1972 *Ferrous metallurgy: Express-information* **33** (146) 1–30
[13] Spirin N A et al 2011 *Model of Decision Support System in the Process Control System of Blast Furnace Smelting* (Ekaterinburg: UrFU) p 462
[14] Junichiro Yagi 2012 *6th Int. Con. on Science and Technology of Iron Making – ICSTI* (Rio de Janeiro, RJ, Brazil) pp 1660–74
[15] Jian Xu, Sheng-li Wu et al 2012 *Tech. Contrib. to the 6th Int. Cong.on Science and Technology of Ironmaking – ICSTI* (Rio de Janeiro, RJ, Brazil) pp 414–424
[16] Hatano M, Misaka Y, Matoba Y and Otsuka K 1982 *Trans. Iron and Steel Inst. Japan* **22** (7) 524–533
[17] Spirin N et al 2014 *Trans. on Ecology and the Environment* (Ekaterinburg: UrFU) **190** (1) pp 467–478
[18] Bokovikov B A et al 2015 AISTech 2015 *Iron and Steel Technology Conf. and 7th Int. Conf. on Science and Technology of Ironmaking, ICSTI 2015* (Cleveland) **1** pp 237–244

Types of greenhouse gas emissions in the production of cast iron and steel

V G Lisienko, Yu N Chesnokov, A V Lapteva and V Yu Noskov

Department of Automation and Control in Technical Schemes, Ural Federal University, 19 Mira Street, Yekaterinburg, 620002, Russia

E-mail: annalapteva@mail.ru

Abstract. Types of carbon dioxide emissions in iron and steel production are indicated. Production processes have been classified according to mechanisms of carbon dioxide formation. Mathematical models for calculation of carbon dioxide emissions for each type of process are found. Calculations results of carbon dioxide emissions of coke (BF + EAF) and cokeless processes (Corex, Midrex, HyL-3, Romelt) in combination with EAF are provided.

1. Introduction

Greenhouse gases (GHG) are responsible for temperature increase on earth. Ferrous metallurgy produces a lot of such greenhouse gas as carbon dioxide (CO2). Methane, the other GHG, is a part of the secondary energy resources (SER) and is burnt to carbon dioxide in metallurgical units.

In addition to blast furnace (BF) process in iron and steel production Corex, Romelt, Midrex, Hyl-3 processes are used currently. Pig iron or sponge iron with scrap iron addition are loaded into electric arc furnaces (EAF) for steelmaking [1-4].

There is a problem of comparative assessment of carbon dioxide through emission (carbon footprint) in the following metallurgical tandem processes: BF + EAF, Corex + EAF, Romelt + EAF, EAF + Midrex, Hyl-3 + EAF. To solve it, we need to analyze the types of carbon dioxide emissions in metallurgical processes in terms of CO2 formation, find formulae (mathematical models) for carbon dioxide emissions calculation in different types of metallurgical processes.

2. Classification of industrial emission of carbon dioxide in metallurgy

Let us reduce worldwide spread concept of carbon footprint in ferrous metals industry to M_{th} - integrated through emission of carbon dioxide, which is the sum of CO_2 emissions, consequently appearing in all processes of technological chain, starting with raw materials extraction and ending with the product for which emission is provided. In addition, we distinguish a process integrated emission M_p from a transit emission M_{tr} determined by share of total mass of carbon dioxide emission, generated in previous process, which has been transferred to the analyzed process. The through M_{th} obey the formula:

$$M_{th} = M_p + M_{tr}. \qquad (1)$$

Blast furnace gas consists of the CO 25 – 27 % vol., and the CO_2 16 – 23 % vol., depending on content of oxygen in blast [4]. We call this CO_2 direct gas or direct emission M_{dir}. Part of the CO is used directly in blast furnace process, it is oxidized in blast heaters to CO_2. But for the most part it is burnt in boilers of local power plants, and as the simplest CO is burnt in flares or completely burnt in special units. We call the resulting CO_2 indirect gas or indirect emission M_{ind}. Fig. 1 shows CO and CO_2 gas flows in blast furnace operating on flux-bearing iron ore materials (no limestone in charge).

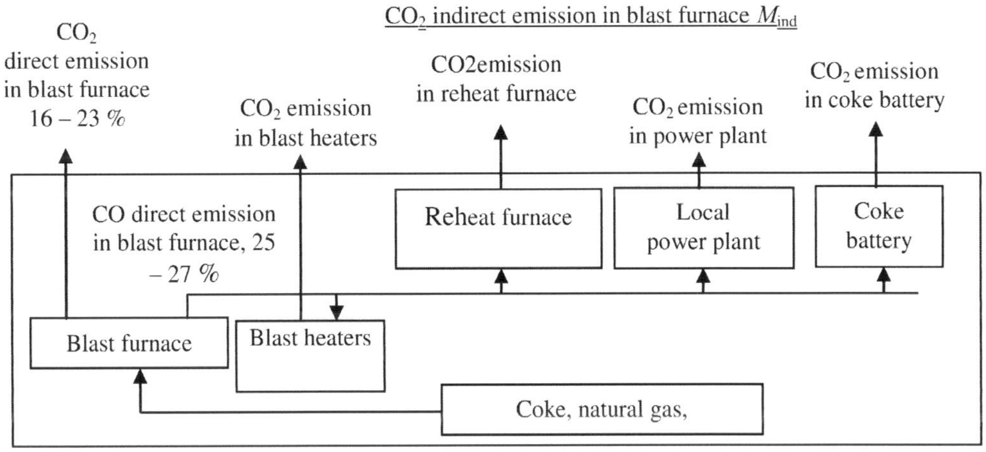

Figure 1. Gas flows during blast furnace operation.

Blast furnace is the cause of total or process integrated emission of carbon dioxide

$$M_p = M_{dir} + M_{ind}. \tag{2}$$

Figure 2 shows CO_2 and CO gas flows for Corex process. It is obvious, that it would not be correct to compare blast furnace emission with Corex process units only by values of direct emissions, since their work causes formation of additional CO_2 mass during CO burning in the other technological units. Different processes should be compared by carbon dioxide emissions by total (2) of their direct and indirect CO_2 emissions, i.e., by values of CO_2 process integrated emissions M_p.

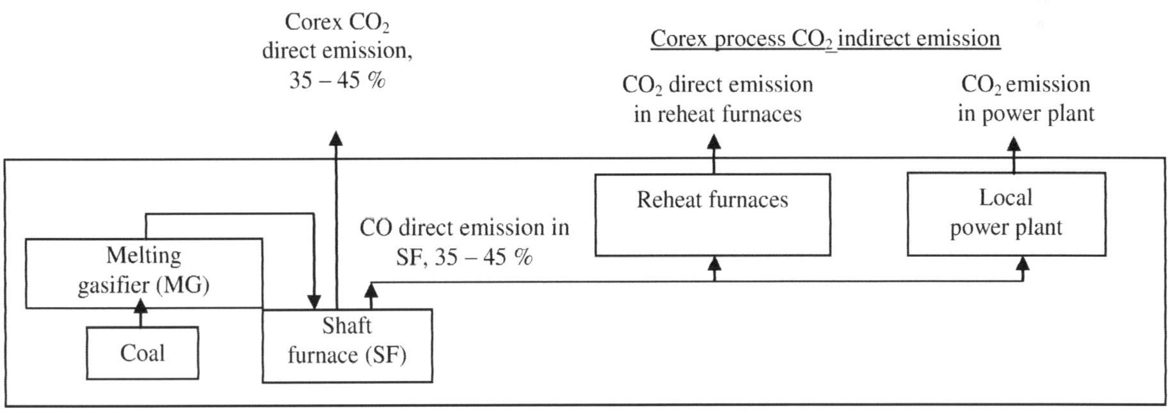

Figure 2. Gas flows in Corex process.

It is not possible to use formula from [5-8] to calculate the process direct and indirect emissions, as there is no data on SER consumption in reheat furnaces, boilers of local power plant, etc.

Thus, CO_2 mass produced in blast furnace, coke, electric arc and open-hearth processes will be determined by total carbon mass containing in raw fuels, taking into account CO reburning. In the

BOF process with reburning CO_2 mass [4] will be proportional to the mass of carbon burnt from charge.

It is interesting to estimate the so-called through emission of carbon dioxide throughout all the production chain to the final product. To calculate value of through emission we will present processes and their relations as a directed weighted graph with marked nodes - a sort of signal graph. In order to visualize formula derivation for through emissions calculation let us look at an example of general directed graph of emission (Figure 3).

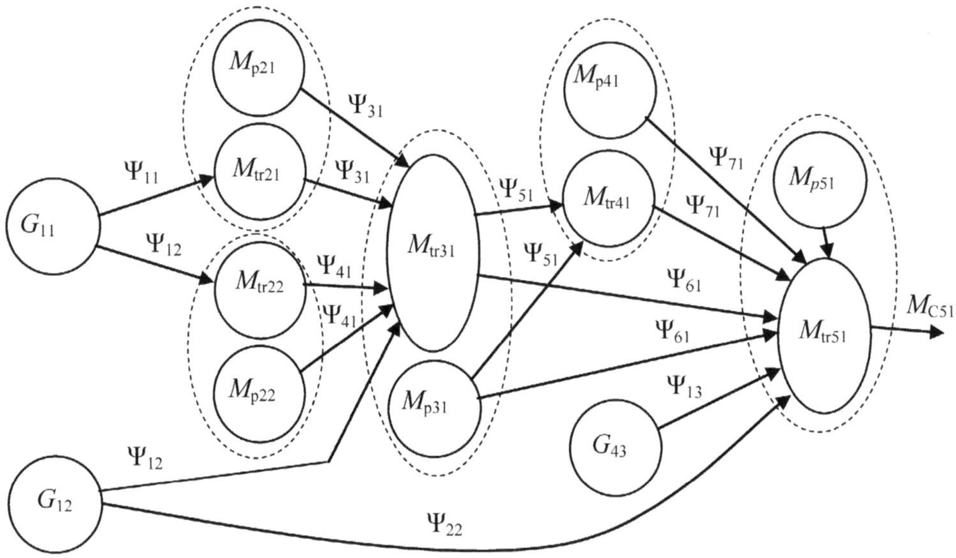

Figure 3. Graph of carbon dioxide emissions.

Weights of Ψ_{ik} directed lines, drawn from the node k to the node i, correspond to specific consumption of resources in tons or cubic meters, depending on emission dimension in the node, from which the line proceeds (kg/t of product or m^3/ton of product). Inside the twin nodes of complete graph (Fig. 3) values of transit emission M_{tr} and process emission $M_p$are indicated. The first index indicates the number of process stage, the second - number of process in the stage. Let us call sources the nodes from which arrows only proceed. Variety of processes correspond to these nodes (mining, transportation, crushing, screening, preparation, etc.), which are difficult to split. The value of their trough emission we denote as G_{ik}, where the first index indicates the number of process stage, the second index – the source number.

For signal graphs signal value at the node totals to the sum of signals entering from the other nodes, taking into account the directed lines transmission factor. Then, through emission for the node 51 can be found by using the following calculations:

$$M_{th51}=M_{p51} + G_{43}\Psi_{13} + (M_{p31}+ M_{tr31})\, \Psi_{61} + (M_{p41}+ M_{tr41})\, \Psi_{71} + G_{12}\Psi_{22}.$$

After simple transformations we obtain a formula from which follows general formula for through emission calculation:

$$M_{Cik} = M_{Pik} + \sum_{i=1}^{L}\left(G_{ik} \cdot \sum_{j=1}^{N} P_{jk}\right) + \sum_{i=1}^{M}\left(M_{Pil} \cdot \sum_{j=1}^{P} P_{jl}\right), \tag{3}$$

where L – number of sources (graph nodes, from which lines only come out); N - number of paths from the source node to the analyzed node; P_{jk} - transmission of the appropriate way, the product of directed lines weights, for example, $\Psi_{11}\Psi_{31}\Psi_{51}\Psi_{71}$; M - number of nodes, relevant to processes that have not zero process emission; P - number of paths from such nodes to the analyzed node. The length of the paths is different. Here index i corresponds to the number of the process stage in technologic chain, and index k marks emission of carbon dioxide from the the kth source in the ith process stage, l - number of node in the ith process stage, which has not zero process emissions.

For example, per 1t of agglomerate CO_2 emission is 319kg, per 1t of coke – CO_2 emission is 392kg, per 1t of pig iron CO_2 emission is 1.551kg. To produce 1t of pig iron 1.8t of agglomerate (Ψ_1 = 1.8, k index is not applied, as on each process stage one source of carbon dioxide is considered) and 0.6 t of coke (Ψ_2 = 0.6) should be used. Through emission per 1ton of pig iron M_{th} = 1.8·319 + 0.6·392 + 1 551 = 2 360 kg/t of pig iron. Considering injection of natural gas in blast furnace, CO_2 emission for blast furnace would have two components.

The integrated process emissions of carbon dioxide distinguish specific coefficient of carbon dioxide emission

$$\beta_{CD} = \frac{m_{CD}}{m_{prod}}, \qquad (4)$$

which is produced by the resulting mass of carbon dioxide per 1t of product (sinter, coke, iron, steel, etc.). Value of this coefficient depends on many factors. For this reason, we can speak of the set of its values for variety of technological parameters. Average values of this coefficient are analyzed further on.

3. Classification of metallurgical processes

In this paper, for the purpose of deriving formulas for calculation of integrated emission of carbon dioxide in production of pig iron and steel these processes have been analyzed in terms of distribution of carbon between gases and products, as well as carbonates content in charge. Here are the types of processes division into types that are different by ways of carbon dioxide formation that defines the methods of generated carbon dioxide weight or volume calculation. Division is performed by the product (columns) and mechanism of carbon dioxide formation (lines). Types of processes are sequentially numbered in all columns in order to eliminate unnecessary complexity in their listing.

Type 1.Processes in which volatization of volatile fractions occurs, which are used as fuel in the same processes (coking).

Type 2. In processes of this type carbon dioxide is emitted due to oxidation of carbon fuel and decomposition of charge components (limestone, dolomite) under affect of high temperature (sintering machines and fluxed iron ore pellets firing machines).

Type 3.Processes, in which fuel is burnt, part of carbon transfers to the final product and flux carbonates decompose (blast furnace process, Romelt, Corex).

Type 4. Processes in which charge metal carbon is burnt, but fuel is not used (oxygen converter production).

Type 5. The processes in which carbon from metal charge is burnt and fuel is burnt (electric furnace production).

Type 6. The process type in which fuel is burnt (HyL-3, Midrex).

4. Mathematical models of the carbon dioxide emissions in metallurgy

There is specific mathematical model (MM) for calculation of carbon dioxide emissions corresponding to each process type.

4.1. MM of the type 1 technological process

Processes in which volatization of volatile fractions from charging feed without access of air occurs and combustion of the purified of these fractions for the purpose of these processes (coking) takes place. During all produced coke oven gas combustion the whole carbon is oxidized in a way of direct and indirect emissions to CO_2. Carbon mass in coke oven gas is calculated based on data on mass or volume of coke oven gas outlet, its composition, its density.

$$w_c = M_c / M_{cc} = (0{,}65 - 0{,}75);\ w_{cog} = M_{cog} / M_{cc} = (0{,}12 - 0{,}15),$$

where M_{cc} - mass of coking coal; M_c - mass of coke produced from this coal; M_{cog} - mass of dry or reverse coke oven gas; w_{cog} – reverse coke gas yield coefficient of coal; w_c- coke yield coefficient of coal. CO_2 mass produced in combustion of M_F = M_{cog}mass of fuel gas mixture - coke oven gas, is determined by formula:

$$M_{G5} = 3{,}667 \cdot w_C \cdot M_F = 3{,}667 \cdot w_C \frac{w_{cog}}{w_c} M_{cog}. \quad (5)$$

Per 1,000kg of coke oven gas the amount of CO_2 is produced, mass of which is determined by formula

$$\beta_{DC5} = 3667 \cdot w_C \frac{w_{cog}}{w_c}, \text{ kg of } CO_2\text{/t of coke} \quad (6)$$

4.2. MM of the type 2 technological process

Specific weight of CO2 produced in type 2 technologic process is determined by mass of carbon completely oxidized during use of N types of fuel and masses of carbonates:

$$M_{G1} = 3{,}667 \sum_{1}^{N} C_I^P M_{FI} + 0{,}44 \cdot m_L + 0{,}47 \cdot m_D \quad (7)$$

where M_{GI} - mass of gas formed during total fuel combustion of the, t; 3.667 ≈ 44/12 - coefficient that determines amount of mass of generated gas per unit of combusted carbon (at CO combustion, this coefficient is 1.571); C_I^P – mass volume of carbon content (concentration) in the Ith fuel - ratio of carbon mass M_{CI} to M_{FI}; M_{FI} - mass of burnt I^{th} fuel, t; M_L - limestone mass, t; M_D – dolomite mass, t.

4.3. MM of the type 3 technological process

Representatives of this process type are blast furnace, Corex, Romelt processes. Formula for calculation of carbon dioxide mass formed in the type 3 process is as follows:

$$M_{G1} = 3{,}667 \left(C_{F1} M_{F1} + C_{F2} M_{F2} + 0{,}44 \cdot m_L + 0{,}47 \cdot m_D - C_1 m_1\right) \quad (8)$$

where C_{F1} - mass fraction of carbon in coke; M_{F1} - mass of burnt coke; C_{F2} - mass fraction of carbon in injection fuel; M_{F2} - mass of burnt injection fuel; C_1 - mass fraction of carbon in cast iron; m_1 - mass of cast iron; m_L - limestone mass, t; m_D - dolomite mass, t.

4.4. MM of the type 4 technological process

In addition to carbon dioxide in BOF carbon monoxide - combustion gas is formed due to oxygen blowing. Carbon monoxide goes with flue gases to purification and post-combustion system. In this respect, assume that the entire carbon of initial charge being burnt forms carbon dioxide. Integrated CO_2 emissions in type 4 process is determined by the following formulas

$$M_{G2} = 3{,}667\ \Delta m_C \,, \tag{9}$$

$$\Delta m_c = m_{ch}(C_i D_i + C_s D_s) - C_{st} m_{st} = m_{st}\left(\frac{C_i D_i + C_s D_s}{K_{loss}} - C_{st}\right), \tag{10}$$

$$K_{loss} = \frac{K_B[D_i(1 - C_i - Si_i - Mn_i - P_i - S_i) + D_s(1 - C_s - Si_s - Mn_s - P_s - S_s)]}{1 - C_{st} - Si_{st} - Mn_{st} - P_{st} - S_{st}} = \frac{K_{Tr}\sum D_l(1 - C_l - Si_l - Mn_l - P_l - S_l)}{1 - C_{st} - Si_{st} - Mn_{st} - P_{st} - S_{st}} \tag{11}$$

where M_{G2} - mass of CO_2 formed at carbon burning, t; Δm_C - amount of carbon burnt, t; C_I - mass fraction of carbon in cast iron; m_I - mass of iron in metallic charge, t; C_S - mass fraction of carbon in scrap; m_S - weight of scrap in metal charge, t; C_{St} - mass fraction of carbon in produced steel; m_{St} - mass of produced steel, t; D_I - mass fraction of iron in metal charge; D_S - mass fraction of scrap in metal charge; $m_{Ch}=m_I+m_S$ - mass of metal charge, t. $K_{loss}=m_S/m_{Ch}$ - the loss factor, which accounts to the loss of initial mass of charge due to carbon, iron and other charge raw components burnout; K_{Tr} – coefficient of charge pure iron transition into steel; D_l, C_l Si_l Mn_l P_l S_l - mass fractions of charge, carbon, silicon, manganese, phosphorus, sulfur in the lth component.

4.5. MM of the type 5 technological process

In this process, carbon dioxide is produced in fuel carbon oxidation M_{G1} (7) and in burning of initial materials carbon (9) M_{G2}:

$$M_{G3} = M_{G1} + M_{G2} \,. \tag{12}$$

This type includes melting in electric arc furnaces. Carbon in charge is in metal scrap and cast iron.

4.6. MM of the type 6 technological process

Specific gravity of CO_2 formed in the type 6 process, M_{GI}, is determined by the mass of completely oxidized carbon during using N types of fuel (7).

5. Results and discussions

Based on the above noted calculations are performed, the results of which are presented in Tables 1 and 2.

Table 1. Values of carbon dioxide emissions in various metallurgical processes.

Process stages	Through emission, kg per 1 t of product	Process emission to through emission ratio	Process range by CO_2 emission
EAF on scrap	1021	0.087	1
HyL-3	1044	0.593	2
Midrex	1211	0.524	3
EAF with iron	1434	0.062	4
BOF	2147	0.067	5
Blast furnace	2148	0.655	6
Corex	3475	0.768	7
Romelt	3925	0.954	8

IOP Conf. Series: Materials Science and Engineering **150** (2016) 012023 doi:10.1088/1757-899X/150/1/012023

Table 2. Values of carbon dioxide emissions in different metallurgical tandem processes of steel production.

Process stages	Through emission per 1 t of product		Process range by emission
	Mass, kg	Volume, m^3	
EAF on scrap	988	500	1
HyL-3 + EAF	1171	592	2
Midrex + EAF	1226	620	3
BF + EAF	1401	709	4
Corex + EAF	1832	926	5
Romelt + EAF	1980	1001	6

6. Conclusions

1. Processes at EAF scrap, HyL-3, Midrex and HyL-3+EAF, Midrex+EAF tandems have the lowest through emission (carbon footprint).
2. Corex, Romelt processes and Corex+EAF, Romelt+ EAF tandems have the highest carbon footprint.
3. BF + EAF tandem by its carbon footprint is at intermediate position.

7. References

[1] Cvinolobov N P and Brovkin V L 2004 *Furnaces of Ferrous Metallurgy: Learner's Guide for Univirsities* (Dnepropetrovsk: Porogi) p 154
[2] Yusfin Yu S and Pashkov N F 2007 *Metallurgy of Iron: Textbook for Universities* (M.: IKTs "Akademkniga") p 464
[3] Romenets V A et al 2005 *Romelt Process* (M.: MISiS, Publishing House "Ore and Metalls") p 400
[4] Voskoboynikov V G, Kudrin V A and Yakishev A M 1998 *General Metallurgy* (M.: Metallurgiya) p 768
[5] Eggleston H S, Buendia L and Miva K 2007 *Program MGEIK on National Cadasters of GHG: Guiding Principles of National GHG Inventories* (MGEIK 2006) (Japan: IGES)
[6] Potapochkin A N et al 2004 *Steel* **9** 69–72
[7] Shevelev L N 2007 *Steel* **4** 97–102
[8] Shevelev L N 2008 *Bulletin of Scientific, Technical and Economic Information "Ferrous Metalurgy"* **8** (1304) 3–8
[9] Chesnokov Yu N, Lisienko B G and Lapteva A V 2011 *Steel* **8** 74–77
[10] Chesnokov Yu N, Lisienko B G and Lapteva A V 2012 *Metallurg* **12** 23–26
[11] Lisienko B G, Chesnokov Yu N and Lapteva A V 2015 *Metallurg* **5** 18–24

IOP Conf. Series: Materials Science and Engineering **150** (2016) 012024 doi:10.1088/1757-899X/150/1/012024

Modification of gray iron produced by induction melting with barium strontium

O G Modzelevskaya[1], A V Feoktistov[1], I F Selyanin[2], A I Kutsenko[3] and A A Kutsenko[4]

[1] Department of Quality Management, Siberian State Industrial University, 42 Kirova Street, Novokuznetsk, 654007, Russia
[2] Department of Material Science, Foundry and Welding Technologies, Siberian State Industrial University, 42 Kirova Street, Novokuznetsk, 654007, Russia
[3] Department of Business and Economics, Siberian State Industrial University, 42 Kirova Street, Novokuznetsk, 654007, Russia
[4] Department of Research Projects, Siberian State Industrial University, 42 Kirova Street, Novokuznetsk, 654007, Russia

E-mail: aak_mail@mail.ru

Abstract. The article provides analysis of results of gray iron experimental melts in induction furnace and the following melt modification with barium-strontium carbonate (BSC-2). It is shown that modification positively affects mechanical and casting properties and as-cast iron structure. It was established that BSC-2 granulated immediately prior to use has greater impact on melt than BSC-2 of the same faction, supplied by the manufacturer.

1. Introduction

Recently, in iron-carbon alloys casting significant attention is paid to liquid melt modification, namely to use of relatively cheap natural materials, represented by complex carbonate ores containing barium and strontium, as modifiers. One of the named advanced materials are barium-strontium carbonate ores, which deposit is being developed in the north-east of the Irkutsk region in Russia, based on these ores the BSC-2 barium-strontium carbonate modifier is industrially produced.

BSC-2 is supplied by "Metalltehprom" SEE Ltd. (Irkutsk) in two main classes by size: 1 class – 10...70mm; 2 class - from powdered to 10mm. Chemical composition of the supplied BSC-2, % is: CaO – 21.5; SiO_2 – 24.8; BaO – 16.0; SrO – 5.5; MgO – 0.9; K_2O – 3.0; NaO – 3.0; Na_2O – 1.5; Fe_2O_3 (FeO) – 4.0; MnO – 0.2; Al_2O_3 – 2.9; TiO_2 – 0.9; CO_2 – 18.0.

This modifier contains barium and strontium, active elements, which by their deoxidizing and modifying effect on steels and cast irons are significantly superior to conventionally used silicocalcium and ferrosilicon.

Industrial testing of BSC-2 modifier is launched 2004 at many machine-building enterprises of "Kamaz" JSC [1, 2], "AvtoVAZ" JSC [3], "Universal" JSC (Novokuznetsk) [4-6]. Results of experimental introduction at metallurgical plant of "Uralvagonzavod Production Association" FSUE of technology of 20GL open-hearth steel modifying with BSC-2 indicates possibility of significant improvement in the range of foundry and mechanical properties of steel and reduction of castings defects along with actually achieved high technical and economic effect for the enterprise [7].

Modifier can form more stable compounds with sulfur and phosphorus than classic fluxing materials used in metallurgy, taking them out of metal into slag, and is more oxygen hungry [8]. Barium and strontium have a modifying effect on ferrite-perlite basis and on large group of non-metallic inclusions, sulphides and phosphides.

Research is fulfilled in accordance with the task No. 2014/213 to perform state scientific research within the base part of the state task of the Ministry of Education and Science of the Russian Federation (Project N2447).

2. Results and discussion

The experiments were conducted to study BSC-2 effect on castings structure and mechanical properties of SCh 15 gray iron. Experiments were performed in the induction casting furnace (ICF) with capacity of 0.06t. The charge of 50 kg weight, consisting of iron LK (40%), casting redistribution (10%), scrap iron (50%), was put into the induction furnace. Chemical composition of the metallic charge is shown in Table 1.

Table 1. Chemical composition of metal charge.

Mass fraction of elements, %											
C	Si	Mn	Cr	Ni	Cu	Ti	V	Mo	S	P	Fe
2.92	2.53	0.46	0.16	0.09	0.15	0.04	0.03	0.006	0.050	0.25	the rest

After melting the charge temperature was driven to 1500°C. The temperature was fixed by tungsten-rhenium thermocouple connected to a digital measuring system [9]. Two meltings we conducted with three options of ladle melt processing:

- without adding the modifier;
- adding the BSC-2 modifier in fraction from pulverized to 10mm, as supplied by manufacturer;
- adding the BSC-2 modifier granulated to the size of pulverized fraction to 10mm instantly before the experiment.

In each variant, 15 kg of liquid metal was consumed. Both meltings were performed in identical conditions with different methods of modifier input. In the first melting the modifier was introduced in the heated ladle under the heel of metal with the help of alloying basket, in the second melting – at the bottom of the ladle. Temperature of metal tapping into the ladle in both cases was 1450±10°C.

BSC-2 was added at the rate of 0.3kg per 50kg of liquid metal, i.e. 0.1kg per ladle. Ladle modifying using alloying basket gave considerable pyroeffect. When modifier was introduced into the bottom of the ladle under metal stream pyroelectric effect was not observed.

From each ladle the following samples were poured at 1350±5°C:

- fluidity sand spiral sample in accordance with GOST 16438, to study melt fluidity;
- V-shape sample for mechanical tests according to State Standard (GOST) 24648, to determine microstructure and hardness;
- 3 cast samples with diameter of 30mm and length of 300mm according to GOST 24648, to study mechanical properties, followed by testing on IK-500 pull test machine;
- wedge sample, to determine cast iron chilling tendency after modification;
- thermal curves were taken during pouring molten iron into die casting sample, to carry out differential thermal analysis (DTA) [10-12].

Study of iron structure was performed on prepared thin sections of samples on Olympus GX-71 optical microscope programmed with Siams Photolab 700.

For metal and slag element composition analysis XRF-1800 wavelength dispersive X-ray fluorescence spectrometer (Shimadzu, Japan) was applied.

Analysis of the results of experimental melts of gray cast iron in induction furnaces with subsequent modification of melt with barium-strontium carbonate showed BSC-2 effectiveness in obtaining fine-grained structure, improvement of casting and mechanical properties of cast iron. Important factor is the state of modifier.

The X-ray spectral analysis showed no Ba and Sr in the samples modified with BSC-2, probably because of their low content. Introduction of modifier did not affect significantly chemical composition of cast iron. Cast iron modification enables reduction of silicon and harmful impurities – i.e. sulfur and phosphorus content.

At cast iron treatment with BSC-2 modifier, an increase of its fluidity (L) is observed. Experimental data on cast iron spiral sample fluidity are shown on Figure 1.

Fluidity of the 1st melt of cast iron modified with BSC-2 granulated as-prepared is increased by 1.7 times compared to unmodified cast iron, it is increased by 1.4 times in comparison with cast iron, modified with BSC-2 as-supplied by manufacturer Fluidity of the 2nd melt of cast iron modified with BSC-2 granulated as-prepared is increased by 1.5 times compared to unmodified cast iron, and by 1.07 times in comparison with cast iron, modified by BSC-2, as-supplied by manufacturer.

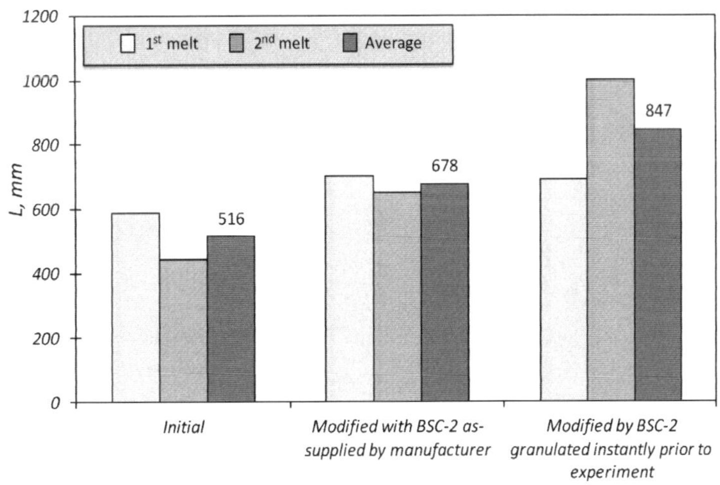

Figure 1. Values of cast iron fluidity for the 1st and 2nd melts.

In all the experiments, absence of chill is typical for wedge samples, due to high carbon content in initial cast iron.

In cast iron microstructure metal base and graphite inclusions were explored. These data are summarized in Table 2 and shown on Figure 2.

Table 2. Analysis of cast iron microstructure.

Melt No.	Sample	Structure by State Standard 3443-87		Size of graphite flakes, micron	
		Metal base	Graphite	min	max
1	Initial without introduction of BSC-2	Pt1 (lamellar perlite) –P (perlite) 6–F (ferrite) 94–PD (perlite, dispersion) 1.0–1.4	PGf (flake graphite inclusions size)1–PGr (flake graphite distribution) 9–PGd (flake graphite inclusions size) 180–PG (flake graphite) 6	5-7	145-155
1	Modified with BSC-2 as-supplied by	Pt1 (lamellar perlite)–P(perlite)45–	PGf (flake graphite inclusions size) 2– PGr (flake graphite	2.5-3.5	80-90

	manufacturer	F(ferrite)55– PD(perlite, dispersion)1.0	distribution)4– PGd(flake graphite inclusions size)90– PG(flake graphite)10		
1	Modified by BSC-2 granulated instantly prior to experiment	Pt1(lamellar perlite)– P(perlite)45– F(ferrite)55– PD(perlite, dispersion)1.0	PGf(flake graphite inclusions size)2– PGr(flake graphite distribution)6– PGr(flake graphite distribution)9– PGd(flake graphite inclusions size)45– PG(flake graphite)10	1.8-2.0	55-65
2	Initial without introduction of BSC-2	Pt1(lamellar perlite)– P(perlite)92– F(ferrite)4– PD(perlite, dispersion)1.4	PGf(flake graphite inclusions size)2– PGr(flake graphite distribution)3– PGd(flake graphite inclusions size)25– PGd(flake graphite inclusions size)90– PG(flake graphite)6	10-14	110-120
2	Modified with BSC-2 as-supplied by manufacturer	Pt1(lamellar perlite)– P(perlite)92– F(ferrite)8– PD(perlite, dispersion)1	PGf(flake graphite inclusions size)2– PGr(flake graphite distribution)8– PGr(flake graphite distribution)9– PGd(flake graphite inclusions size)45– PG(flake graphite)12	3-4	50-60
2	Modified by BSC-2 granulated instantly prior to experiment	Pt1(lamellar perlite)– P(perlite)85– F(ferrite)15– PD(perlite, dispersion)1	PGf(flake graphite inclusions size)2– PGr(flake graphite distribution)4– PGd(flake graphite inclusions size)15– PGd(flake graphite inclusions size)45– PG(flake graphite)12	1.5-2.0	45-55

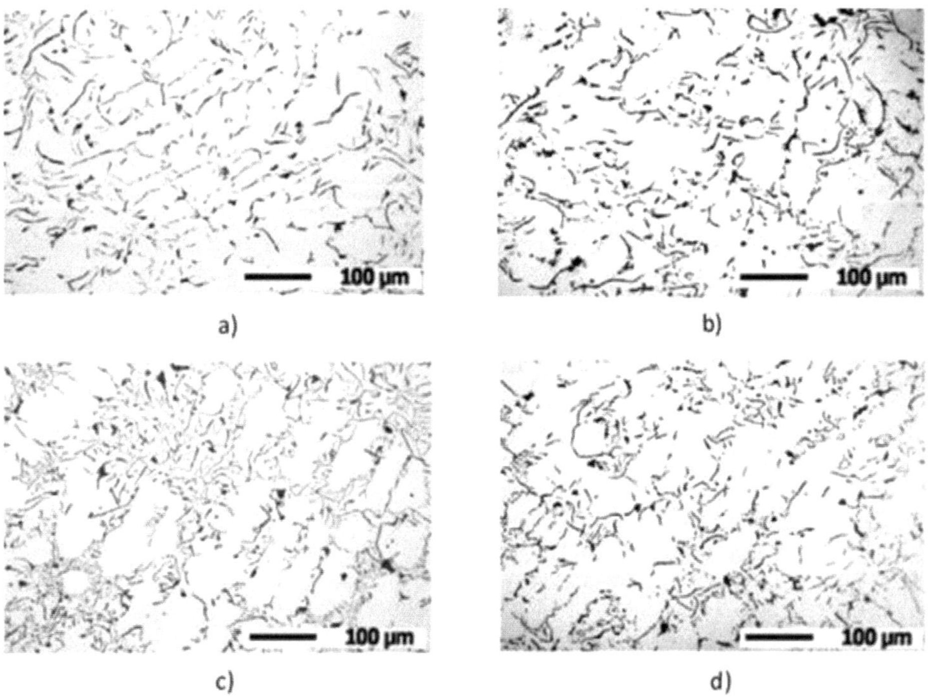

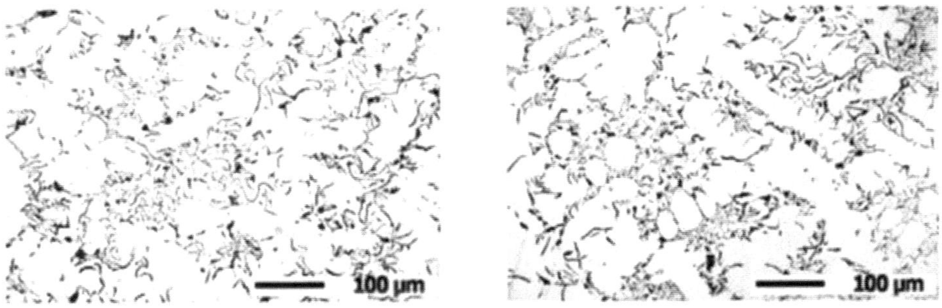

Figure 2. Structure of cast iron, obtained from two melts. Left a, c, e – samples obtained from the 1st melt, x100: a – initial, without introduction of BSC-2; c – modified with BSC-2, as-supplied by manufacturer; d – modified with BSC-2 granulated before the experiment; right b, d, f – samples obtained from the 2nd melt, x100: b – initial, without introduction of BSC-2; d – modified with BSC-2, as-supplied by manufacturer; f – modified with BSC-2 granulated before the experiments.

Microstructure analysis of three samples of cast iron from the 1st melt has shown that modifying leads to grain refinement of metal matrix and change in ferrite-perlite ratio. Refining of graphite inclusions takes place also, graphite is changing its form into vermicular. Greater refining of grains and graphite inclusions is observed when using BSC-2 granulated before the experiment.

Based on the Brinell hardness test results, average hardness value HBav, dispersion D (HB) and mean-square deviation σ(HB) were calculated. 22 measurements of metal hardness were carried out for each sample. Calculation results are presented in Table 3.

Table 3. Results of statistical parameters of cast iron hardness (HB) calculation.

Melt No.	Sample	Specification	Average hardness value HB_{av}	Dispersion$D$ (HB)	Mean-square deviation $\sigma(HB)$
1	Initial without introduction of BSC-2	1-1	240	56	7.5
	Modified with BSC-2 as-supplied by manufacturer	1-2	232	40	6.3
	Modified by BSC-2 granulated instantly prior to experiment	1-3	219	14	3.2
2	Initial without introduction of BSC-2	2-1	219	11	3.3
	Modified with BSC-2 as-supplied by manufacturer	2-2	217.7	10.2	3.2
	Modified by BSC-2 granulated instantly prior to experiment	2-3	221.4	6.1	2.5

Study of samples hardness has shown that cast iron mechanical properties after modification are more uniform along the body of casting. This is indicated by decrease in values of dispersion D(HB) and mean-square deviation σ (HB).

Reduction of average hardness HBav value is associated with increase in plasticity of cast iron modified with BSC-2. This is also evidenced by the data obtained as a result of tensile test carried out in accordance with GOST 1497-84 (Figure 3). Tensile strength σ of cast iron from the 1st melt modified with freshly granulated BSC-2 has increased by 32% compared to unmodified iron, and by 15% compared to cast iron modified with BSC-2 as-supplied by manufacturer.

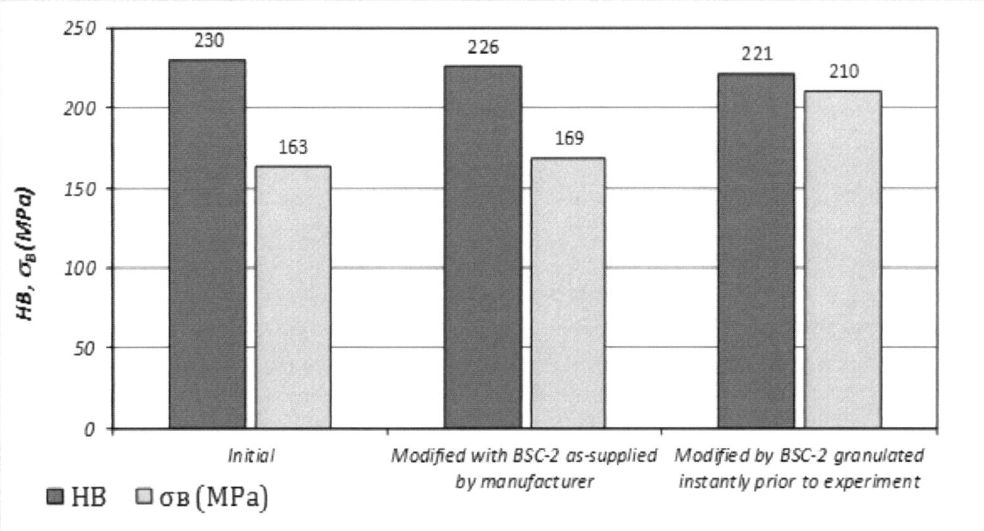

Figure 3. Average values of the Brinell hardness (HB) and tensile strength σ_B (MPa) of cast iron for the 1st and the 2nd melts.

Emission of latent crystallization heat during eutectic transformation is characterized by an increase in melt temperature during eutectic transformation and appeared to be eutectic recalescence (R) [10]. High values of R may indicate uncontrolled formation of graphite inclusions at the early stages of cast iron crystallization process, increasing risk of mold cavity geometry perturbation and incipient shrinkage occurrence. Furthermore, graphite inclusions formation at the early stages of melt crystallization and their growth may lead to a shortage of carbon needed for graphite growth at the end of crystallization cycle, which increases risk of internal shrinkage at microlevel.

Experiments have shown that introduction of BSC-2 modifier into the melt reduces the recalescence value, and modifier granulated before the experiment has lower R values, than the one as-supplied by manufacturer.

Under the influence of BSC-2 modifier temperature range between the lower temperature of eutectic transformation interval (point in which heat loss in the sample are compensated by the heat generated during dendrites growth) and cast iron solidus temperature in cast irons decreased, that indicates reduction of the risk of graphite, carbides and shrinkage defects inter-dendrite distribution in cast iron body.

These are witnessed by the results of metallographic study of samples structure.

3. Conclusions

Analysis of the results of gray cast iron experimental melts in induction furnace with subsequent modification of melt with barium-strontium carbonate have shown the BSC-2 effectiveness in terms of obtaining fine-grained structure, improving casting and mechanical properties of cast iron.

As a result of modification of induction melt cast iron with BSC-2 observed are refining of metal matrix grains and change in ferrite-perlite ratio. Refining of graphite inclusions is also observed, form of graphite becomes more vermicular.

Fluidity of cast iron, modified by freshly granulated BSC-2, has increased by 1.7 times compared to unmodified cast iron. Tensile strength limit σ of cast iron modified with freshly granulated BSC-2 rise by 32% compared to unmodified cast iron. Mechanical properties of cast iron after modification are more uniform along the casting body, hardness HB reduces by 4% correspondingly along with cast iron plasticity increase.

Eutectic recalescence R value reduction for modified cast irons decreases risk of shrinkage and formation of incipient shrinkage microporosity in cast iron body.

Significant factor is the state of modifier. BSC-2 granulated instantly prior to the experiment is has greater effect on melt than BSK-2 of the same fraction, as-supplied by manufacturer.

4. References

[1] Panov A G and Gurtovoy D A 2012 *Russian Foundrymen* **8** 25–27
[2] Panov A G et al 2010 *Russian Foundrymen* **12** 40–42
[3] Mukhayev O N and Panov A G *Russian Foundrymen* **10** 23–25
[4] Modzelevskaya O G et al 2015 *Liteinoye Proizvodstvo* **6** 6–8
[5] Feoktistov A V et al 2015 *Metallurgia Mashinostroyeniya* **4** 7–9
[6] Selyanin I F et al 2015 *Izvestiya Visshikh Uchebnykh Zavedenii. Chernaya Metallurgiya* **58** (10) 749–755
[7] Filippenkov A A et al 2015 *Russian Foundrymen* **9** 23–25
[8] Ryabchikov I V et al 2013 *Siliceous Ferroalloys and Modifiers of New Generation. The Production and Application* (Chelyabinsk: CSU Press) pp 295
[9] Kutsenko A I et al 2012 *"Tempol" Software for Collection, Primary Processing and Numerical Differentiation of the Data of Full Scale Experiment* patent RF No. 2012610147
[10] Boldyrev D A et al 2012 *Russian Foundrymen* **7** 15–20
[11] Byalik O M et al 1981 *Liteinoye proizvodstvo* **5** 2–3
[12] Snigir A N 1987 *Liteinoye proizvodstvo* **10** 3–4

Regularities of bainitic steel deformation transition

V E Gromov[1], E N Nikitina[2], Yu F Ivanov[3] and K V Aksenova[1]

[1] Department of Physics n.a. Prof. V.M. Finkel, Siberian State Industrial University, 42 Kirova Street, Novokuznetsk, 654007, Russia
[2] Vice-Rector for Assets Development, Siberian State Industrial University, 42 Kirova Street, Novokuznetsk, 654007, Russia
[3] Institute of High Current Electronics, Siberian Brunch of the Russian Academy of Sciences, 2/3 Akademichesky Ave, Tomsk, 634055, Russia

E-mail: gromov@physics.sibsiu.ru

Abstract. Quantitative analysis of defect and carbide subsystems evolution in medium-carbon bainitic steel subjected to compressive strain up to 36% was performed by means of transmission electron diffraction microscopy. Dislocation substructure and carbide phase parameters dependence on degree of deformation are identified, possible reasons of staging in their changes are discussed. It is suggested that the reason for bainitic steel softening at high (over 15%) degrees of deformation is activation of deformation microtwinning process.

1. Introduction

Attention of physical material science researchers is recently drawn to study of bainitic transformation in steels [1-3]. Steels with bainitic structure, due to the set of their characteristics (good welding properties, high creep-resistance and relatively high strength), are widely used in automobile production, energy production, production of rails, high-strength pipes for oil and gas industry, etc. [4, 5].

Products use often leads to deformation effects, resulting in plastic deformation. In turn, understanding of structure formation regularities and steel properties under plastic deformation is essential for mechanical hardening process control. To identify promising areas for application of technologies, based on plastic deformation after heat treatment, it is necessary to investigate hardening effect dependence on structural state of material and parameters of deformation processing mode, to establish cause-and-effect relationships between phenomena that determine comprehensive improvement of properties.

The aim of present work is to study formation and evolution of phase composition and defect substructure of bainitic steel under active compressive plastic strain.

2. Materials and methods

$30Cr_2Ni_2MoV$ structural steel [6] was used as a research material. Austenitization of steel was carried out at temperature of 960 °C, during 1.5 hours, with subsequent air cooling. Deformation of steel was carried out by axial compression at rate of $\sim 7\cdot 10^{-3}$ c^{-1} in 4x4x6 mm^3 columns in "Instron-1185" type test machine. Compression is more convenient to use as a mean of deformation, since it enables to

IOP Conf. Series: Materials Science and Engineering **150** (2016) 012025 doi:10.1088/1757-899X/150/1/012025

achieve more severe deformation than under tension. Study of structure and phase composition of steel was carried out by means of electron diffraction microscopy of thin foils.

3. Results and discussion

Typical profile of steel strain hardening curves is shown on Figure 1a. Mathematic processing of strain hardening curves shows that σ-ε dependence is parabolic and is described by polynomial of the fourth degree. Differentiation of σ-ε dependence curve allows to determine the coefficient of steel strain hardening

$$\theta = \frac{\partial \sigma}{\partial \varepsilon}$$

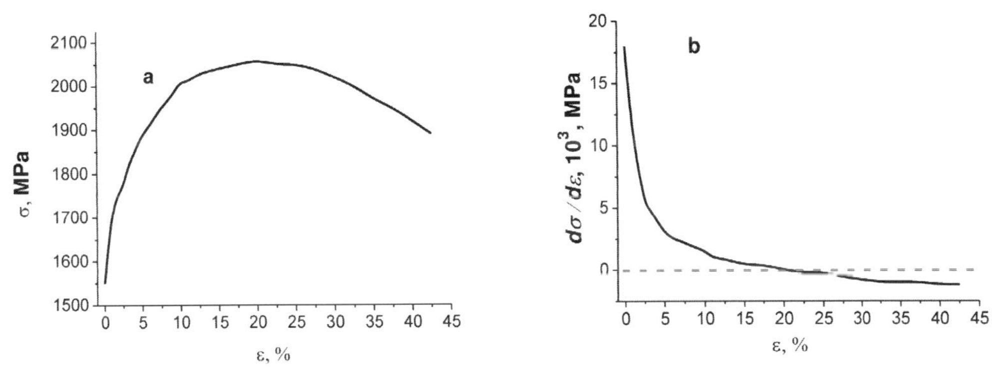

Figure 1. Curve of strain hardening (a) and strain hardening coefficient dependence from degree on deformation (b) for bainitic steel.

Analyzing results shown on Fig. 1b, two stages of strain hardening can be distinguished: the stage with parabolic σ-ε dependence or decreasing hardening coefficient Θ, and a stage and with slightly varying negative value of hardening coefficient. Transition from the first stage to the second one occurs in 19 ... 25% interval of strain degree. Destruction of test samples of steel occurred at $\varepsilon \approx 0.4$ through cleavage generating several large fragments. It is obvious that samples deformation behavior is determined by change in material phase composition and defect substructure. Consider this question in more detail.

Bainite transformation of steel leads to formation of multiphase structure [7, 8]. The main is α-phase (solid solution based on body-centered cubic lattice), additional phases are γ-phase (solid solution based on face-centered cubic lattice) and iron carbide (cementite).

Martensite (shear) mechanism of ferrite formation results in generation of dislocation substructure of reticulate type with relatively high scalar density of dislocations in bainite plates, that comprises $\approx 7*10^{10}$ cm^{-2} in investigated steel, indicating shear mechanism of $\gamma \rightarrow \alpha$ transformation. Plastic deformation of steel increases scalar density of dislocations (Fig. 2a). Dislocation substructure does not change - reticulate substructure remains unchanged. Scalar density of dislocations depends on degree of deformation. Namely, at 0 % $<\varepsilon<$ 18 % scalar density of dislocations increases monotonically. With increase in degree of deformation (18 % $<\varepsilon<$ 36 %) scalar density of dislocations remains practically unchanged.

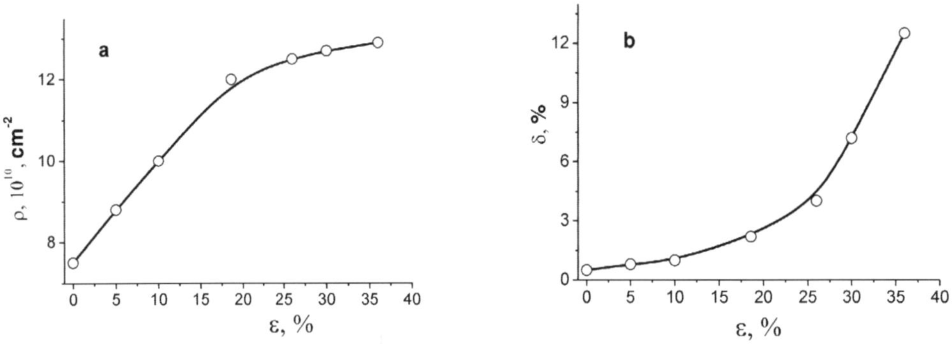

Figure 2. Dependence of dislocations scalar density (a) and volume of material containing microtwins, (b) on degree of deformation.

One possible reason for a significant reduction in rate of dislocations concentration at the late stressing stages is switching in additional mechanism of material deformation. Twinning may be one of such mechanisms, performed under deformation. Objectively, study performed within this research revealed substantial increase in volume of material containing deformation microtwins, at deformation degree exceeding ≈18 % (Figure 2b). Typical image of steel volume containing micro-twins of deformation origin is shown in Figure 3.

Elastic stress occurring due to shear mechanism of $\gamma \rightarrow \alpha$ transformation not only leads to formation of substructure with high dislocation scalar density, but also to fragmentation of bainite plates, i.e. to segmentation of plates into areas with small-angle disorientation.

Increase in deformation degree leads to decrease in average longitudinal dimensions of the fragments. Whereas transverse fragments dimensions are limited by bainite plates boundaries and during deformation stay almost unchanged. Decrease in average size of the fragments occurs against the backdrop of increase in degree of disorientation. Similar to dislocation scalar density behavior, change in size of the fragments shows up certain staging: at deformation degree of ε> 26 % change in average size of fragments practically stops. We can assume that fragments size reaches certain critical value (≈200 nm). It is shown [9] that such structure is not capable of further evolution during deformation process (so-called critical structure), it is preferential place for micro-cracking.

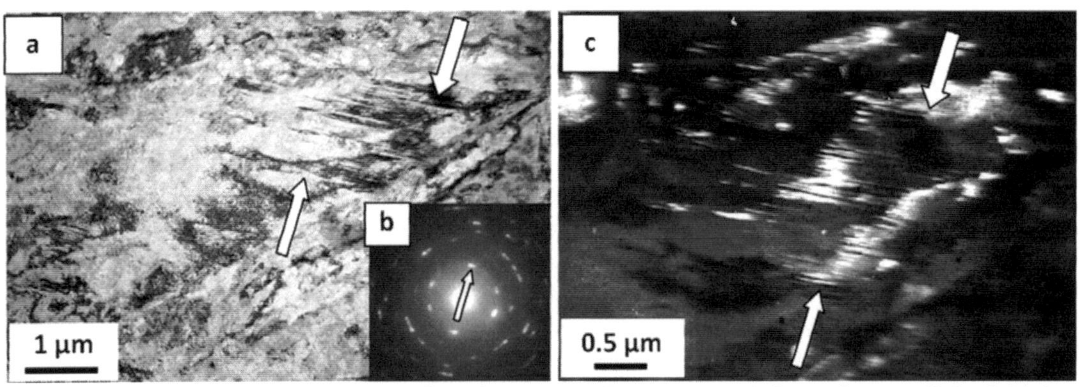

Figure 3. Electron microscope image of steel structure after ε = 36 % deformation; a – light field; b – micro diffraction pattern; c – dark field obtained in reflection [101] α-Fe; arrows indicate: on (a) and (c) – microtwins of deformation origin; on (b) – a reflection, in which dark field is obtained.

Electron-microscopic micro-diffraction analysis of steel by means of thin foils method has identified bend contours of extinction (Figure 4). Presence of these contours indicates bend and twist of the material crystal lattice [10-12]. It is found that in steel before deformation (initial state) and in steel after deformation of small extent (≈5%) contours are mostly arranged crosswise bainite plates, crossing the plate from one edge to the other (Fig. 4b). After large deformation (ε≈18% or more) ring contours are formed in material embracing certain areas of the plates (Figure 4c).

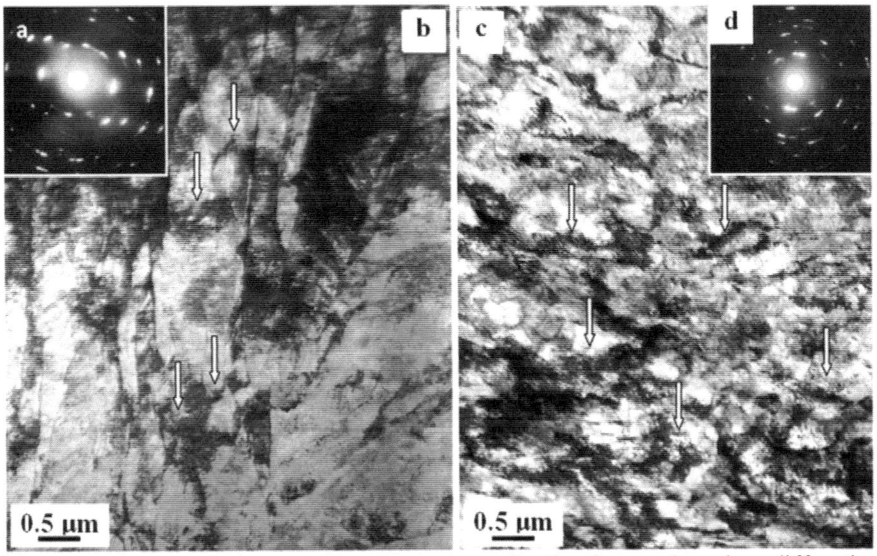

Figure 4. Electron microscopic image of bend contours of extinction; a, d – micro diffraction patterns; b, c - light field; extinction contours are indicated by arrows; b – ε = 5%; c - ε = 30%.

Quantitative analysis of steel structure demonstrated that with increase in extent of deformation, number of contours per a unit of image area (contours surface density) increases (Figure 5, curve 2). At the same time average crosswise dimensions of contours (Figure 5, curve 1) reduce. The first fact indicates increase in amount of stress raisers in material with the rise in extent of deformation, the second one points on increase of material crystal lattice bend and twist amplitude [10-12].

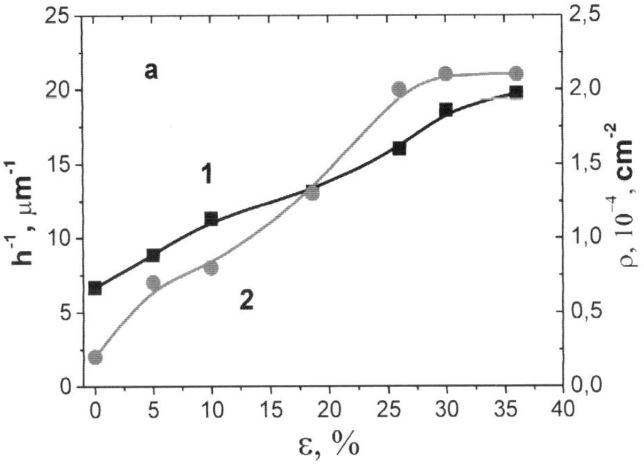

Figure 5. Dependence of average transverse dimensions of contours h (a, curve 1), and their surface density ρ (a, curve 2) on extent of deformation.

Deformation of steel is accompanied by significant transformation in material carbide subsystem. Initially, lamellar particles (ratio of longitudinal size (L) to transverse size (d) L/d ≈ 8) are transformed into ellipsoidal at the last stage of deformation (L/d = ≈5). At the same time inside bainite crystals (on dislocations and boundaries fragments) particles of round shape are found, their number is increasing with increase in extent of deformation.

Location of cementite particles is changing: with increase in deformation extent volume fraction of particles located on boundaries of bainite plates is increasing, and by the time of sample fracture almost all cementite is located on intraphase boundaries (boundaries of grains and plates of ferrite) (Figure 6a, curve 1). Increase in volume fraction of cementite particles located on intraphase boundaries proceeds non-monotonically: it is very rapid in deformation extent range of 5 % <ε< 10 % and is significantly slower at deformation of higher extent. The reason for an abrupt increase in fraction volume of cementite particles located on intraphase boundaries can lay in transformation of residual austenite and origination of α-phase and cementite, initiated due to steel deformation. Genuinely, micro diffraction electron microscopic studies revealed the fact of rapid decrease in volume fraction of retained austenite at steel deformation of low extent (ε≈10 %) (Figure 6b).

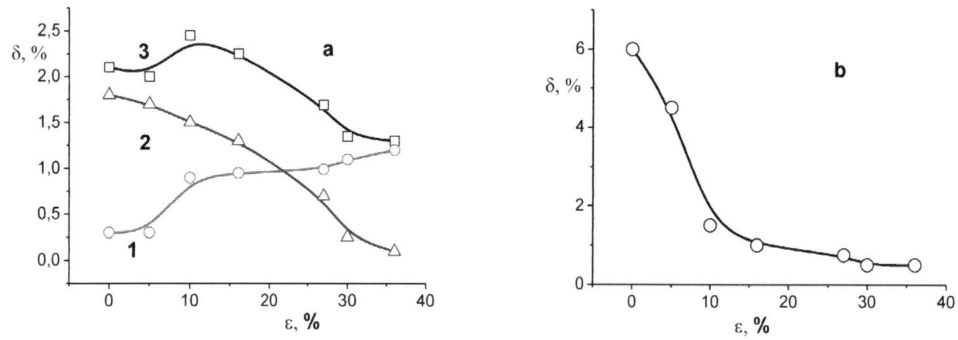

Figure 6. Dependence of volume fraction of cementite particle (a) and residual austenite layers (b) on ε extent of steel deformation; curve 1 - cementite particles located on ferrite crystal boundaries, curve 2 - cementite particles located in ferrite crystals; curve 3 - change in cementite total volume fraction in steel.

Initial stage of deformation (ε≈10 %) is accompanied by an increase in total proportion of cementite particles in steel (Figure 6a, curve 3); at deformation of high extent volume fraction of cementite in steel is reduced. This means that atomic carbon transfers to defects of steel crystal lattice (dislocations, sub-boundaries and boundaries) and to solid solution based on α-phase.

Stated facts in total allows for assumption that there are two concurrent processes taking place in steel during deformation: (1) dissolution of cementite particles formed during bainitic transformation in ferrite plates and (2) emission of cementite particles on dislocation substructure elements ("strain age") and on intraphase boundaries (subsidiary transformation of residual austenite).

Comparison of the results shown on Figure 1, Figure 2, Figure 5 and Figure 6, brings us to the statement that transition from the first stage of steel deformation to the second one is conditioned by the following change in material structure-phase state: first, by completion of intensive accumulation of dislocations, secondly, by initiation of deformation microtwinning, thirdly, by completion of bainite plates fragmentation, fourthly, by reaching maximum density of bend extinction contours, fifthly, by significant increase in steel solid-solution strengthening. Considering together, those processes result in propagation of areas with critical substructure in material having a potential of micro-cracking with subsequent destruction of sample.

4. Conclusions

IOP Conf. Series: Materials Science and Engineering **150** (2016) 012025 doi:10.1088/1757-899X/150/1/012025

It is shown that $30Cr_2Ni_2MoV$ bainitic steel plastic deformation by means of uniaxial compression is followed by, first, increase in dislocations scalar density and volume of material with deformation microtwins, secondly, decrease in average longitudinal size of fragments and increase in degree of their disorientation, thirdly, increase in number of stress raisers and material crystal lattice bend and twist amplitude. Revealed are stages of changes in steel structure parameters. Change of deformation mechanism is suggested as follows: on the first stage of stressing (0 % $<\varepsilon<$ 18 %) deformation is mostly conditioned by dislocations move; on the second stage (18 % $< \varepsilon <$ 36 %) by of dislocations move and twinning.

Quantitative electron microscopic micro diffraction analysis of steel phase composition evolution under plastic deformation by means of uniaxial compression is performed. It is shown that carbide transformations in bainitic structures occur within two concurrent processes – dissolution of cementite particles formed during bainitic transformation in ferrite plates, and emission of cementite particles on dislocation substructure elements as part of "strain aging". Concurrently subsidiary transformation of residual austenite initiated by steel deformation is observed.

5. References

[1] Quidort D and Bréchet Y 2002 *Scripta Materialia* **47** 151–156
[2] Clarke A J et al 2008 *Acta Materialia* **56** 16–22
[3] Borgenstam A et al 2009 *Acta Materialia* **57** 3242–52
[4] Matrosov Y I et al 1989 *Steel for Main Gas Pipelines* (M.: Metallurgy) p 288
[5] Pavlov V V et al 2007 *Metallurgist* **4** 51–53
[6] Pridantsev M V et al 1980 *Structural Steel: Reference Book* (M.: Metallurgy) p 288
[7] Kurdyumov V G et al 1977 *Transformations in Iron and Steel* (M.: Science) p 236
[8] Bhadeshia H K D H 2001 *Bainite in Steels* (London: Institute of Materials) p 460
[9] Rybin V V 1986 *Large Plastic Deformations and Fracture of Metals* (M.: Metallurgy) p 224
[10] Hirsch P et al 1977 *Electron Microscopy of Thin Crystals* (Huntington. NY: Krieger) p 574
[11] Konyeva N A and Kozlov E V 1982 *Soviet Physics Journal* **8** 3–14
[12] Ivanov Y F et al 2010 *Hard Construction Steel: The Structure and Hardening Mechanisms* (Novokuznetsk: Siberian State Industrial University) p 174

Detection of cold cracks in the cast-steels by the methods of ultrasonic and eddy-current infrared thermography

A I Cheprasov[1], S V Knyazev[2], A A Usoltsev[2], A E Dolgopolov[2] and R O Mamedov[2]

[1] Department of Physical Methods and Quality Control Units, National Research Tomsk Polytechnic University, 30 Lenina Street, Tomsk, 634050, Russia
[2] Department of Materials Science, Foundry and Welding Engineering, Siberian State Industrial University, 42 Kirova Street, Novokuznetsk, 654007, Russia

E-mail: krookia@mail.ru

Abstract. The aim of this study was to investigate the possibility of detection of cold cracks in the massive steel products using ultrasonic and eddy-current excitation, as well as the thermal imaging method of temperature recording, that in the perspective should be completed by the development of requirements for the monitoring equipment.

1. Introduction

The essence of the proposed methods is as follows. Ultrasound (US) infrared (IR) thermography presuppose local excitation of pulse or continuous ultrasonic vibrations in the control object with the registration of the temperature field of a large part of the control object by the infrared imager. Surface and subsurface cracks appear in the form of local temperature anomalies caused by the effect of internal friction and mechanical hysteresis. The basis for the development of this method was positive experience in control of composite materials, as well as foreign experience in metals control. Eddy-current infrared thermography uses the induction excitation of eddy-currents in metal, and the temperature rises in the cracks zone as a result of the crowding of the electromagnetic field lines.

2. Discussion and results

Ultrasonic IR thermography. Famous classical methods of thermal loading under thermal nondestructive control (TNC) are applicable only to a certain category of detected defects. Typically, these techniques require thermal stimulation of all control objects which is not always acceptable. Moreover, certain categories of defects can not be detected in principle by conventional thermal stimulation, e.g., via optical heaters. The optimal stimulation method, apparently, could be the one which would not lead to the change in the temperature of defect-free zone.

At the present time this kind of stimulation of internal defects is carried out by thermo-mechanical methods.

In the 90's of the last century the term "vibrothermography" became popular, indicating the thermal control process (TC) of the hidden structural heterogeneities of materials on their surface temperature fields under cyclic mechanical loading. A similar procedure can be performed with the acoustic or ultrasonic excitation of materials as the source of a significant increase in the temperature is the internal friction of the defects walls under their irradiation by mechanical vibrations, i.e. a thermo-mechanical heat takes place. This method is based on the analysis of absorption and scattering of acoustic (ultrasonic vibrations) in the solid media. Ultrasonic stimulation is most frequently used, realizing the principle of optimal heating and advantages of the synchronization method of heating function and the recorded temperature readings. The method is being developed by an independent group of G. Busse (University of Stuttgart, Germany) under the name "Ultrasonic lock-in thermography" [1-4], as well as by R. Tomasom, R. Favro and others (Wayne State University, USA) [5], who use the term "Sound IR imaging", or "Thermosonics". In recent years in the United States to refer to the appropriate procedure the term VibroIR or SonicIR is used (Lawrence Livermore National Laboratory) [6].

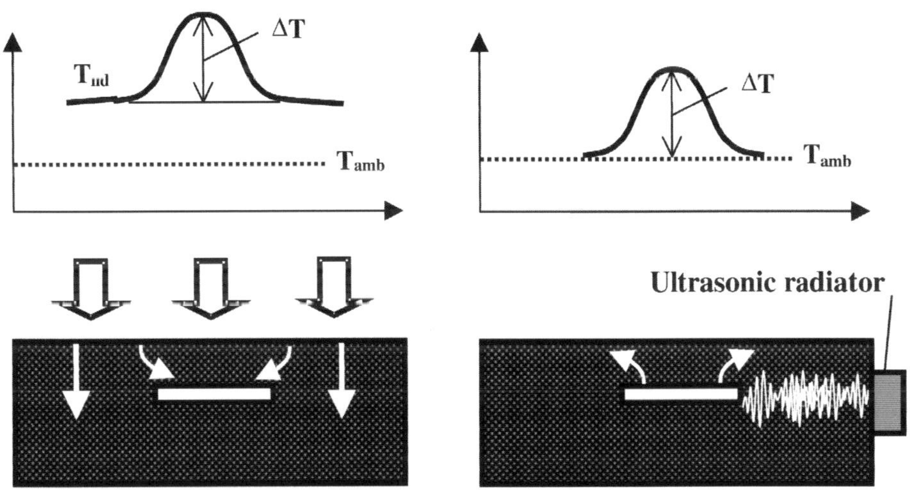

Figure 1. Method of heat waves (ultrasound stimulation).

The difference in surface optical methods and volumetric ultrasonic heating is illustrated in Figure 1. During the surface heating the excess temperature signal occurs not only above the defect, but also in the defect-free zones, which leads to a current noise contrast caused by the surface heterogeneities. Ultrasound stimulation produces mainly temperature signal only in the area ofdefect, and the temperature of defect-free zones remains close of the temperature of the environment, which was called the "dark field principle". As a result, ceteris paribus the probability of the correct defects detection increases (the requirements for the qualification of operators are reduced). Importantly, form of the temperature signal during at ultrasonic stimulation does not always coincide with the form of a signal appearing during surface heating: in many cases only a few zones of internal defects generate significant amount of heat due to wall friction, therefore, the topography of the surface temperature signal may be associated with the zones of developing cracks and do not reflect the already "formed" defects.

The authors of the method [3, 4] describe the experimental setup in which the ultrasonic excitation of products was carried out by elastic waves of frequency 20 kHz and a power of several hundred watts, the amplitude of which was modulated with a frequency of 1 Hz. For temperature recording was the thermal imager Jade II of company CEDIP (now FLIR Systems) was used, which provided a record of thermal images of format 320x240 at a frame rates up to 200 Hz in the spectral range 3 ... 5 μm. Duration of a single test reached 3 minutes.

The objects of study were composite materials (carbon and glass-fibre plastics) and ceramics, which demonstrated such advantages of ultrasonic stimulation as selectivity for defects and the effective use of excitation energy. In particular, detection of cracks and stratifications in the stringer ailerons, made of carbon fiber, on the modulating frequency of 0.4 Hz proved to be very effective. Another controlled material was silicon carbide reinforced with carbon fibers, which is used in space shuttles, high speed train brakes, etc., where it is necessary to resist the fast and powerful changes in temperature. The authors of study [3] suggest that ultrasound stimulation can detect those zones, where cracks develop, while the surface optical excitation reveals relatively large areas with anomalous TPC. With respect to the riveted joints of aluminum sheets, widely used in aviation, it was found that the conventional thermal images reflect the influence of the rivets on the heat-transfer and are insensitive to cracks. The temperature field pattern under a modulated ultrasonic load is more bind to the relative movement of interconnected sheets under load. In particular, when checking the data of the eddy-current control detecting a crack along a row of rivets, the ultrasound thermography showed that the same crack is detected at a frequency of 0.06 Hz (power of ultrasonic stimulation 600 W), and compared to the eddy-current control its mark has a greater length on the phasegram and on the ampligram.

A further increase in the competitiveness of the ultrasound thermography method is associated with the increased energy input and, consequently, the expansion of the control zone, which is currently about 1 m^2. Choosing the ultrasonic heater parameters it is necessary to prevent the formation of standing waves, in the nodes of which the defects can be overlooked (on the thermograms the standing waves can be seen as regular strips).

One of the main drawbacks of the periodic ultrasound stimulation method is a need to maintain good contact between the product and the ultrasonic stimulator for a few minutes required to reveal sufficiently deep defects. To solve this problem, in analogy to optical excitation it is proposed to use for products heating short ultrasound pulses (with a certain carrier frequency), and the surface temperature field to investigate using the thermal imager at the cooling stage (phase angle thermography with ultrasound burst excitation) [1, 4].The experimental results were obtained during detection of a crack in the row of rivets in the aluminum aircraft stringer panel. The duration of the ultrasonic pulse with power 2 kW was 100 ms. The cooling stage was investigated for 3 seconds at a frequency of thermal images recording of 15 Hz. The crack was visible in 270 ms due to its walls friction of its walls, but the interpretation of the initial thermograms was hampered by the presence of noise, as well as product heating near the ultrasound system.

In [4] a comparative data on the optical and pulse ultrasonic excitation during testings of aerospace products are provided. Ultrasonic pulses with the power up to 600 W and duration 100 ms (the carrier frequency 20 kHz) were used. The results of ultrasound thermography were recognized as the most informative in the control of composite materials, ceramic coatings on the turbine blades and riveted joints.

Research at Wayne State University is focused on the detection of cracks under pulsed ultrasonic excitation [5]. The authors use the "ultrasonic gun" for welding produced by the company "Branson", which generates a pulse with power of 1 kW and a duration of 30...200 ms with a carrier frequency of 20 kHz. The pulse is introduced by the direct contact through a steel horn of diameter 1.3 cm, using an immersion paste; the temperature development process is observed for a few seconds. The sub-surface cracks appear in the form of local temperature increase in the background of almost "cold" product (Figure 2a). Most clearly the effects of heat generation in the crack zone are revealed during recording the sequences of thermal images and viewing them in the form of a IR film (example of thermograms of a growing crack 1 mm in length is given in Figure 2b). The example of detection of a fatigue crack in the turbine blade is shown in Figure 3.

The Lawrence Livermore National Laboratory performed studies on the use of ultrasound vibrothermography for detecting defects of welded joints in the products of complex shape, in particular, in the zone of tubes and plates joints. It was found that the defective welded joints exhibit specific signatures of a thermal field under loading the product for 0.1 sec by a package of ultrasound pulses with power 300 W at a frequency of 30 kHz [6]. The advantage of the method is the weak dependence of the results on the ultrasound entry point. The disadvantage is the possibility of inducing new defects at sufficiently high power of ultrasonic pulses. In the same laboratory the comparative tests of acoustic microscopy, radiography, ultrasonic method and vibrothermography for detection of various defects in the nose panel of a space shuttle, made of 36-layer carbon fiber and coated with silicon carbide, were conducted. The results of vibrothermography turned out to be very promising.

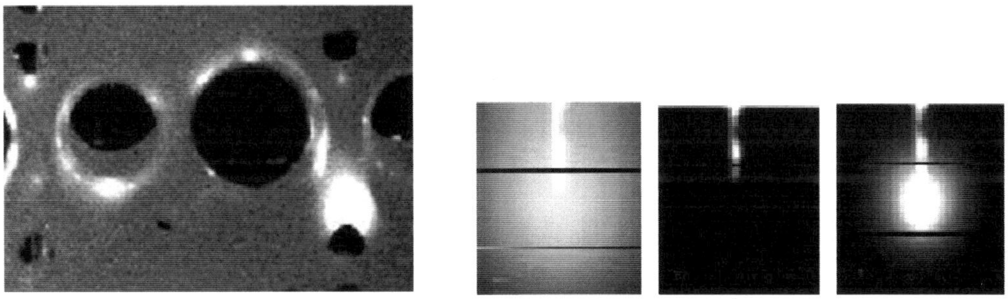

Figure 2. Pulse ultrasonic infrared thermography (results obtained by the foreign partner of the Contractor – Wayne State University, Detroit, USA): a – thermogram of numerous cracks (bright zones) in the aluminum panel of the internal combustion engine cylinder; b – development of cracks in the cutting zone in the aluminum plate 3 mm thick (the crack length is less than 1 mm.

In general, it should be stated that the infrared thermography method with ultrasonic stimulation is promising, but its features are still poorly understood; for example, in many cases, where intuitively the emergence of dynamic temperature signals were anticipated, such signals were not registered. The dynamic response of the control object depends on the excitation energy, the method of fixing the control object, in particular, thin products, and material properties. The frequency of the ultrasonic stimulation needs to be close to the natural frequencies of products in order to increase the heat generating efficiency.

Eddy-current infrared thermography. Method of non-destructive testing that combines thermal imaging principle of temperature recording and eddy-current (induction) heating principle, has recently acquired a certain popularity in the metal control. Metals often have a low coefficient of electromagnetic radiation absorption, which hampers their optical heating by lamps and lasers. During induction stimulation in the surface layer of the control object the eddy-currents are induced, and material heating is due to Joule heating. Stimulation is performed either by pulsed or by periodically-modulated waves, and the test results are analyzed using the principle of synchronous detection or phase analysis.

The eddy currents are generated only in the surface layer of the material (skin layer), the thickness of which depends of the excitation frequency, magnetic conductivity, and electrical conduction of the material. In magnetic materials, such as steel, the penetration depth is very small, for example, about 0.03 mm for the excitation frequency of 200 kHz. The eddy-currents in the skin layer, round the crack and concentrate on its corners increasing the Joule heating rate and, hence, the temperature by several degrees. The temperature is higher as deeper the crack. In the non-magnetic materials the penetration depth of eddy-currents is much higher than in the magnetic ones, thus, in the crack zones their density can decrease leading to the temperature drop. The effectiveness of the eddy-current simulation depends on the orientation of the magnetic field relative to the revealed cracks, so, it is necessary to revolve the control object during practical control.

For example, the Austrian researchers [7] investigated cast steels having surface defects with the depth of several millimeters. For heating the high frequency (HF) generator with power 10 kW, pulse duration 0.1-1 s was used, and the control object was placed inside the coil. A uniform magnetic field was created by the Helmholtz coils placed apart at a distance of the radius. The texture of the temperature field can be made sharper using Fourier analysis. It was found that in the phasegram the defective marks weakly depend on the depth of cracks, so to assess the parameters of defects the initial data of infrared thermograms and phasegram should be jointly analyzed. The advantage of phase analysis is that the phase of the signal varies considerably in the zone of cracks and weakly depends on the geometry of the control object, even if it leads to significant temperature anomalies.

In [8] the eddy-current thermal control of castings is described. It is shown that in the thin-walled pipes the through cracks are characterized by the increase in temperature only in the mouths of cracks. The scheme of full automatic control is offered, in which the product falls through the heating coil, and three infrared cameras record the product temperature field in the process of their falling. For the analysis of the test results a set of algorithms, eliminating the distortions of thermograms due to falling and segmenting the image of defects, was developed.

3. Conclusions

In the field of ultrasonic and eddy-current infrared thermography to date in the TPU the following studies have been performed:

- Experimental studies on continuous ultrasonic excitation (ultrasound frequency 22 kHz, electric power up to 300 watts have been carried out.
- Taking into account the international experience an pulsed ultrasound device for ultrasound materials stimulation was assembled (ultrasonic frequency of 22 kHz, pulse duration 0.1 s, electric power up to 2000 W).
- Due to the fact that for the input of ultrasound into the massive steel products, which are the parts of the bogies, special equipment is required, for the present moment, a number of preliminary experimental studies on the samples was conducted. These studies have shown that the main problem is the low efficiency of the ultrasound input into the metal, that does not provide the necessary temperature signals in the zone of cracks.
- For further experiments aiming at more efficient input of powerful ultrasonic vibrations into the product the equipment was designed.
- To perform an eddy-current heating of metal products with cracks a standard unit for induction hardening of metals is used, operating at a frequency of 30 kHz at an electric power of 10 kW or more.
- In the initial IR thermograms the cracks are not found on the background of a significant heating of the sample directly under the inductor coil. During the processing of the recorded sequence of infrared images, in particular, according to the so-called method of principal component analysis, the cracks are weakly detected at any mutual orientation of the cracks and coil. Apparently, in the experiments the IR camera was placed too far from the sample, as a result, a crack appears in the form of a very narrow strip.
- The research in the field of ultrasonic and eddy-current thermography should be continued. Further research is needed to optimize the frequency ranges for the two methods, the method of powerful ultrasound input into the metal and the choice of an effective inductor form.

4. References

[1] Dillenz A, Zweschper Th and Busse G. 2000 *Insight* **42** (12) Dec. 815–817
[2] Zweschper Th, Dillenz A and Busse G. 2001 *Insight* **43** (3) March 173–179
[3] Zweschper Th, Dillenz A and Busse G 2000 *Proc. Eurotherm Seminar No. 64 on Quant. IR Thermography* (Reims, France) pp 212–217
[4] Dillenz A, Zweschper Th and Busse G 2000 *Proc. Eurotherm Seminar No. 64 on Quant. IR Thermography* (Reims, France) pp 247–252
[5] Favro L D, Han X, Ouyang Z et al. 2000 *Proc. SPIE "Thermosense-XXII"* **4020** pp 182–185
[6] Burke M W and Miller W O 2004 *Proc. SPIE "Thermosense-XXVI"* **5405** pp 313–321
[7] Oswald-Tranta B and Sorger M 2010 *Proc. QIRT'2010 Conference* pp 162–167
[8] Oswald-Tranta B, Sorger M and O'Leary P 2010 *J. Electronic Imaging* **19** (3) 1–6

Vibration influence on structure and density of aluminum alloys

A A Usoltsev[1], S V Knyazev[1], A I Kutsenko[2], A E Dolgopolov[1] and R O Mamedov[1]

[1] Department of Materials Science, Foundry and Welding Engineering, Siberian State Industrial University, 42 Kirova Street, Novokuznetsk, 654007, Russia
[2] Department of Business and Economics, Siberian State Industrial University, 42 Kirova Street, Novokuznetsk, 654007, Russia

E-mail: a.us@rambler.ru

Abstract. The results of study on aluminum alloys of grade AK9M2 AK12M2 are provided. Alloy crystallization time for alloys AK9M2 AK12M2 decreases, the intensity of reduction depends on the vibration amplitude. For alloys AK9M2 and AK12M2 the optimal amplitude is 2÷2.2 mm, allowing a dense cast alloy with a fine grain structure to be obtained. Density of samples from AK12M2, cut from the bottom part, is slightly increases with the rise of the vibration amplitude.

1. Introduction

Improvement of the cast alloys density and creation of a fine-grained structure have always been important tasks. The structure and density of the cast alloys, largely determine the basic properties of metal in the casts. One of the trends in the industry is a fundamental improvement in the metal quality, improvement of their technical and economic characteristics. There are many ways to increase the density of cast alloys. The solution of this problem is possible b active interference in the crystallization process using a vibration treatment. The defining parameters influencing the melt crystallization during vibration treatment are frequency and amplitude of vibration exposure.

2. Discussion of the problem and results

According to the data in [1, 2] and our own research the optimum vibration frequency for aluminum alloys is 50 Hz [3, 4]. Vibration loads were observed only in vertical planes. The amplitude of vibration was varied in the range from 0.5 to 5 mm. Studies were carried out on aluminum alloy of grade AK9M2 and AK12M2. Alloy overheating temperature remained constant at 835÷850 °C. The samples casting temperature also remained constant and was 750÷760 °C. The melt was poured into a cylindrical mold of the following dimensions: inner diameter 64 mm and height 145 mm. The mold was mounted on vibration unit, firmly fixed with special ties to the vibration exciter [4].

To study the effect of vibration impact on the crystallizing melt the unit was designed the diagram of which is shown in Figure 1.

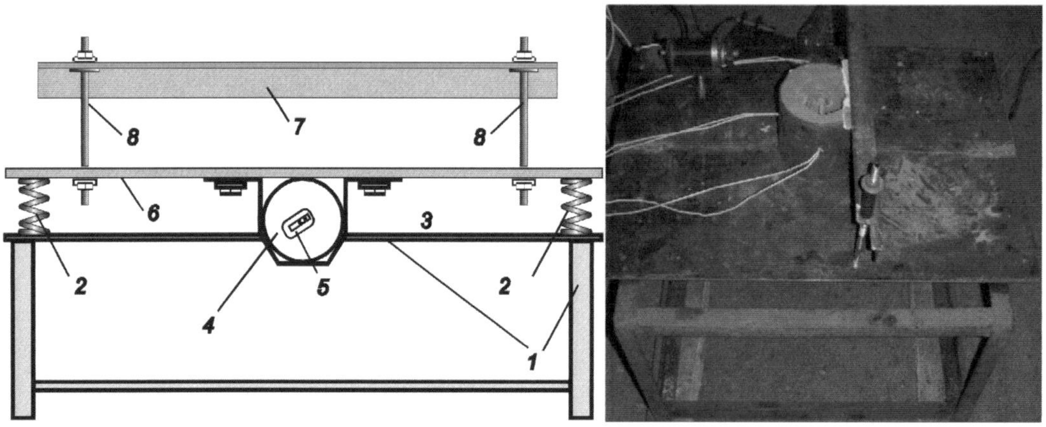

Figure 1. Diagram of vibration unit: 1 – frame; 2 – springs; 3 – bolting; 4 – the electric motor; 5 – eccentric; 6 – vibration platform; 7 – metal clamp for fixing the mold; 8 – bar with thread).

Starting from casting and to the complete solidification the melt was subjected to vibration with different amplitudes. Further, from the obtained ingots the samples were cut from their center at the same distance from the bottom and from their bottom part. Samples density was measured by hydrostatic weighing. In addition to the determination of cast aluminum alloys density the temperature dependences were recorded using thermocouples installed in the center of the ingot and on its surface.

Metal quality depends on the alloys composition and features of their crystallization, which is characterized by crystallization heat and melt supercooling value. One of the main methods for determining the crystallization parameters is the differential thermal analysis. To perform this analysis the temperature dependences were recorded using thermocouples installed in the center of the ingot, on the contact boundary – mold, and the external temperature of the mold. For the continuous recording of the temperature dependence $T = f(\tau)$ the automated measuring system was used on basis of a PC and controllers of type ICP CON 7018, 7520 [5].

The measured temperature was smoothed by the method of piecewise exponential spline-approximation [4] and $dT/d\tau$ was determined to obtain differential thermal curves. Besides, the heating rate of the ingot mold surface was determined.

It is known that the change of the temperature gradient in the ingot body (between its center and the surface) changes the length of the characteristic structural zones in the ingot macrostructure.

The main reason for the destruction of the dendritic crystals is the imposition of periodically changing during the wave motion of the liquid the pressure and depressions causing the elastic deformation and stresses.

In addition, the wave process also contributes to the activation of insoluble impurities already present in the melt, which serve as a substrate for the emergence of such centers. The effectiveness of crystals refinement during solidification of the molten metal depends on the type of vibration applied. In the process of crystalline structures formation the vibration energy is consumed not only for the destruction dendritic branches and creation in the system of additional crystallizing nuclei, but also for enhancement of the heat removal rate from the melt to the mold walls. This leads to an increase in the rate of melt crystallization and significant (by 25 ... 30%) acceleration of the ingot solidification. Besides, the source of more intensive crystallization during vibration is the formation of the supercooled micro volumes of melt not only on the boundary of vertical front of solidification, but also around each of the growing floating crystal. The structure refinement, increase in the gas yield, decrease in porosity during vibration treatment of the melt influences on the alloy density making it increase. The forced vibrations of the melt very effectively refine the crystals and influence the cast macrostructure. The destruction of dendritic crystals takes place with the formation of additional nuclei formation. To control the size and shape of the grain it is necessary to know the mechanism of the influence of vibration on the growing crystal and the places of crystals origin in the solidified cast.

The refinement degree depends on the type of vibrations, which are located in the sequence: no vibration, vibration with amplitude orientation in the vertical plane, vibration with the amplitude orientation in the horizontal plane, vibration with circular amplitude in the horizontal plane. The most effective in terms of structure refinement is vibration with circular amplitude in the horizontal plane.

In the analysis of the impact of dynamic pulses of inertial forces on the mass of the liquid alloy we need to consider not only the variation range of their acceleration values, but the average value of these accelerations acting during each half-cycle of oscillations.

The most likely places of crystals separations are the mold surfaces and the opened surfaces of the liquid metal subjected to vibration. The increased concentration of segregates at the crystallization front contributes to obtaining narrow neck-shaped crystals on the mold surface, untimely formation of a stable crust of solidified metal, crystals uncoupling with the mold wall and prevention of remelting of the already separated crystals in the inner mold volumes. Vibration is significantly affects the processes of destruction and dispersion of dendrites branches growing on the boundary of crystallization.

Reduction in the temperature gradient between the center of the ingot and its surface increases the tendency of the alloy to the volumetric crystallization, reduces the zone of columnar crystals and, accordingly, increases the equiaxed crystal zone.

The ingots subjected to vibration have hardening boundaries of two types: at the directional heat sink to the bottom part of the mold it has a parabolic shape, and with the full heat sink – tapered. The disperse crystalline structure of the ingots is formed, primarily due to the breaking under vibration impact of peaks of columnar crystals, growing from the cooled walls and the mold bottom under the vibration impact – further about 20% (of the total) of disperse crystals are formed.

The nature of the vibration effects on the crystallization process and the formation of the alloy structure should also be linked to the thermal conditions of its solidification.

Thus, upon solidification of the ingot from the melt in the conditions of heat sink the intense peaks breaking of dendrites are observed, growing from the bottom and side walls of the mold under the influences of vibration. The intense flow of crystal fragments along the vertical solidification boundaries is noted. This movement is caused by the combined action of vibration and gravitational forces. Sinking debris of dendrites branches along the vertical solidification boundary become the centers of crystallization and provide an ingot with a fine crystalline structure.

3. Conclusions

The heating rate of the mold surface shows an increase in the crystallization rate of alloys subjected to vibration. It is found that the maximum surface temperature of the mold surface heating is observed in alloys AK9M2 AK12M2 subjected to vibration with an amplitude of 7 mm.

In the studied alloys AK9M2 AK12M2 subjected to vibration with an amplitude of 0.5 mm at the initial moment of time of cooling the maximal temperature gradient and its rate of change was low.

The analysis of differential thermal curves allowed us to establish the following. Vibration treatment of the alloys shifts the critical points of the crystallization onset (liquidus) and the end of crystallization (solidus).

For hypoeutectic alloy AK9M2 at the increase in the vibration amplitude the liquidus temperature drops from 624.8 °C to 621.7 °C, and the solidus temperature decreases from 576.9 °C to 570.5 °C.

For hypereutectic alloy AK12M2 it was found that as the vibration amplitude increases the liquidus temperature increases from 605 °C to 607 °C, and the solidus temperature decreases from 576.4 °C to 574.3 °C.

Changes in critical points, the crystallization rate and interval affect the structure of aluminum alloys.

The following regularities of the influence of vibration modes on the structure were revealed. For AK9M2 composition, with the increase in the vibration amplitude up to 1 mm, the grain refinement in

1.33 times is observed. Further, when the vibration amplitude increases from 0.5 to 2 mm, the grain size increases in 1.9 times. This is apparently explained by the fact that the maximal amount of solid phase begins to be formed closer to the solidus line.

With the vibration from 2 to 3 mm the grain refinement in 2 times takes place. With the amplitude more than 3 mm there is a slight increase in the grain size –in 1.36 times. For the composition of AK9M2, with the vibration amplitude from 0.5 to 2 mm the decrease in the eutectic phase content occurs in 1.2 times. With the increase in the amplitude from 2 to 3 mm, the increase of the eutectic phase formation is in 1.4 times. With the amplitude higher than 3 mm a slight increase in eutectic can be seen.

For the composition of AK12M2 the following conclusions can be drawn. With the amplitude of 2 mm the smallest grain size is observed. The increase of the amplitude from 2 to 3 mm or decrease from 2 to 1.2 mm there is an increase in the grain size in 1.4 times. With the increase in the amplitude from 0.45 mm to 2 mm a reduction in α-phase content is registered. With the increase in the amplitude higher than 2 mm there is a slight decrease in α-phase. The degree of crystals refinement depends on the type of vibration: vibrations with the amplitude oriented in the vertical or horizontal planes, and a vibration with circular amplitude in the horizontal plane. With the increase in the vibration amplitude from 0.5 mm to 2÷2.15 mm there is an increase in the density in 1.2 times and in 1.15 times for AK12M2 and AK9M2 alloys respectively. Increase in the density of samples, cut from ingots, is due, primarily, with vigorous stirring of the molten metal layers, improvement of gas release during vibration, increase in the zone of more dense columnar crystals. A further increase in the vibration amplitude leads to a decrease in the density of the cast alloy (Figure 2), which can be explained as follows. At this mode of vibration the molten alloy mass partly (or completely) moving away from the solid phase surface, captures the gas particles into the intercrystalline space overlapping them above with the dropping from the liquid phase fine crystals. Samples density from alloy AK12M2, cut from the bottom part, with the increase in the vibration amplitude slightly rises. The increase in the amplitude up to 2 mm increases the samples density from 2.68 g/cm^3 to 2.71 g/cm^3. Further increase in the vibration amplitude slightly lowers the density of the samples.

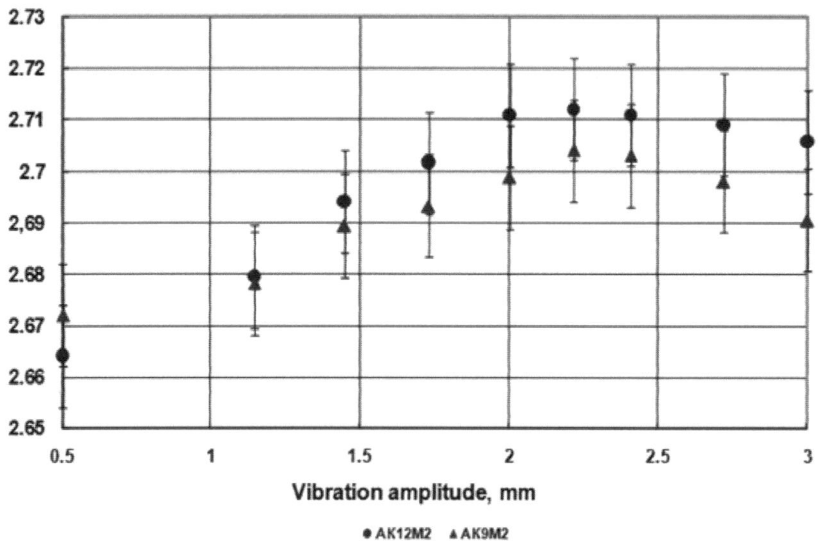

Figure 2. Effect of vibration amplitude on the density of aluminum alloys.

During formation of crystalline structures the vibration energy is consumed not only for the destruction of the dendrites branches and creation in the system of additional crystallization nuclei, but also for the enhancement of heat removal rate from the melt to the mold walls. This leads to the increase in the rate of crystallization and acceleration of the ingot solidification. The rate of crystallization for AK12M2 and AK9M2 alloys increases with the amplitude of 0.5 mm by 1.1 times and by 1.26 times, respectively, with the amplitude of 3 mm by 1.8 times and 2.25 times. The crystallization time for alloy AK12M2 decreases in 1.1 times at the vibration amplitude of 0.5 mm and in 1.43 times with the vibration amplitude of 3mm.

Time for the crystallization of alloy AK9M2 decreases in 1.35 times with the vibration amplitude of 0.5 mm and in 2.34 times with the vibration amplitude of 3 mm. For alloys AK9M2 and AK12M2 the optimal amplitude is 2÷2.2 mm, allowing the dense cast alloy with a fine grain structure to be obtained.

4. References

[1] Ivantsov A A and Krushenko G G 2003 *Foundry Production* **2** 29–31
[2] Gladkov M I et al 2003 *Poc. Higher Educ. Inst. Ferrous Metallurgy* **9** 56–60
[3] Usoltsev A A, Korotkikh I K and Kutsenko A I 2008 *Proc. IV Int. Conf. on Technological Ensurance of Quality of Machines and Devices* (Penza) pp 22–24
[4] Usoltsev A A et al 2009 *Study of Foundry Processes: Textbook* (Novokuznetsk: SibSIU) p 194
[5] Korotkikh I K et al 2005 *Proc. V All-Russian Conf. on Automation Systems in Education, Science and Industry* (Novokuznetsk: SibSIU) pp 424–426

Improvement of rolling modes of long length rails on the universal rail and structural steel mill "EVRAZ ZSMK"

A V Golovatenko[1], A A Umansky[2] and V N Kadykov[3]

[1] JSC "EVRAZ Consolidated West Siberian Metallurgical Plant", Novokuznetsk, 654000, Russia
[2] Department of Metallurgy of Ferrous Metals, Siberian State Industrial University, 42 Kirova Street, Novokuznetsk, 654007, Russia
[3]Department of Metal Forming and Material Science. "EVRAZ ZSMK", Siberian State Industrial University, 42 Kirova Street, Novokuznetsk, 654007, Russia

E-mail: umanskii@bk.ru

Abstract. Using the results of theoretical and experimental research a new rolling mode of long-length rails on the universal rail-and-structural steel mill JSC "EVRAZ Consolidated West Siberian Metallurgical Plant" ("EVRAZ ZSMK") was developed with a reduced number of passes in the breakdown stands. Experimental and industrial testing and implementation of the new rolling mode showed a high technical and economic efficiency of its use – reduction in the specific energy consumption was registered, as well as decrease in rejection of rails due to their surface defects, consumption of mill rolls and increase in mill productivity. The cost-effectiveness after implementation was more than 98 million RUB per year.

1. Introduction

The improvement of rolling modes, including grooving of rolls, is the most important reserve for the energy efficiency increase in the rolling production and quality improvement of the finished sections [1]. To the maximum extent this statement is valid with respect to the shaped sections of a complex shape, including rail sections [2-9].

In 2013 "EVRAZ ZSMK" was the first in Russia to put into operation an universal rail-and-structural steel mill. Its main product is differentially hardened long-length rails (up to 100 m long) [10]. In accordance with the contract grooving of the equipment supplier (company "SMS Meer" Germany) the rolling scheme includes the following steps (Figure 1). Deformation in the first breakdown stand BD1 is done in 7 passes, of which the first 6 passes are carried out in the box groove, and the last one – in the groove "recumbent trapeze".

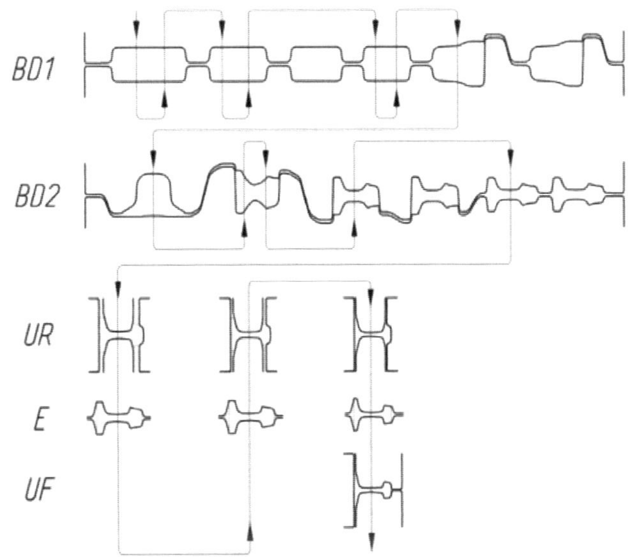

Figure 1. The contract rolls grooving for rolling rails R65 on the on universal rail-and-structural steel mill "EVRAZ ZSMK".

Rolling in the second breakdown stand BD2 was performed in 5 passes: the first pass – in the trapezoidal groove without cutting from the side of the future rail base; the second and third passes – in the first closed rail groove of "beam-type"; the fourth pass – in the second closed rail groove of "beam-type"; the fifth pass – in the open symmetrical rail groove. After rolling in the breakdown stands the rolled piece is forwarded for the further deformation into a continuous group of stands, which is carried out in three passes: the first pass – continuous rolling in the first universal stand (UR) and subsidiary stand (ER), the rolls of the second universal stand (UF) are driven apart; the second pass – in the first universal stand (UR) after its reversing with the driven apart rolls of the subsidiary stand (ER); the third pass – continuous rolling in all three stands of the groups (UR, ER, UF).

The start-up and initial period of operation of the universal rail-and-structural steel mill revealed the significant deficiencies in the contract grooving of company "SMS Meer" and the significant reserves for improving the rolling mode.

In the usage of the contractual grooving the bending the front end of the feed took place during rolling in the second rail groove (the second and third passes in stand BD2) and in the symmetrical rail groove (the fifth pass in the stand BD2), which led to the need for rolling with "starting" of the stock in the indicated passes. As a result of this mode the formation of "rolling skin" on the rails surface occurred.

The mechanism of this defect consists in the rolling of the existing mechanical damages of the stock surface during rolling [11]. Further, rolling with "priming" resulted in the decrease of rolls durability due to the increased mill rolls production and the increase in the rolling time in the breakdown stand BD2. Thus, in practice, it was shown that the use of the contract mode of rails rolling was low-tech, and led to a significant deterioration of the technical and economic operational performance of the rail-and-structural steel mill and the decline in the quality of finished products.

2. Results and discussion

In order to eliminate these disadvantages of the contract mode of rails rolling on the universal rail-and-structural steel mill "EVRAZ ZSMK" a new intensified rolling mode was developed with a reduced number of passes in the second breakdown stand BD2 (Figure 2).

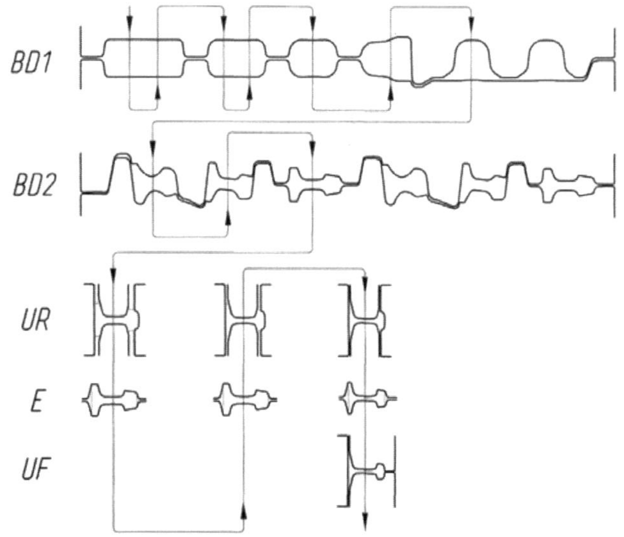

Figure 2. Intensified mode of rails rolling R65 on the universal rail-and-structural steel mill "EVRAZ ZSMK".

Using the new rolling mode the deformation in the stand BD1, as well as using the contract scheme, is carried out in 7 passes. However, the rolling in the box grooves is performed only in the first 5 passes, in the sixth pass the deformation is achieved in the groove "recumbent trapeze", and in the seventh pass – in the trapezoidal groove.

Rolling in the stand BD2 is done in 3 passes: the first and second passes are carried out in the "oblique located" rail grooves with slopes of the side walls up to 18% and use of thrust cones with slopes up to 25%, and the last pass – in the open symmetrical rail groove. After BD stands the stock, as in the contract rolling scheme, is forwarded for the further deformation in the stands of continuous group. The rolling scheme in the continuous group of stands remained unchanged.

The possibility for reducing the number of passes was proved by the calculations of the rolling force in BD stands, which were done according to the previously developed calculation methods of the deformation resistance [12]. Based on the obtained data using the new rolling mode the rolling force will not exceed 61% of the permissible values (Figure 3).

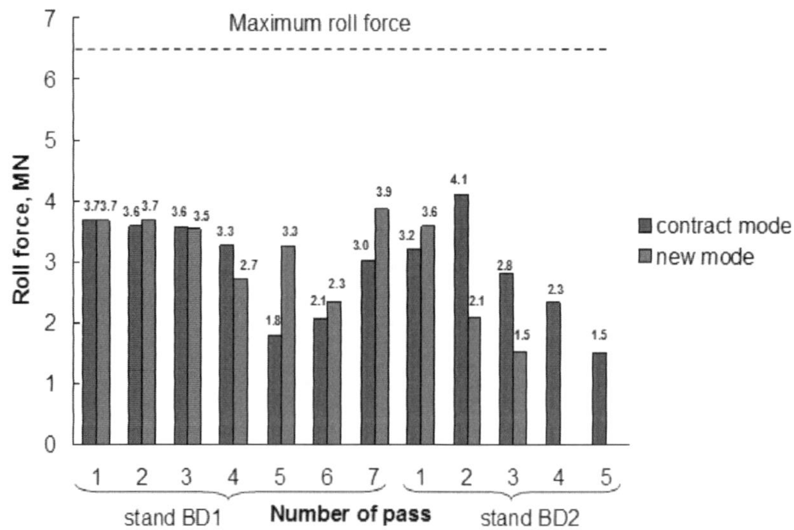

Figure 3. The calculation date on the rolling force while using various modes of deformation.

To determine the producibility of the new rolling mode the pilot industrial tests of the engines load in the BD stands of the universal rail-and-structural steel mill "EVRAZ ZSMK" were performed. Three rolling modes were used:

1. Contract mode (7 passes in the stand BD1 and 5 passes in the stand BD2) – mode No. 1;
2. Mode with redistribution of breakdowns between the stands (9 passes in the stand BD1 and 3 passes in the stand BD2) – mode No. 2;
3. The developed intensified mode (7 passes in the stand BD1 and 3 passes in the stand BD2) – mode No. 3.

The research was conducted by the oscillograph method measuring parameters of the engine operation of stands drives.

Both stand BD1 and BD2 are equipped with the same type of synchronous engines of type AMZ 0900LT08 LSB, the technical characteristics of which are summarized in Table 1.

Table 1. Specifications of engines in the stands BD1 and BD2.

No.	Parameter	Value of the parameter
1	Rotation nominal speed	310 rot./min
2	Rotation maximal speed	982 rot./min
3	Current nominal speed	744 A
4	Nominal voltage of starter	3165 V
5	Nominal power	4000 kW
6	The nominal moment at the motor shaft - at the rotation nominal speed - at maximum speed	 123 kN·m 39 kN·m

For these engines, the following restrictions on overloading of the nominal value in the process:

- Not more than 115% of the nominal value during continuous overload;
- Not more than 225% of the nominal value at overload for 30 seconds;
- Not more than 250% of the nominal value at overload for 2 seconds.

The results obtained after oscillograms processing are shown in Table 2.

On the basis of the conducted analysis we can say that in the process of use of all three rolling schemes the modes of engines operation of stands BD1 and BD2 are within the permissible.

To compare the efficiency of different rolling schemes the specific energy consumption per tonne of the rolled stock is identified. The results are provided in Table 3 and Figure 4.

Table 2. Static moments of BD stands' engines.

Stand	Pass No.	Static moments on the engine shaft (SM) and rotation speed (n) of the engine shaft for different rolling schemes					
		Mode No.1		Mode No.2		Mode No.3	
		SM, % from nom.	n, rot./min	SM, % from nom.	n, rot./min	SM, % from nom.	n, rot./min
BD1	1	100	370	135	315	110	210
	2	82	370	110	368	100	370
	3	105	475	135	485	110	415
	4	77	463	95	465	80	476
	5	35	504	40	485	80	476
	6	30	463	40	465	55	420
	7	55	525	70	420	76	580
	8	-	-	55	355	-	-
	9	-	-	35	537	-	-

	1	66	495	85	366	105	355
	2	20	455	70	344	68	315
BD2	3	92	377	35	484	38	485
	4	70	344	-	-	-	-
	5	33	385	-	-	-	-

Table 3. Specific energy consumption in different rolling schemes in the stands BD1 and BD2.

Stand	Specific energy consumption for different rolling schemes, kW·h/tonne		
	Mode No.1	Mode No.2	Mode No.3
BD1	4.23	5.63	5.19
BD2	5.43	4.38	3.96
Total	**9.66**	**10.01**	**9.15**

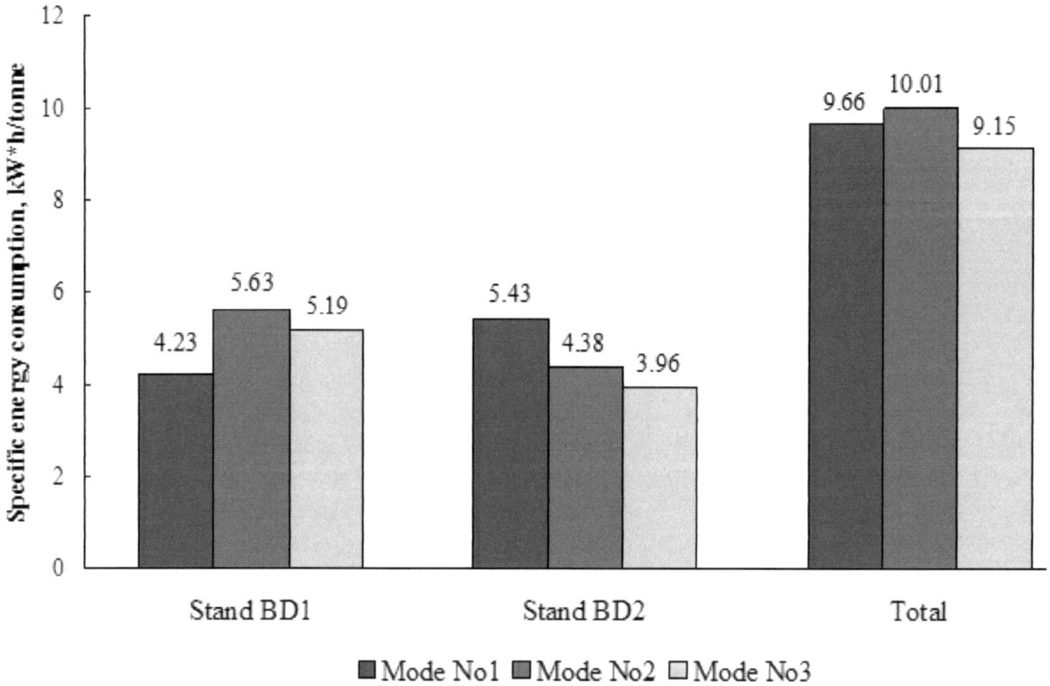

Figure 4. Specific energy consumption for different rolling modes of rails R65 in the BD stands.

According to the data rolling mode No.3 is the most energy efficient: in relation to the contractual scheme of rolling (mode No.1) the decrease of the specific electricity consumption 0.51 kW·h/tonne was registered, and in comparison with the rolling scheme No.2 the specific energy consumption is 0.86 kW·h/tonne less.

During the pilot-industrial approbation and implementation of the intensified mode of breakdowns in the BD stands (mode No.3) the decrease of finished rails rejection due to surface defects us recorded, compared with periods when rolling mode No.1 was used (Figure 5).

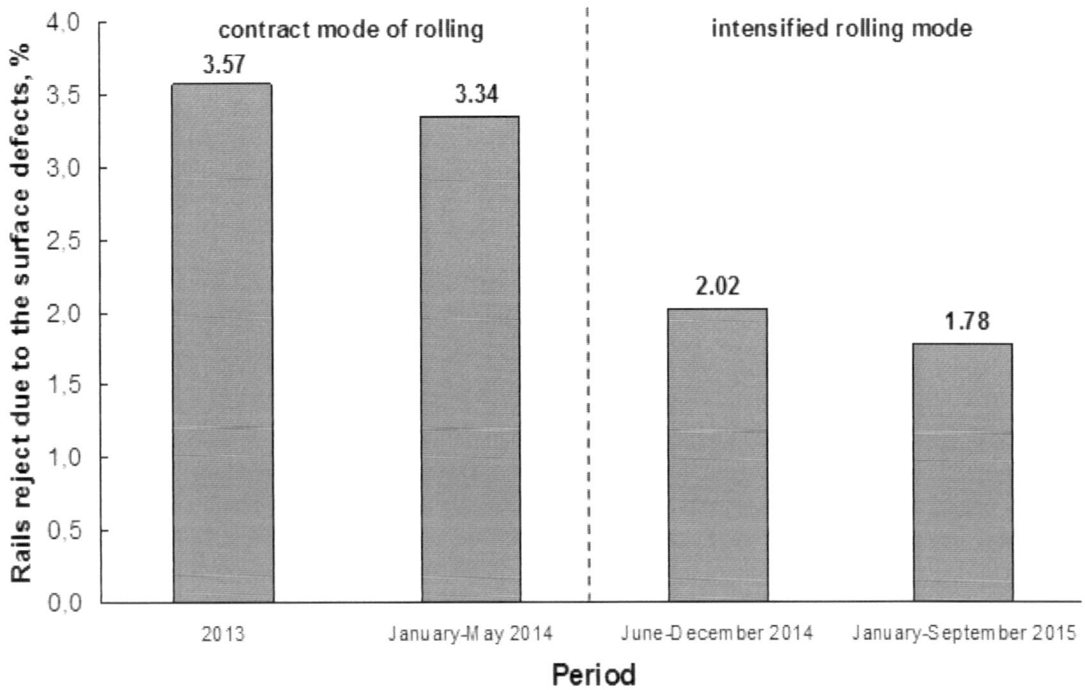

Figure 5. Dynamics of rails rejection due to surface defects at the rail-and-structural steel mill "EVRAZ ZSMK".

The conducted evaluation of the impact of change in the rolling mode on rejection due to the defect "rolling skin" showed that application of the intensified rolling mode reduces rejection for this defect by 0.5%. This is due to the fact that when using the new rolling mode the technological possibility appeared to abandon rolling with "starting" of the stock with the absence of risk for bending of the strip after its leaving grooves of BD stands.

Based on the positive results of pilot testing the rolling mode with a reduced number of passes in the BD stands was adopted as a primary mode for the production of long-length rails on rail rolling mill "EVRAZ ZSMK".

The introduction of this regime in the rolling production allowed, except for the above-mentioned reduction in specific energy consumption and improvement of the rails surface quality, reduction in specific consumption of rolls for stand BD2 to be reduced by 0.51 kg/tonne, as well as reduction in the cycle on rolling stands by 10 seconds (which led to an increase in productivity of the mill up to 146.8 tonnes/h) and the increase in the rolled products volume between repairs from 3.5 to 6.0 thousand tonnes. The cost-effectiveness of implementation was 98.588 million RUB per year.

3. Conclusions

Based on the results of theoretical and experimental studies developed an intensified mode of rolling of long rails in the rail-and-structural steel mill "EVRAZ ZSMK". Its main distinctive features are: reduction of passes number in the BD stages (by 2 passes) and substitution of the closed grooves of "beam-type" by "oblique located" rail grooves with the slope of the side-walls up to 18%. The pilot testing of the new rolling mode showed the producibility, technical and economic efficiency of its use, which was a prerequisite for its implementation into production. The cost-effectiveness from its implementation is more than 98 million RUB per year.

4. References

[1] Umanskii A A and Mart'yanov Y A 2014 *Metallurgist* **58** (5–6) 516–523
[2] Shilov V A, Shvarts D L and Litvinov R A 2008 *Univ. Proc. Ferrous Metallurgy* **3** 51–54
[3] Shvarts D L 2015 *Steel in Translation* **45** (6) 430–435
[4] Shvarts D L 2015 *Steel in Translation* **45** (5) 526–530
[5] Shilov V A, Shvarts D L and Litvinov R A 2008 *Rolled Metal Production* **1** 29–32
[6] Stalinskii D V and Rudyuk A S 2011 *Steel in Translation* **41** (5) 73–77
[7] Rudyuk A S, Azarkevich A A, Sidorchuk R S, Karmazin A V and Koshulé I M 2014 *Metallurgist* **57** (9) 845–848
[8] Sinel'nikov V A, Filippov G A 2001 *Metallurgist* **45** (9) 403–407
[9] Smirnov V K, Bondin A R and Mikhaylenko A M 2002 *Rolled Metal Production* **12** 24–30
[10] Golovatenko A V, Volkov K V, Aleksandrov I V, Kuznetsov E P, Dorofeev V V and Sapelkon O I 2014 *Ferrous Metallurgy* **6** (1374) 32–38
[11] Trotsan A I, Kaverisky V V, Nosochenko A O and Koshulé I M 2012 *Proceedings of PGTU. Series of Tech. Sc.* **25** 106–114
[12] Umansky A A, Golovatenko A V, Kadykov V N and Dumova L V 2015 *Proc. XIX Int. Conf. on Metallurgy: Technologies, Innovations, Quality* (Novokuznetsk: SibSIU) part 2 pp 110–115

Development of mathematical models and methods for calculation of rail steel deformation resistance of various chemical composition

A A Umansky[1], A V Golovatenko[2], V N Kadykov[3] and L V Dumova[4]

[1,4] Department of Metallurgy of Ferrous Metals, Siberian State Industrial University, 42 Kirova Street, Novokuznetsk, 654007, Russia
[2] JSC "EVRAZ Consolidated West Siberian Metallurgical Plant", Novokuznetsk, 654000, Russia
[3] Department of Metal Forming and Material Science. EVRAZ ZSMK", Siberian State Industrial University, 42 Kirova Street, Novokuznetsk, 654007, Russia

E-mail: umanskii@bk.ru

Abstract. Using the device of the complex "Gleeble System 3800" the physical experimental studies of deformation resistance of chrome rail steel at different thermo-mechanical deformation parameters were carried out. On the basis of mathematical processing of experimental data the statistical model of dependence of the rail steel deformation resistance on the simultaneous influence of deformation degree, rate and temperature, as well as the steel chemical composition, was developed. The nature of influence of deformation parameters and the content of chemical elements in steel on its resistance to plastic deformation is scientifically substantiated. Verification of the adequacy of the proposed model by the comparative analysis of the calculated and actual rolling forces during passes in the universal rail-and-structural steel mill JSC "EVRAZ Consolidated West Siberian Metallurgical Plant" ("EVRAZ ZSMK") showed the possibility of its use for development and improvement of new modes of rails rolling.

1. Introduction

Steel resistance to plastic deformation (deformation resistance) is one of the most important characteristics of metal strength during its forming in the rolling process. According to the accepted concepts by deformation resistance the stress intensity is understood sufficient for the plastic deformation of metal under specified deformation conditions [1-5].

Deformation resistance directly determines the power parameters of rolling as the magnitude of this parameter conditions the metal pressure on rollers, and as a result the rolling enforcement. Thus, use of incorrect data on the magnitude of deformation resistance during designing the rolling modes might lead either to overloads of the main equipment (in the case of underestimating of deformation resistance in comparison with its real values) or to underutilization of the equipment and, as a consequence, to the decrease in mill productivity (using the overstated deformation resistance values in relation to the actual values). One of the most important tasks in the development and improvement of rolling modes on the operating and especially on the newly commissioned rolling mills is to obtain reliable data on the magnitude of the steel resistance to plastic deformation under specified conditions of rolling.

Analysis of the literature shows that the following physical and chemical parameters influence the deformation resistance during rolling: thermo-mechanical parameters of deformation (temperature, rate and degree of deformation) and the chemical composition of steel.

Currently used methods for calculating the deformation resistance are based on the presence of the predetermined basic value of this parameter under standard conditions (temperature, rate and degree of deformation). Calculation of the deformation resistance for the specific conditions of rolling in this case is done by multiplying the basic value of deformation resistance by the thermo-mechanical coefficients, values of which are determined experimentally.

A significant disadvantage of this group of methods is the limited field of their application – they provide with the adequate to the actual values results only for the increasing curves, characterizing the dependence of the steel resistance to plastic deformation on the degree of deformation [6, 7]. Furthermore, in the literature there is information about the basic value of deformation resistance only for a limited number of steel grades [8]. In particular for rail steels such data are not available.

The most reliable way to obtain information about the true values of deformation resistance for today is the experimental studies by the methods of hot mechanical testings of samples for tension, torsion or compression. The first two methods have significant drawbacks. When performing hot pulling test even at relatively small deformations (20-30%) there is a high degree of deformation heterogeneity and a change in the deformation rate in the metal volume takes place, which is explained by formation of a web in the sample during testing.

Method of hot torsion is characterized by a considerable deformation heterogeneity over the cross-section: the degree and rate of deformation vary from the maximum value on the surface to the minimum (up to zero) values of true deformation in the center (along the rotation axis) of the cylindrical sample.

The advantage of the compression test is the possibility to apply high deformation degrees. Although it should be noted that at the degree of deformation more than 60-70% the sample takes "barrel" shape, which significantly increases inaccuracy of the research. This phenomenon is due to the fact that during the upset of the metal billet the metal sliding on the contact surface is delayed by friction forces.

Thus, nowadays, in order to develop a method for calculating the deformation resistance experimental studies of deformation resistance of rail steel should be performed. The most preferred method of mechanical testing of samples during the studies is a hot compression.

2. Experimental research

Experimental study of deformation resistance of rail steel E78KhSF was carried out using the apparatus "Hydrawedge II", which is one of four possible changeable modules for the complex of physical modeling of thermo-mechanical processes "Gleeble System 3800" [9]. "Hydrawedge II" consists of the following subsidiary devises (Figure 1): instruments for control of temperature, force, displacement; the mobile module for creation of an artificial atmosphere in the working chamber; deforming bars with a water cooling system, necessary for elimination of the undesirable effect of heating on tools and measuring instruments.

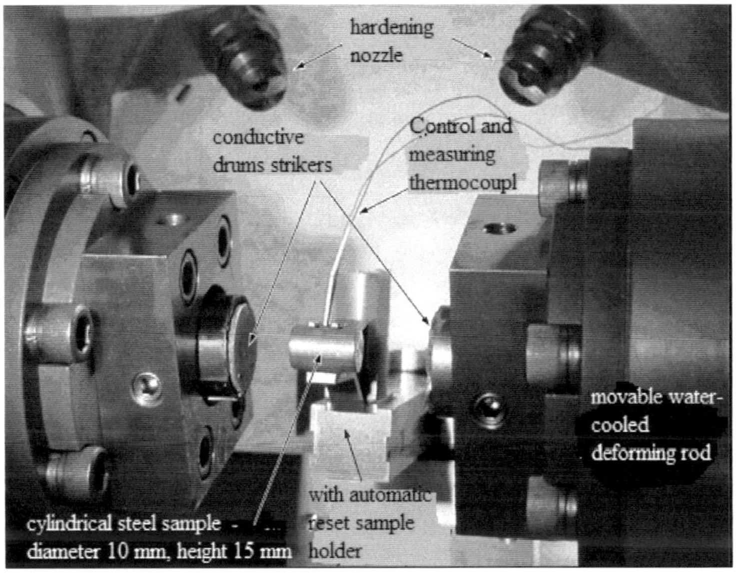

Figure 1. Module scheme of "Hydrawedge II" for the complex of physical modeling "Gleeble System 3800".

During hot compression tests rectangular and cylindrical samples were used, obtained from cast billets of steel produced by JSC "EVRAZ Consolidated West Siberian Metallurgical Plant" ("EVRAZ ZSMK").

Testing mode included the following steps (Figure 2): heating of the template in vacuum at the rate 5°C/sec to a temperature of 1200 °C for 240 sec; isothermal exposure for 600 sec at 1200 °C; steel cooling-down to deformation temperature at 10 °C/sec; deformation at a predetermined temperature; steel cooling at rates from 50 °C/sec to 10 °C/sec. Samples were deformed at rates 0.1, 1 and 10 sec-1 with temperatures 1150, 1100, 1050, 1000, 950 and 900 °C.

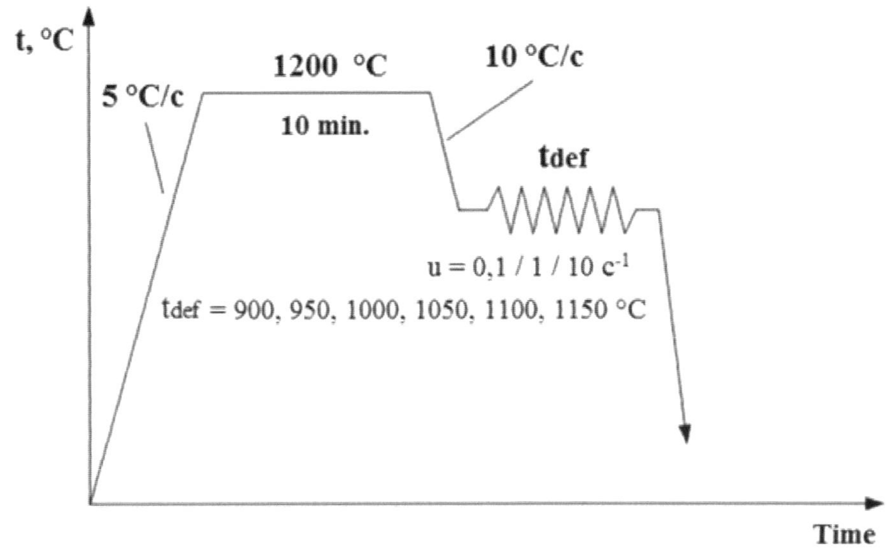

Figure 2. Test scheme of rail steel samples E78KhSF.

3. Results and discussion

Generalization and mathematical processing of the experimental characteristic curves, the examples of which are shown in Figure 3, allowed us to obtain a statistical model of dependence of deformation resistance of rail steel E78KhSF on the simultaneous influence of the degree, rate and deformation temperature:

$$\sigma = A \cdot e^{m_1 \cdot t} \cdot \varepsilon^{m_2} \cdot e^{\frac{m_3}{\varepsilon}} \cdot (1+\varepsilon)^{m_4 \cdot t} \cdot e^{m_5 \cdot \varepsilon} \cdot u^{m_6 \cdot t} \quad (1)$$

t – rolling temperature, °C;
ε – deformation degree;
u – deformation rate, sec^{-1}.

Analysis of the experimental nature of dependencies (Figure 3) showed that in areas where the degree of deformation is less than 0.7 the curves have a pronounced maximum.

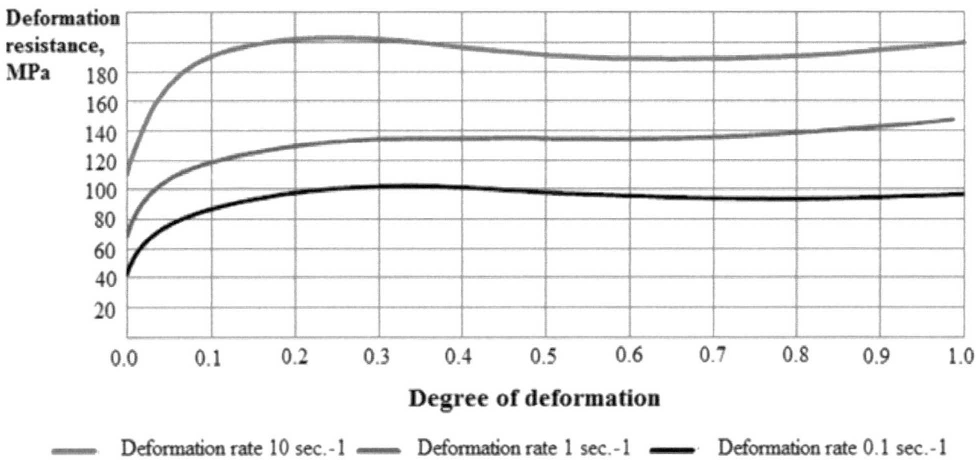

a

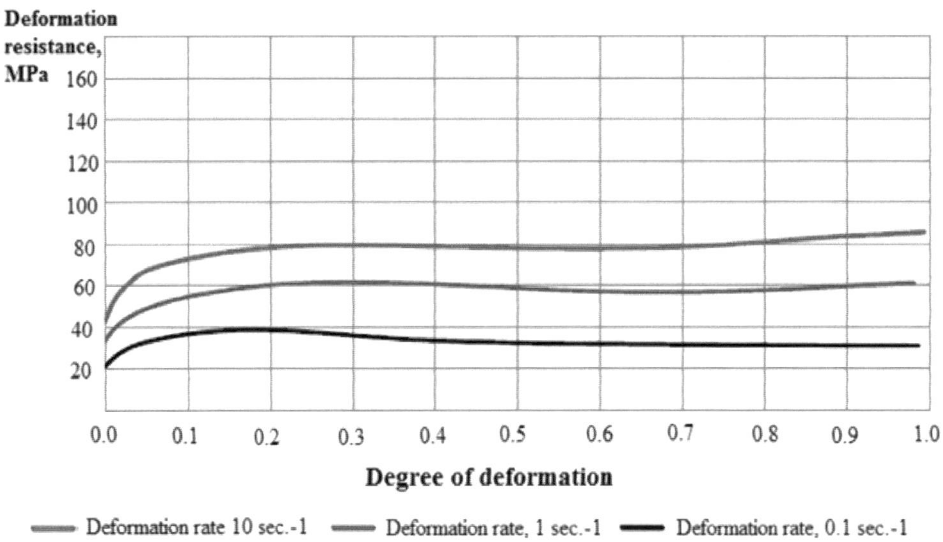

b

Figure 3. Experimental dependence of the deformation resistance of steel E78KhSF on the rate and degree of deformation (a – deformation temperature 900 °C; b – deformation temperature 1150 °C).

That is, after a certain deformation degree is achieved, in the steel in addition to the dynamic recovery and polygonization a dynamic recrystallization takes place. The indicated maximum shifts towards the increase in the deformation degree when the temperatures go up, shifts backward as the deformation rate increases. Then in the zone corresponding to the deformation degree 0.7-1.0 on some curves there is a repeated increase in deformation resistance, which is obviously conditioned by the error of the experiment due to the increased "barreling" of samples at high deformation rates.

According to the data the increase in the rolling temperature of rail steel E78KhSF steel in the range 900-1150 °C leads to a decrease in deformation resistance due to the weakening of the material bindings of particles and the creation of favorable conditions for their mutual displacement. Increase in the deformation rate within the range 0.1-10 sec^{-1} on the contrary increases the deformation resistance, since in that case the hardening processes begin to prevail over the competing processes of steel softening.

As a result of the statistical analysis of the influence of steel chemical composition on the coefficients of equation (1) the regression equations in natural scale are obtained:

$$A = 4365.4 + 69118 \cdot S \tag{2}$$

$$m_1 = -0.0033 - 0.0043 \cdot V \tag{3}$$

$$m_2 = 0.2607 - 5.7663 \cdot P \tag{4}$$

$$m_3 = -0.0025 + 0.00308 \cdot C + 0.00025 \cdot Mn \tag{5}$$

$$m_4 = -0.0015 + 0.0475 \cdot P \tag{6}$$

$$m_5 = -0.407 + 0.655 \cdot Mn \tag{7}$$

$$m_6 = 0.0002 - 0.0012 \cdot V, \tag{8}$$

where S, V, P, C, Mn – content of sulfur, vanadium, phosphorus, carbon and manganese in the steel respectively, %.

Summarizing the findings about the influence of chemical composition of rail steel E78KhSF on the coefficients of equation (2) and, consequently, on the deformation resistance, it can be stated that the increase in the content of manganese, carbon, sulfur and phosphorus in these limits increases the deformation resistance. The increase in vanadium content from 0.04% to 0.07% reduces the steel resistance to plastic deformation due to significant influence of vanadium additives even in small quantities on improvement of steel ductility.

Effects of chromium and silicon in the considered range of their variation were not found, which is consistent with data from other sources [3]. There is evidence that silicon increases the deformation resistance of steels only when its concentration is at the level of 1.5-2.0%, and for the high-carbon chromium steels (with chromium content more than 0.45%), a decisive influence on the deformation resistance has carbon but not chrome.

In order to assess the quantitative impact of changes in the content of chemical elements on the deformation resistance of chrome rail steel the computational experiment was carried out. The essence of the experiment was to do calculations of the deformation resistance of rail steel E78KhSF with different chemical composition, the variation of which was carried out under the condition of alternate change in the content of each considered chemical element and constant content of other elements. The calculation was performed for the deformation conditions, typical for rolling in roughing and finishing stands of rolling mills: t = 1150°C, u = 5 sec-1 (roughing stands); t = 1050 °C, u = 15 sec-1 (finishing stands); (for both groups of stands).

According to the results of numerical experiment it was shown that the most significant influence on the deformation resistance of chrome rail steel had vanadium and sulfur (Figure 4).

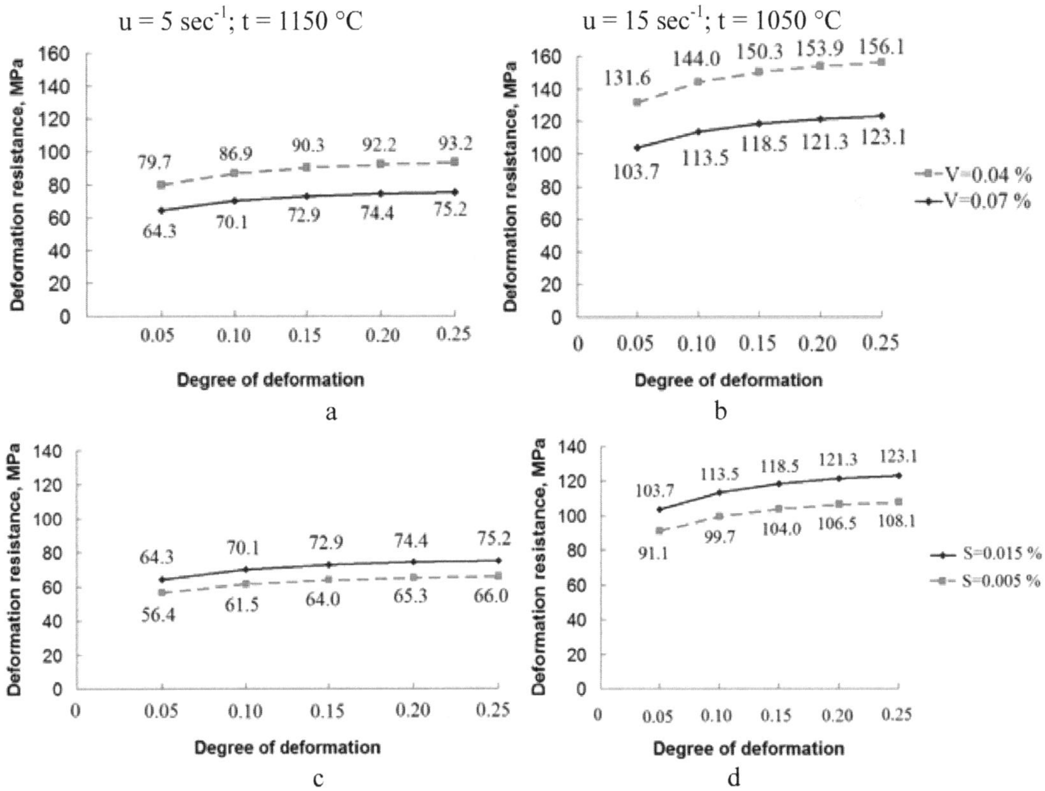

Figure 4. Dependence of deformation resistance on the chemical composition of rail steel E78khSF (a, b – dependence of the deformation resistance on vanadium content; c, d – dependence of the deformation resistance on sulfur content).

Increase in vanadium content from 0.04 to 0.07% leads to a decrease in the deformation resistance 1,24-1,27 times (Figure 4 a, b) and increase in the concentration of sulfur from 0.005 to 0.015% causes an increase in deformation resistance by 1.14 times (Figure 4 c, d). This temperature reduction and increase of deformation rate enhances the effect of vanadium on the steel resistance to plastic deformation. The increased content of phosphorus, manganese, and carbon in the steel within the considered ranges leads to the increase in the deformation resistance up to 10%, 5% and 1% respectively.

On the basis of experimental and statistical studies the method for calculation of deformation resistance was developed, which takes into account the impact of the thermo-mechanical deformation parameters and chemical composition of steel [10, 11].

Application algorithm of this method includes the following steps:

- Calculation of the equation coefficients (1) for a given chemical composition of steel using the regression equation (2-8).
- Calculation of the actual deformation resistance values for given parameters of thermo-mechanical rolling (the degree of deformation, temperature and deformation rate) using equation (1).

In order to test the adequacy of the developed method for determining the deformation resistance of rail steels, a comparative analysis of the calculated and the actual rolling forces in passes in the first breakdown stand of the universal rail-and-structural steel mill “EVRAZ ZSMK” was performed.

During calculations of the deformation resistance two chemical compositions of steel E78KhSF were used: standard chemical composition corresponding to the average content in the melts of current production (average content values of carbon, manganese, silicon and chromium according to State Standards (GOST) R 51685-2000, vanadium content – 0.07%; sulfur and phosphorus – 0.015% for each element); chemical composition similar to the first variant except for the sulfur content and vanadium (0.04% and 0.005% respectively).

The calculation of rolling force in the box grooves and the final horizontal T groove of the breakdown stand was performed using standard methodology proposed by the authors [12]. According to the obtained data the deviations of the calculated data from the actual ones do not exceed 10% (Figure 5).

Thus, it can be concluded that the developed method for deformation resistance calculation can be used in the design of new rolling modes of rail steels. Since no significant effect on the deformation resistance of the silicon and chromium content was revealed, the field of application of this technique can be extended not only on E78KhSF steel, but also on the rail steels not alloyed with these elements, in particular E76F and E76.

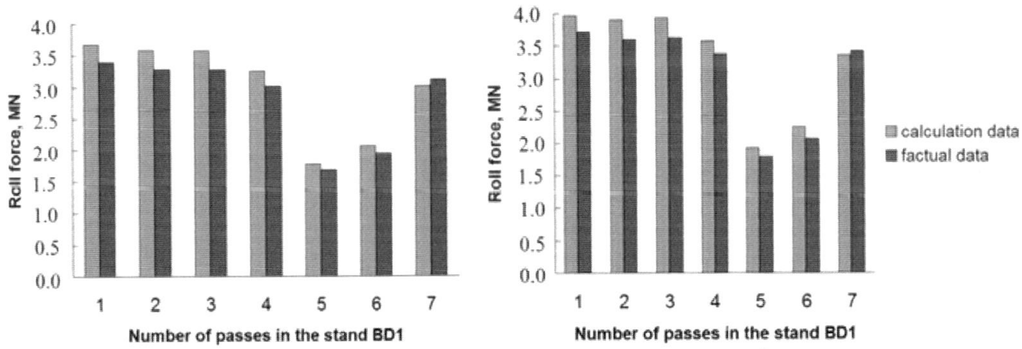

Figure 5. Comparative analysis of the calculated and actual data for the rolling force in the breakdown stand of the universal rail-and-structural steel mill "EVRAZ ZSMK" (a – standard chemical composition of steel E78KhSF; b – E78KhSF steel with a lowered content of sulfur and vanadium).

4. Conclusions

Based on the experimental research of deformation resistance of chrome steel rail, conducted on the facility for physical modelling of thermo-mechanical processes "Gleeble System 3800", a comprehensive statistical model of deformation resistance dependence on the deformation parameters (degree, rate and temperature of deformation) and steel chemical composition was developed. The nature of the influence of deformation parameters and the content of chemical elements in steel on its resistance to plastic deformation is scientifically substantiated. Verification of the adequacy of the developed model by the comparative analysis of calculated and actual rolling forces in the passes in breakdown stand of the universal rail-and-structural steel mill JSC "EVRAZ ZSMK" showed the possibility for its use in the development and improvement of new rolling modes.

5. References

[1] Peretyatko V N, Temlyatsev M V and Filippova M V 2010 *Development of the Theory and Practice of Metallurgical Technologies* (M.: Teplotekhnik) vol 2 p 352

[2] Kalpin Yu G, Perfilov V I, Petrov P A, Ryabov V A and Filoppov Yu K 2011 *Deformation Resistance and Ductility of Metals by Pressure Treatment* (M.: Mashinostroyeniye) p 244

[3] Efimov V N and Brovman M Ya 1996 *Resistance to Deformation in Rolling Processes* (M.: Metallurgiya) p 254

[4] Zyuzin V I, Brovman M Ya and Melnikov A F 1964 *Deformation Resistance of Steel during Hot Rolling* (M.: Metallurgiya) p 270

[5] Polukhin P I, Gun G Ya and Galkin A M 1983 *Resistance to Plastic Deformation of Metals and Alloys* (M.: Metallurgiya) p 352
[6] Karpov S V, Banshchikov A A and Karpova A S 2008 *Polzunov's Almanac* **3** 123–126
[7] Prokhodko I Yu, Sergeenko A A and Raznosilin V V 2007 *Proc. on Fundamental and Applied Problems of Ferrous Metallurgy* (Dnepropetrovsk: IChM NAN of Ukraine) **14** 187–193
[8] Peretyat'ko V N, Pyataikin E M, Dorofeev V V, Karetnikov A Yu and Dobryanskii A V 2011 *Steel in Translation* **41** (8) 649–651
[9] Golovatenko A V, Dorofeev V V, Volkov K V, Dobryanskii A V and Trusov V A 2014 *Metallurgist* **58** (5–6) 528–534
[10] Golovatenko A V, Kadykov V N and Umansky A A 2014 *Proc. of Mining and Metallurgy Section of the RANS* (Moscow – Novokuznetsk) **33** 64–71
[11] Golovatenko A V, Umansky A A and Kadykov V N 2014 *Proc. of XVIII Int. Conf. on Metallurgy: Technology, Control, Innovations, Quality* (Novokuznetsk: SibSIU) pp 160 – 165
[12] Tselokov A I, Nikitin G S and Rokotyan S E 1980 *Theory of Longitudinal Rolling* (M.: Metallurgiya) p 320

Rational chemical composition and heat treatment models of rails made from E76HF steel using the heat of rolling heating

E V Polevoy[1] and M V Temlyantsev[2]

[1] JSC "EVRAZ Consolidated West Siberian Metallurgical Plant", Novokuznetsk, 654000, Russia
[2] Institute of Metallurgy and Material Science, Siberian State Industrial University, 42 Kirova Street, Novokuznetsk, 654007, Russia

E-mail: uchebn_otdel@sibsiu.ru

Abstract. On the pilot plant the experiments in the differentiated heat treatment by air using the heat of rolling heating of rails produced from steel E76HF with three chemical compositions comprising various contents of chromium, manganese and silicon were carried out. The effect of heat treatment parameters on the structure and properties of rails was examined. According to the results of experiments the rational chemical composition was determined, as well as the heat treatment modes recommended for a mass production of differentially thermo-strengthened rails of DT350 category.

1. Introduction

Thermal hardening is one of the effective ways to enhance the performance properties of metal, so at the moment most of the rails produced in the world are subjected to a hardening heat treatment [1-5]. Until recently, the process of production of rail products by the domestic enterprises included a separate heating of the rails before their volumetric oil hardening [5-7]. Compared with the rails hardening using heat of the rolling heating this technique is more expensive. However, as it is known, the rail steel possesses high flake sensitivity [8], and for prevention of flakes formation at the time of technology development for volumetric hardening, rails after rolling were subjected to slow cooling.

After introduction in the production of the steel vacuumization the main reason of flakes formation – high hydrogen content has been eliminated. This allowed without losses in quality of rails the heat treatment technology from the rolling heat to be applied, which was implemented by several foreign rails producers. In 2005 at JSC "NKMK" was put into operation the vacuum vessel, and in 2007 it was decided to conduct a staged reconstruction of rail production, with the introduction of a differentiated line of hardening by air from the rolling heat. In order to identify promising chemical compositions of the rail steel and the modes of heat treatment of rails by air from the rolling heat, a series of experiments was conducted in heat treatment of rail samples on the pilot unit using a separate and rolling heating.

The description of a unit, methods and results of the experiments are presented in [9]. The experimental results of the treated rail samples produced from steel of grade E76F are presented in [10]. In our earlier studies [10] we found that during hardening rails, produced from steel E76F, by rolling heating the necessary level of strength properties and hardness is achieved by hardening at a sufficiently high rate – not less than 4°C/sec (about 4000 mm H_2O), and the plastic properties are at a quite low level and do not meet the standard requirements. To improve the complex of properties and determine the optimal in terms of resource saving heat treatment parameters we conducted experiments of heat treatment of rail samples from carbon steel alloyed with manganese, silicon, chromium and vanadium.

2. Material and research methods

The study object of this paper are thermally hardened by compressed air on the pilot unit sample of length ~ 400 mm, selected from hot-rolled rails of type R65, steel E76HF according to the State Standards (GOST) R 51685. The rolling of the rails was carried according to the current technology. From the bottom end of the feed on the hot cutting saws 2-3 samples were selected. Reaching the temperature of the hardening start one sample was cooled in the pilot unit, the rest all the time remained in the bucket with hot rail crop for prevention of a significant temperature drop.

Samples, selected on the hot cutting saws at a temperature of 900-940°C, were subjected to subcooling to temperatures 750-900°C and hardening with the rate up to 8°C/s at different modes.

The temperature during the experiments was recorded by manually infrared pyrometer of type Raynger MX.

After experiments from the upper part of the head of each probe in accordance with requirements of GOST 51685 samples were cut to determine by Brinell method the hardness on the rolling surface and along the head cross-section, to test for tension and the impact bending, microsections for microstructure control. The hardness test was conducted by the method of Brinell on the hardness tester of TSh-2M type by the ball with diameter of 10 mm under load of 3000 kgf according to GOST 9012-59. Tension properties were determined on the tension testing machine EU-40 with the force of 10 tonnes on the cylindrical samples of 6 mm in diameter and the initial effective length of the working part 30 mm, prepared in accordance with the requirements of GOST 51685 and GOST 1497.

Impact bending test was performed on the impact pendulum-type testing machine MK-15 in accordance with GOST 9454 on the standard samples of sized 10x10x55 mm with U-shape cut of radius 1 mm and depth 2 mm at temperatures of +20°C and -60°C. The metal microstructure was revealed by electrolytic polishing of the microsection surface in 5% acetic solution of perchloric acid and etching in 4% alcoholic solution of nitric acid.

3. Development of a chemical composition for research

The development of advanced chemistry compositions was performed on the basis of rails with the category T1 (NE), micro-alloyed with vanadium in the amount of 0.07-0.08%. Traditionally manganese and silicon are introduced into the rail steel that enhance the strength properties of the steel.

However, studies showed [10] that air hardening due to its low cooling capacity, containing ~ 0.30% of silicon and up to 0.90% of manganese, failed to achieve a balanced set of properties meeting the regulatory documents requirements. Therefore, to provide the desired standard properties an additional alloying with austenite-stabilizing elements was needed.

One of the most common and widely available elements enhancing depth of hardening and hardening capacity of steel, is chrome. In order to achieve the desired set of properties under moderate conditions of heat treatment the minimal chromium content was limited at the level of 0.35-0.40%. It should be noted that chromium has a significant influence on the weldability of steel, which is especially important due to the annual increase in the length of the welded rail joints on the network of JSC "Russian Railways", so during development of the chemical composition the upper limit of the chromium content was limited at level 0.50-0.55%.

Based on the above considerations variants for experimental steels No.1 and No.2 were proposed which are similar in chemical composition of the rails of T1 (NE) category of the current production and differed in chromium content, the quantity of which amounted to 0.42% and 0.55% respectively.

On the basis of previous experiments [9] on air hardening from a separate heating, it was found that the optimum ratio of strength and plastic properties is achieved in the steels rails with high silicon content, with a moderate content of manganese and chromium. On this basis we developed an experimental variant of chemical composition No.3 differing from No.1 and 2 in the lower content of manganese (0.78%), higher silicon content (0.55%) and moderate amounts of chromium (0.46%).

For experiments on heat treatment on the basis of the offered recommendations three experimental melts were performed O74, O77, O76. The smelting of the experimental material was carried out in the electric arc furnaces of 100 tonnes. Casting of the vacuum degassed metal took place in CCM in the form of billets with section 300x330 mm. Heating of continuous cast steel billet in the WB furnace and rolling were carried out according to the current rails production technology of R65 type. The content of chemical elements in the metal of experimental melts defined by the chemical and spectral method is presented in Table 1.

Table 1. Contents of chemical elements in the experimental metal.

Variant No.	Mass concentration of chemical elements, %									
	C	Mn	Si	Cr	P	S	Al	V	Ni	Cu
O74	0.76	0.87	0.32	0.55	0.015	0.006	0.002	0.07	0.08	0.13
O77	0.77	0.91	0.31	0.42	0.015	0.008	0.003	0.08	0.07	0.12
O76	0.79	0.78	0.55	0.46	0.014	0.015	0.002	0.07	0.08	0.14
GOST requirements R 51685-2013 for steel E76HF	0.74-0.82	0.75-1.15	0.20-0.80	0.20-0.60	not less 0.025	0.025	0.005	0.03-0.15	not less 0.20	0.20

3.1. Heat treatment of rail steel – melt O74

The heat treatment of rails square. O74 chromium content of 0.55% was carried out by temperature 705 - 850° C at the rate of 2.4 - 2.8 0C / s during 105 - 125 s. Modes of heat treatment and mechanical test results are presented in Table 2.

Table 2. Heat treatment parameters using heat of rolling heating and mechanical properties of rails R65 from steel grade E76HF – melt O74.

No.	Heat treatment parameters			Mechanical properties				KCU, at temperature		Hardness over the head cross-section, HB				
	T, °C	Cooling rate °C/sec	Time, sec	σ_{yield} N/mm²	σ_{ten} N/mm²	δ_5 %	ψ %	+20°C J/cm²	-60°C J/cm²	HB_{rail} tread top	HB_{10}	HB_{22}	HB gage corner 1	HB gage corner 2
1.1	750	2.8	125	960	1370	10.5	36	23	8.5	406	406	398	415	415
1.2	780	2.8	125	1040	1400	9.4	35	20	9.7	415	415	395	415	415
1.3	815	2.8	125	1000	1410	14	30	16	4.9	420	415	406	415	415
1.4	845	2.8	125	1020	1410	11	27	14	6	429	406	393	415	415
1.5	700	2.4	105	760	1230	10	18.5	12	3.6	380	380	388	359	361
1.6	780	2.4	120	970	1380	11.5	27	8.4	6	411	390	385	395	393
1.7	850	2.4	120	950	1380	15	31	22	6	415	415	404	409	409
GOST requirements R 51685-2013 for rails of category DT350				not less 800	1240	9.0	25.0	15.0	-	363-401	not less 341	341	341	

The presented in Table 2 data show that all test samples except sample No. 1.5, showed high values of ultimate strength and yield stress and plastic properties of satisfactory values. Sample No. 1.5 was treated by heat from 700 °C. During tensile test the metal showed the values of mechanical properties which do not meet the standard requirements for heat hardened rails.

With the increase in the cooling rate the strength properties and hardness improve over the head cross-section. Among the samples No. 1.1-1.4, hardened at 2.8 °C/sec, there is a tendency to increase the strength properties and reduce specific elongation and impact viscosity with the temperature of the start of heat treatment.

The impact viscosity of the rails hardened at lower rate (2.4 °C/sec) showed lower values not satisfying the standard requirements. The experimental rails are characterized by high values of hardness over the head cross-section, and except for the rail, strengthened at the temperature from 700°C, exceed the set by the standard maximal allowable hardness values on the surface of the head rolling.

Impact tests at temperatures below freezing showed a low level of impact viscosity of all experimental metal not exceeding 10 J/cm2.

The metal microstructure of experimental rails is a thin-plate and sorbitic pearlite. From the rolling surface of the head in the head centre and from the fillets to a depth of 11 mm in samples No. 1.1-1.4, hardened at 2.8 °C/sec, we observed the intermediate transformation products, unacceptable according to the requirements of normative and technical documentation on rail products. As the temperature of the beginning of heat treatment lowers the tendency appears to increase the magnitude of the layer with the presence of a bainite structure.

With a decrease in the cooling rate to 2.4°C/sec the layer depth with the areas of bainitic structures is reduced to 0.3-0.5 mm. Metal sample No. 1.5 does not contain bainite; pearlite has a rough, lamellar structure.

3.2. Heat treatment of rail steel – melt O77

The rails heat treatment of type R65 melt O77 with chromium content of 0.42% was carried out at the rate 2, 2.4, 2.8 °C/sec, from the temperatures 750-860°C during 120-125 sec. The heat treatment parameters and results of mechanical tests parameters are given in Table 3.

From the data presented in Table 3 it can be seen that the metal of this group is characterized by high values of hardness on the rolling surface and over the head cross-section, very high strength properties and relatively low level of plastic properties, mainly the relative contraction. Fully complies with the standard for rails DT350 category only metal sample No. 2.4, heat-treated for 125 sec with a cooling rate 2.8°C/sec from 810°C.

The temperature of the beginning of hardening at a given cooling rate or decrease in the cooling rate ceteris paribus leads to a decrease in the values relative of relative contraction.

Table 3. Parameters of heat treatment and mechanical properties of rails R65 from steel grades E76HF – melt O77.

No.	Heat treatment parameters			Mechanical properties				KCU, at temperature		Hardness, HB				
	T, °C	Cooling rate, °C/sec	Time, sec	σ_{yield}	$\sigma_{tensile}$	δ_5	ψ	+20°C	-60°C	HB_{rail} tread top	HB_{10}	HB_{22}	HB gage corner 1	HB gage corner 2
				N/mm^2		%		J/cm^2						
2.1	780	2.8	125	980	1380	13	30	21	10	420	404	390	404	401
2.2	780	2.8	125	940	1360	13	25	8.4	8.4	423	409	395	401	401
2.3	802	2.8	125	980	1380	9.3	26	19	9.7	415	415	401	415	415
2.4	810	2.8	125	940	1370	9	26	15	6	398	409	393	415	415
2.5	830	2.8	125	1010	1400	11.5	22.5	16	8.4	415	409	395	415	415
2.6	860	2.8	125	990	1400	11.5	21.5	22	9.7	404	404	398	415	415
2.7	760	2.4	120	880	1350	9.9	21.5	17	6	423	404	398	404	401
2.8	800	2.4	120	930	1230	8.9	23.5	19	12	409	409	401	401	401
2.9	870	2.4	120	950	1360	9.2	24	19	6	409	409	409	409	404

2.10	750	2	120	900	1350	12.5	23	11	12	401	390	390	383	383
2.11	790	2	120	900	1350	12.5	23.5	17	8.4	409	395	383	395	395
2.12	845	2	120	920	1340	11.5	24	18	6.1	404	404	393	409	409
GOST requirements R 51685-2013 for rails of category DT350				not less					-	363-401	not less			
				800	1240	9.0	25.0	15.0			341			

In the microstructure of samples No. 2.1, 2.2 and 2.7 with the temperature of the beginning of heat treatment 780-760°C, from the rolling surface, in the centre and from the fillets side, to a depth of 3 mm and 1.5 mm respectively, we observed areas of bainite.

3.3. Heat treatment of rail steel – melt O77

In view of the fact that melt O76 was made in accordance with the recommendations [9] obtained earlier, based on the positive results in experiments on differential hardening of rail samples by furnace heating, then the metal of this melt was used for the largest number of experiments, at different modes about 50 samples of rails R65 were hardened. The heat treatment was carried out rails at a cooling rate from 2 to 3.3°C/sec for 60-160 seconds. Below we summarize the results of the tests.

Hardening at 2°C/sec

Table 4 contains the data on heat treatment of rails melt O76 at a cooling rate by air 2°C/sec.

From the data presented in Table 4 it can be seen that with the temperature increase of the beginning of hardening and cooling duration the mechanical properties are improved. At lower temperatures of the beginning of hardening, a relatively low level of plasticity properties are observed.

The highest properties are obtained by hardening 800°C/sec.

Hardening at 2.4°C/sec

The test results of metal hardened at a rate 2.4°C/sec are given in Table 5, from which it follows that for all samples relatively high properties were obtained. With the temperature increase of the beginning of hardening the yield strength rises at a comparable level of ultimate tensile strength and plastic properties. The best level of impact viscosity we obtained in the samples hardened from the temperature ~750°C.

Hardening at 2.8°C/sec

Test results of the metal hardened at a rate 2.8°C/sec are shown in Table 6.

From the data presented it can be seen that at hardening rate of 2.8°C/sec for 60 sec the requirements of mechanical properties of GOST R51685-2013 are meet, but at a depth of 22 mm a low hardness was found. With the increase of the cooling duration the level of strength properties and hardness increases as well. For all samples except for sample No. 1.13, hardened from the temperature of 700°C, a satisfactory level of impact strength at a test temperature of + 20°C was observed.

Table 4. Parameters of heat treatment and mechanical properties of rails from melt O76 at a cooling rate by air 2°C/sec.

No.	Heat treatment parameters			Mechanical properties				KCU, at temperature		Hardness, HB				
	T, °C	Cooling rate, °C/sec	Time, sec	σ_{yield}	$\sigma_{tensile}$	δ_5	ψ	+20°C	-60°C	HB_{rail} tread top	HB_{10}	HB_{22}	HB gage corner 1	HB gage corner 2
				N/mm²		%		J/cm²						
2.1	780	2.8	125	980	1380	13	30	21	10	420	404	390	404	401
2.2	780	2.8	125	940	1360	13	25	8.4	8.4	423	409	395	401	401
2.3	802	2.8	125	980	1380	9.3	26	19	9.7	415	415	401	415	415
2.4	810	2.8	125	940	1370	9	26	15	6	398	409	393	415	415
2.5	830	2.8	125	1010	1400	11.5	22.5	16	8.4	415	409	395	415	415
2.6	860	2.8	125	990	1400	11.5	21.5	22	9,7	404	404	398	415	415
2.7	760	2.4	120	880	1350	9.9	21.5	17	6	423	404	398	404	401
2.8	800	2.4	120	930	1230	8.9	23.5	19	12	409	409	401	401	401

2.9	870	2.4	120	950	1360	9.2	24	19	6	409	409	409	409	404
2.10	750	2	120	900	1350	12.5	23	11	12	401	390	390	383	383
2.11	790	2	120	900	1350	12.5	23.5	17	8.4	409	395	383	395	395
2.12	845	2	120	920	1340	11.5	24	18	6.1	404	404	393	409	409
GOST requirements R 51685-2013 for rails of category DT350				not less						363-401	not less			
				800	1240	9.0	25.0	15.0	-		341			

Table 5. Parameters of heat treatment and mechanical properties of rails from melt O76 at cooling rate by air 2.4°C/sec.

No.	Temp at the beginning, °C	Time, sec	Mechanical properties				KCU at temperature		Hardness, HB				
			σ_T	σ_B	δ	ψ	+20°C	-60°C	HB_{rail} tread top	HB_{10}	HB_{22}	HB gage corner 1	HB gage corner 2
			N/mm²		%		J/cm²						
1.7	735	105	935	1360	10	29	12	8.4	415	401	383	395	393
1.8	745	135	940	1375	10.5	32	20.5	7.3	409	401	383	395	401
1.9	750	125	960	1360	11	34	20	6.1	415	401	378	406	398
1.10	810	125	960	1375	10	32	16	6.2	415	409	390	406	415
1.11	830	135	975	1375	10	34	16.5	7.4	415	415	388	415	388
1.12	850	125	970	1370	9.8	30	6.1	7.2	415	401	375	401	398
GOST requirements R 51685-2013 for rails of category DT350			not less						363-401	not less			
			800	1240	9.0	25.0	15.0	-		341			

Hardening at 3.3°C/sec

Data from test samples hardened at a rate 3.3°C/sec are shown in Table 7.

The presented data show that during hardening at a rate of 3.3°C/sec sufficiently high properties throughout the temperature range of the hardening start – from 700 to 880°C were obtained. The increase in the heat treatment duration leads to a significant increase of strength properties and hardness; with the reduction in cooling duration the impact viscosity increases.

Metal microstructure of all rails melt O76 is a thin-plate pearlite with scattered ferrite formations on the grain boundaries. Bainite is not detected in the microstructure of the rails. Some increase in the volume fraction of the structure-free ferrite with the decrease of the temperature of the hardening start.

Table 6. Parameters of heat treatment and mechanical properties of rails from melt O76 at a cooling rate by air 2.8°C/sec.

No.	Temp at the beginning, °C	Time, sec	Mechanical properties				KCU at temperature		Hardness, HB				
			σ_{yield}	$\sigma_{tensile}$	δ	ψ	+20°C	-60°C	HB_{rail} tread top	HB_{10}	HB_{22}	HB gage corner 1	HB gage corner 2
			N/mm²		%		J/cm²						
1.13	700	125	925	1350	12.5	29	10.1	7.7	415	412	393	401	401
1.14	760	110	975	1380	10.5	33	17	6.6	415	404	388	415	406
1.15	775	105	940	1370	10	31	21	6.1	409	406	388	409	409
1.16	810	60	850	1260	11.5	25	23.5	8	378	363	333	375	370
1.17	840	70	850	1255	11.5	26.2	23.5	7.9	378	356	337	368	370
1.18	840	125	995	1395	9.8	28	18	7.3	438	420	401	423	417
1.19	865	115	955	1390	10.2	30	21	8.6	412	401	383	409	401
GOST requirements R 51685-2013 for rails of category DT350			not less					-	363-401	not less			
			800	1240	9,0	25	15			341			

Table 7. Parameters of heat treatment and mechanical properties of rails from melt O76 at a cooling rate by 3.3°C/sec.

No.	Temp at the beginning, °C	Time, sec	Mechanical properties				KCU at temperature		Hardness, HB				
			σ_{yield}	$\sigma_{tensile}$	δ	ψ	+20°C	-60°C	HB_{rail} tread top	HB_{10}	HB_{22}	HB gage corner 1	HB gage corner 2
			N/mm²		%		J/cm²						
1.19	700	125	950	1355	10	27	16,5	7,5	426	404	395	406	406
1.20	770	125	960	1365	10	31	15	9,2	426	409	393	409	406
1.21	787	90	895	1290	13,1	25,2	20	8	388	383	370	385	383
1.22	800	110	945	1370	11,3	29,5	15,5	7,2	415	409	388	409	409
1.23	840	105	890	1290	13,5	28	25,5	10	388	383	368	388	385
1.24	850	100	970	1380	9,8	28	15	8	406	404	375	393	398
1.25	880	125	985	1385	10,5	32	16	8	426	409	390	409	409
GOST requirements R 51685-2013 for rails of category DT350			not less					-	363-401	not less			
			800	1240	9,0	25	15			341			

4. Conclusions

The performed studies revealed that:

- The most consistent results for the microstructure, without occurrence of needle structures in the largest temperature range (from 700 to 880°C) and cooling rates (from 2 to 3.3°C/sec) are provided by heat treatment of metal melt O76.
- Hardening of rails from steel of grade E76HF at a temperature below 730-750°C might result on poor plasticity properties.

Thus, the most promising for the commercial production is the rail steel close in its chemical composition to the metal of melt O76 with chromium content in the range of 0.38-0.43%; silicon 0.54-0.60%; manganese 0.80-0.90%; vanadium 0.03-0.04%. This chemical compound has high processability, since it can be processed in the wide temperature range of the beginning of hardening – from 700 to 880°C, and at various cooling rates – from 3.3 to 2°C/sec.

For the industrial implementation it is recommended:

- To provide a high performance it is recommended to perform heat treatment at temperatures above 800°C;
- To improve the efficiency it is recommended to perform heat treatment at a cooling rate of 2-2.4°C/sec;
- Optimal cooling duration 90-110 sec.

5. References

[1] Poukhin P I, Gridina Yu V and Zarvin E Ya 1962 *Rolling and Heat Treatment of Rails* (M.: Metallurgizdat) p 510
[2] Zolatarsky Ya R, Rauzin E A et al 1976 *Heat Hardened Rails (*M.: Transport) p 264
[3] Shur E A 2012*Rails Damag*es (M.: Intekst) p 192
[4] Fedit V M 2002 *Volume and Surface Hardening of Parts of the Rolling Stock and Track Upper Structure* (M.: Intekst) p 208
[5] Pavlov V V, Temlyantsev M V et al 2007 *Advanced Technologies of Heat and Therma treatment in the Production of Rails* (M.: Teplotekhnik) p 280
[6] Mikhaylets N S, Gorelkina A E et al 1964 *Rails Production at Kuznetsk Metallurgical Plant* (M.: Metallurgiya) p 222
[7] Polyakov V V and Velokaniv A V 1990 *Fundamentals of Production Technology of Rails* (M.: Metallurgiya) p 416

[8] Pavlov V V, Temlyantsev M V et al 2006 *Handbook on Defects and Quality of Rail Steel* (M.: Teplotekhnik) p 218
[9] Volkov K V, Polevoy E V et al 2014 *Proc. of SibSIU* **3** (9) 17–23
[10] Polevoy E V, Temlyantsev M V et al 2015 *Proc. of SibSIU* **1** (11) 5–11

New materials for welding and surfacing

N A Kozyrev, G V Galevsky, R E Kryukov, D A Titova and V M Shurupov

Institute of Metallurgy and Materials Science, Siberian State Industrial University, 42 Kirova Street, Novokuznetsk, 654007, Russia

E-mail: kozyrev_na@mtsp.sibsiu.ru

Abstract. The paper provides description of research into the influence of new materials and technologies on quality parameters of welds and deposited metal carried out in the research and production centre "Welding processes and technologies". New welding technologies of tanks for northern conditions are considered, as well as technologies of submerged arc welding involving fluxing agents AN - 348, AN - 60, AN - 67, OK.10.71 and carbon-fluorine containing additives; new flux cored wires and surfacing technologies, teaching programs and a trainer for welders are designed.

1. Introduction

The basic requirement of industrial safety consists in assuring the needed level of mechanical characteristics of a weld. Here, mechanical characteristics are pre-determined by non-metallic impurity degree and total content of gas in a weld. Most non-metallic impurities in welds are oxide ones irrespectively to their exogenous or endogenous nature, the quantitative and qualitative composition of the latter depends on the total concentration of oxygen. The issue of new fluxes and their additives development has been currently attracting much attention, as well as the research into their influence on welding and technological characteristics of a weld and on the concentration of oxygen and non-metallic impurities in a weld [1-5].

2. Materials and methods of research

Most grades of the domestically produced fluxes, applied for welding low-alloyed steels, are oxidizing ones and ground on silicon-manganese oxidation-reduction processes. Here, the products of these reactions are oxide compounds of silicon, manganese, ferrum, aluminum etc., which often cannot surface and assimilate by the slag formed from a welding flux; as a result, the level of impurity of weld metal by non-metallic admixtures increases, causing deterioration of physical and mechanical characteristics. Obviously, reducing agents forming gaseous products of reactions should be applied in order to avoid impurity of weld metal. Carbon can serve as a reducing agent forming gaseous compounds CO_2 and CO when reacting with oxidizers.

Taking into account these pre-conditions a series of research activities has been carried out into the influence of additives to various fused and ceramic fluxes on the quality parameters of welds.

In recent years exploration of oil fields in the North has necessitated manufacturing of wall plates for bulk-oil tanks suitable for northern conditions. At JSC "Novokuznetsk Plant of Reservoir Metalwares" manufacture of tanks for northern conditions has been developed according to "Standards of vertical cylindrical steel tanks for oil and oil products"PB-03-605-03 and Construction norms and regulations (SNiP) II – 23-81.

The technological process of assembling, welding, controlling and rolling of wall plates is performed on special roll facilities. Silicon-manganese steel 09G2S (State Standard (GOST) 6713-91) is used for tanks manufacturing. As a result, an optimal technology has been developed for two-side welding of tank wall plates operating in low temperature conditions: inside the tank it is welded by Sv-08GA wire (on the upper tier) with mixed fluxing agents AN-67Б and AN-348A taken in proportion 1:1, outside the tank it is welded by Sv-10NMA wire (on the lower tier) with welding fluxes AN-60 and AN-348A taken in proportion 1:1. Flay welding was carried out up to 18 mm ofplate thickness. On the upper tiers flat butt welds are made at reduced power modes, which provide penetration into metal up to 0.55 of plate thickness. On the lower tier the welding was carried out in the conditions of high current providing penetration up to 0.7 of plate thickness.

Weld metal consists principally of re-melted base 09G2S and added Sv-10NMA metals. The mechanical properties of joint welds and welds were determined at temperature 20°C, impact strength of weld metal and thermal impact zone KCU were measured at temperature -50°C. Mechanical characteristics and impact strength of a weld and thermal impact zone exceeded the required normalized values.

The developed technology made it possible to obtain a range of required mechanical properties and the impact strength of tank metalwares at temperatures below zero, avoid defects – cracks during manufacturing [6]. This technology was patented in the Russian Federation [7, 8].

The influence of carbon and fluorine containing additive to AN-348, AN-60, AN-67 fluxing agents and imported fluxing agent OK.10.71 on refinement of weld metal was experimentally tested [9,10]. As a result, it was found out, that the growing content of carbon and fluorine-containing additive in various fluxing agents at a permitted slight increase of carbon concentration (Figure 1) causes the drop of total concentration of oxygen in the weld (Figure 2), and considerable improvement of mechanical properties, especially of impact strength at temperatures below zero (Figure 3).

Fraction gas analysis conducted by the method of reduction melting using the gas analyzer "LECO"TC-600 revealed that the redistribution of oxygen in impurities depends on oxidation and basicity of slag system.

The designed submerged arc welding technology of metalwares, operating in the conditions of extremely low temperatures, relies on theoretical pre-conditions and practical try-out of carbon and fluorine-containing additive technology. The flux of FD-UFS additive is produced by JSC "Novokuznetsk Plant of Reservoir Metalwares n.a. N.E. Kryukov". The proposed welding technologies of metalware welding involving flux-additive are patented in the Russian Federation [11], the composition of the new basic welding flux is also patented in the Russian Federation [12].

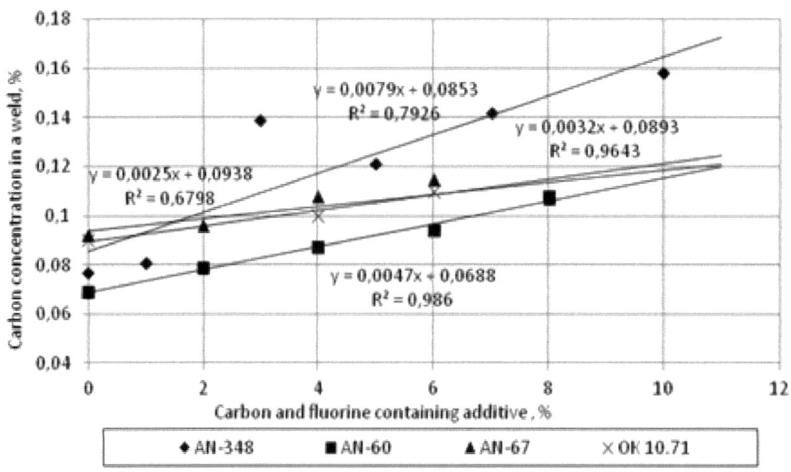

Figure 1. Influence of carbon- and fluorine-containing additive on carbon concentration in a weld.

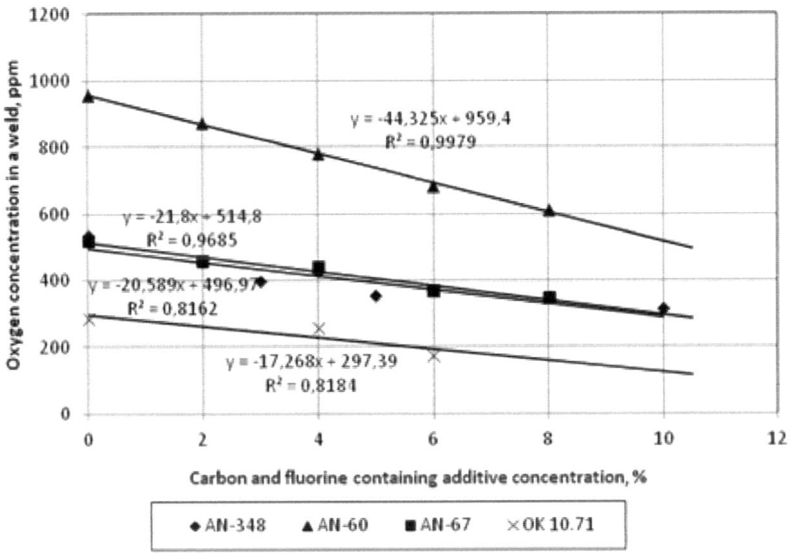

Figure 2. Change in the total oxygen concentration in fluxing agents depending on carbon and fluorine additive.

Surfacing of forming rolls is widely applied at manufacturing enterprises in Russia for restoration and improvement of resistance. Despite a wide range of systems used for surfacing, two systems are applied more often: C–Si–Mn–Cr–W–V and C–Si–Mn–Cr–V–Mo.

One of the elements in the first system is PP-Np-35V9H3SF flux cored wire containing up to 10% of rare and expensive tungsten, applied for wear-resistant surfacing of steel rolls in different hot-rolled mills. The deposited metal is distinguished by a significant abrasion resistance at high temperatures, but its thermal resistance is quite low; therefore, rolls surfaced with the wire often fail because of erosion and chipping cracks. Sometimes stripe wear is caused by areas of deposited metal with inhomogeneous structure and hardness. Such areas can be found after multilayer surfacing of alloyed steels with overlapping of a bead made before.

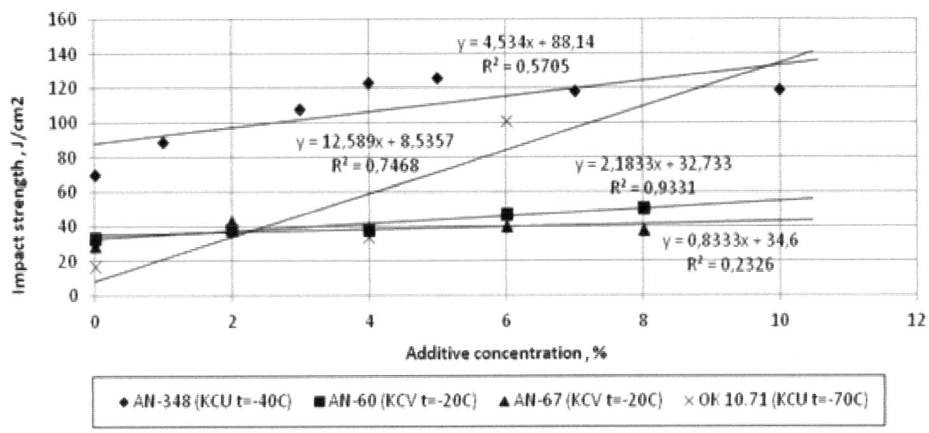

Figure 3. Change in the impact strength depending on carbon and fluorine additive concentration.

The surfaced metal of the same type as economically alloyed chromium-molybdenum and chromium-tungsten-molybdenum steels demonstrated good results after restoration of steel rolls at hot-rolled mills. These steels are not worse than chromium-tungsten ones in terms of heat resistance, and outperform them regarding thermal fatigue resistance. Vanadium alloying is also applied to improve the thermal resistance and abrasion strength of added metal.

Machining of deposited layer is rather difficult when surfacing rolls with complex grooves because of its relatively high hardness. Surfacing materials like martensite-ageing or disperse-hardening steels are promising ones for the above mentioned rolls. After being surfaced such steel have hardness to 28...35 HRC_E, and they are easy to be machined. After tempering the hardness increases up to 48...55 HRC_E, and the deposited metal gains sufficient service qualities. However, special equipment is needed for heat treatment.

The second system provides the highest thermal fatigue resistance of metal. This system represented by PP-Np-25H5FMS flux cored wire makes it possible to gain significant impact strength of the deposited metal at high temperatures.

As a result of laboratory and experimental-industrial research it is recommended to use PP-Np-35V9HSF flux-cored wire for surfacing rolls of continuous bar mills, pipe-rolling, and wire mills; and Pp-Np-25H5FMS flux cored wire on blooming mills (blooming, slabbing), on heavy section and rail and structural steel mills, as well as on medium and small section sheet mills.

The samples of standard PP-Np-25H5FMS flux cored wire were fabricated in the laboratory conditions. Multilayer surfacing of samples was carried out with the preliminary heating of sheets up to 350°C and subsequent slowed cooling (after surfacing). 5-layer surfacing of 09G2S steel sheets was performed by welding tractor ASAW-1250 and fabricated flux cored wire; the surfacing conditions were: welding current I_a=400A, arc voltage U_d = 32W, rate of welding υ_s=0.8 cm/s.

During fabrication of wire the concentration of silicon, manganese, chromium, vanadium was varied wider than the values of these elements required for PP-Np-25H5FMS wire by GOST 26101-84. Nickel was added to chemical composition of some samples, and amorphous carbon was changed into carbon and fluorine containing dust (wastes of metallurgical production), chemical composition was as follows, mass %: Al_2O_3 = 21 – 46; F = 18 – 27; Na_2O = 8 – 15; K_2O = 0.4 – 6%; CaO = 0.7 – 2.3; SiO_2 = 0.5 – 2.5; Fe_2O_3 = 2.1 – 3.3; C_{tot} = 12.5 – 30.2; MnO = 0.07 – 0.9; MgO = 0.06 – 0.9; S = 0.09 – 0.19; P = 0.10 – 0.18. The dispersion of this material enables good mixing with metal constituent of flux cored wire burden. Previously the conducted research into application of this material as an additive to welding flux agents revealed that carbon as a component of carbon and fluorine containing dust is super active, it deoxidizes metal and reduces oxides contained in metal and slag, as the result, carbon oxides are formed. To simulate oxidation and reduction processes, taking place during surfacing, the low oxidizing flux agent AN–67 was used according to GOST R 52222-2004. Manganese was not added to chemical composition of the wire because of its reduction from forming slag. Silicon was partially reduced from slag. The experiments demonstrated that the formed slag can hardly oxidize the alloying elements in the deposited metal.

The chemical composition of the deposited metal was determined by roentgen-fluorescent method with spectrometer XRF-1800 and by atom-emission method with spectrometer DFS-71. Metallographic research into micro-sections was carried out by optical microscope OLYMPUS GX-51 in bright field at different zooming modes after etching in the alcohol solution of nitric acid, as well as in solution of hydrofluoric acid. Hardness was measured by ultrasonic hardness testing instrument UZIT -3. The data on chemical composition of flux cored wires, hardness of the deposited metal and volume concentration of retained austenite are provided in Table 1.

When amorphous graphite was used as an additive, in the structure of deposited metal (samples 1 and 2) martensite and retained austenite in the axle spacing were found. Significant oxygen inclusion lines are found. Vanadium and chromium carbides are distributed over the grain. It should be noted that that introduction of more than 3% of vanadium caused a considerable drop in the deposited metal hardness.

Metallographic analysis demonstrated that the structure of metal surfaced by flux cored wire includes an acicular troostite with martensite and formed along the grain boundaries of separate thin austenite fringes containing some carbide impurities. Carbon- and fluorine-containing additive to wire burden facilitated removing of oxide non-metallic impurities, smoother distribution of carbon in the weld metal and formation of fine-dispersed carbides that in practice should lead to improvement of thermal strength.

Austenite-forming nickel added to the flux cored wire facilitated additional grain refinement. Introduction of the system C–Si–Mn–Cr–V–Mo carbon-fluorine containing additive and nickel into the wire would significantly decrease the level of oxide non-metallic impurities in the deposited metal, refine the grains, and forms fine-disperse carbides, improving thermal resistance of the forming rolls.

As we mentioned above the most loaded hot forming rolls are strengthened by surfacing withPP-Np-35V9H3SF flux cored wire according to GOST 26101-84 with high content of tungsten. From the experience of surfacing forming rolls and their operation we know that their wear of their surface does not occur evenly. In the process the forming rolls are subjected to cyclic thermo-mechanical stresses, corrosion and abrasive wear. Stripe wear, caused by the areas with the deposited metal of different structure and hardness, is possible. Such areas are observed after multilayer surfacing of alloyed steels with overlapping of the previously welded bead. The deposited metal of 35V9H3SFtype has high wear resistance at increased temperatures, but its thermal strength is relatively low, therefore, rolls, surfaced with this wire, can often fail due to heat cracks and chipping.

Table 1. Chemical composition of the tested flux cored wires and hardness of the deposited metal of the system C–Si–Mn–Cr–V–Mo.

№	Mass concentration of elements, %											HRC (HB)	VCA
	C	Si	Mn	Cu	Cr	Mo	Ni	Al	W	V	Ti		
1	0.16	1.77	2.25	0.26	3.51	1.51	0.16	0.044	0.107	3.25	0.018	(186.2)	-
2	0.15	1.59	1.67	0.18	0.35	1.11	0.16	0.032	0.06	0.35	0.006	43.2	4.3
3	0.20	1.33	2.16	0.17	3.52	1.51	0.13	0.04	0.09	0.36	0.020	48.1	5.6
4	0.24	0.93	2.11	0.15	3.18	1.31	0.21	0.054	0.071	0.46	0.021	51.8	15.6
5	0.31	1.19	2.13	0.20	3.50	1.34	0.32	0.064	0.074	0.47	0.022	53.4	14.7
6	0.24	0.92	1.93	0.22	3.00	1.85	0.38	0.034	0.081	0.43	0.014	54.14	8.4

Notes:
1) VCA – volume concentration of retained austenite,
2) samples 1 and 2 include amorphous graphite;
3) samples 3 and 6 include carbon- and fluorine-containing dust at different concentrations.

The samples of standard PP-Np-35V9H3SF flux cored wire, which is used for surfacing hot forming rolls and rolls of roller conveyers were fabricated in laboratory conditions. Multilayer surfacing was carried out as follows: sheets were pre-heated up to 350°C and then slowly cooled down. The surfacing was carried out by the welding tractor ASAW-1250 using the produced flux cored wire on the steel sheets 09G2S in 5 layers; the surfacing mode: welding current I_a=400A, arc voltage U_a = 32W, rate of welding υ_s=0.8 cm/s.

In the process of wire fabrication the concentration of expensive tungsten and chromium was calculated lower than required by GOST 26101-84 for PP-Np-35V9H3SF. A number of wire samples were produced with nickel and amorphous carbon was substituted by carbon- and fluorine- containing dust. Chemical composition of the deposited metal was defined by roentgen-fluorescent spectrometer XRF-1800 and atom-emission spectrometer DFS-71. Metallographic investigation of micro-sections were conducted by optical microscope OLYMPUS GX-51 in bright field in different zooming conditions after etching in alcohol solution of nitric acid, as well as in solution of hydrofluoric acid. The data on chemical composition of flux cored wires, and hardness of the deposited metal are provided in Table 2.

Provided that amorphous graphite is used as an additive (sample 1) the deposited metal gets a dendrite structure (martensite with retained austenite in axial spacing). There are a lot of oxygen inclusion lines, which are stress concentrators and areas of brittle cracks formation. Point tungsten and chromium carbides are distributed over the grains.

Metallographic analysis demonstrated that metal structure, which was surfaced with flux cored wire, includes an acicular troostite with martensite and formed along the grains boundaries separate thin austenite fringes containing some carbide impurities. Carbon- and fluorine-containing additive to wire burden facilitated removing of oxide non-metallic impurities, smoother distribution of carbon in weld metal and formation of fine-dispersed carbides, all this may in practice improve thermal strength. Nickel was added to flux cored wire to improve thermal resistance of metal through stabilization of austenite and grain refinement. The change in volume concentration of retained austenite in the surfaced layer with various concentration of nickel is provided in Table 3.

Therefore, the experiments demonstrated that carbon- and fluorine-containing additive and nickel added to wire of the system C–Si–Mn–Cr–W–V could reduce considerably the degree of the deposited metal impurity by oxide non-metallic inclusions, and facilitated formation of fine-dispersed carbides, thus, thermal resistance of forming rolls can in practice improve. Chemical composition of these flux cored wires and their modifications is protected by patents of the Russian Federation [13, 14].

Table 2. Chemical composition of tested flux cored wires and hardness of the deposited metal of system C–Si–Mn–Cr–W–V*.

Sample No.	Mass concentration of elements, %											
	C	Si	Mn	Cu	Cr	Mo	Ni	Al	W	V	Ti	HRC
1	0.27	1.62	1.42	0.19	2.93	0.14	0.14	0.04	11.39	0.47	0.02	49
2	0.23	0.69	1.04	0.21	1.92	0.09	0.16	0.039	5.58	0.15	0.019	50.8
3	0.24	0.83	1.83	0.24	2.16	0.09	0.29	0.054	7.49	0.27	0.019	53.3
4	0.26	0.77	1.72	0.26	2.17	0.09	0.33	0.042	6.12	0.29	0.016	53.0
5	0.33	1.37	1.06	0.22	2.80	0.10	0.61	0.07	10.71	0.42	0.04	54.8

* sample 1 included amorphous graphite, samples 2-5 were fabricated using carbon- and fluorine-containing dust and nickel powder with different concentrations.

Table 3. Volume concentration of retained austenite in the deposited layer (mean for three fields).

Test	2	3	4
Volume concentration of austenite, %	15.4	5.2	6.8

Most welding defects are caused by low qualification of welders. Taking into account, that the most significant factor is the length of the arc, and constant control results in eye fatigue decreasing training efficiency, various training devices are used at the first stage of welders training.

To develop correct psychomotor skills that would help to meet the standard requirements (length of arc space, incline angle of welding electrode, displacement from the grooving center of work pieces to be welded, welding rate) in different spatial positions, to increase training efficiency through control of the required parameters of welding and reduce costs, a training device is used for training at the department (Figure 4); it consists of the following elements: 1 – screen; 2 – lamp; 3 – simulator of the electrode holder; 4 – headphones; 5 – unit of signal receiving.

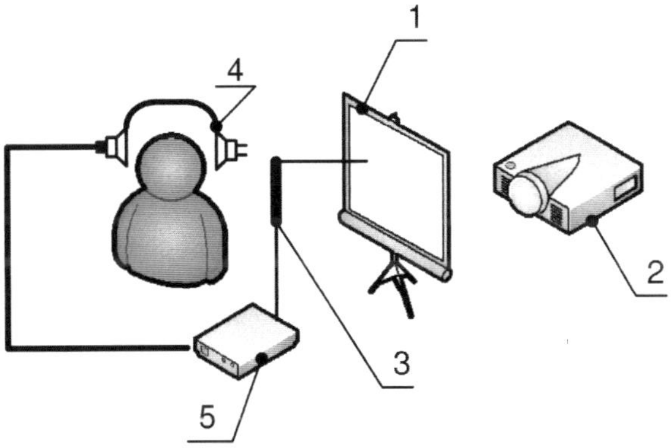

Figure 4. Principal scheme of welder training device.

Training of welders with this training device is performed according to the following method. On the screen there are transparent lines imitating the trajectories of arm movement in conditions of various welding techniques. If the simulator of electrode holder is over the transparent line the light from the source gets into the photoreceiver, which is placed on the end of the electrode model, there is no acoustic signal meaning that the technique of welding is correct. Provided that the end of the electrode model is displaced for a period of time within the limits 0.5 – 1 s, an acoustic signal is given meaning the displacement from the set trajectory, an error, respectively. On the light-proof screen there are various transparent lines, imitating different depths of the base metal fusion and weld formation, depending on transverse vibrations of the electrode. After the movement skills are acquired, the welder improves them through more complicated the trajectories of transparent lines. Training device makes it possible to consolidate and unify the technical skills of electrode movement and create a correct electrode inclination angle towards the product and welding direction, which predetermines the quality of weld formation.

Application of this training device in groups of welders during their preliminary training for certification showed that after training welders groups without using the training device, the number of welds meeting the requirements of quality amounted to 45-50 %, whereas after training on the training device – 87-95%. The training device is protected by the patent of the Russian Federation [15]

3. Conclusions

1) The technologies of new carbon-containing additives to welding fluxes have been developed, which can reduce the level of steel oxide non-metallic impurity, as well as gas content in a weld, and improve the required mechanical properties. These technologies are introduced into production process and used for welding bulk-oil tanks in conditions of low temperatures. Flux-additive FD-EFS is produced and protected by the patent of the Russian Federation.

2) Flux cored wires of the systems C–Si–Mn–Cr–W–V and C–Si–Mn–Cr–V–Mo with the added carbon and nickel containing elements, used for surfacing forming rolls, have been developed and are protected by the patents of the Russian Federation. Application of flux cored wires for surfacing provides efficient removal of oxide non-metallic impurities, even distribution of carbon in weld metal, and improvement of thermal resistance of the surfaced product.

3) For training and formation of correct psychomotor skills in the process of manual arc welding a training device for welding simulation has been developed and introduced into the training process, also patented in the Russian Federation.

4. Acknowledgements
The work was performed in Siberian State Industrial University within the project of State Order of the Ministry of Education and Science of the Russian Federation No. 11.1531.2014/k. For tests, investigations, measuring we used equipment in the Collective Center "Material Science" of Siberian State Industrial University.

5. References
[1] Rafael Quintana Puchola, Jeily Rodríguez Blancoa et al 2009 *Welding International* **23** (2) 132–140
[2] Crespoa A C, Puchola R Q et al 2007 *Welding International* **21** (7) 502–511
[3] Amado Cruz Crespoa, Rafael Quintana Puchola et al 2009 *Welding International* **23** (2) 120–131
[4] Volobueva Yu S, Volobueva O S et al 2012 *Welding International* **26** (8) 649–653
[5] Golovko V V and Potapov N N 2011 *Welding International* **25** (11) 889–893
[6] KryukovN E, Kovalsky I N, Kozyrev N A et al 2012 *Steel in Translation* **42** (2) 118–120
[7] Kryukov N E, Kovalsky I N, Kozyrev N A, Igushev V F and Kryukov R E 2011 *Submerged Arc Welding* patent of the Russian Federation No. 2011123342/02 (034573)
[8] Kryukov N E, Kovalsky I N, Kozyrev N A, Igushev V F and Kryukov R E 2011 *Ceramic Flux-additive* patent of the Russian Federation No. 201112341602/02 (034654)
[9] Kozyrev N A, Igushev V F, Krukov R E et al 2013 *Welding International* **27** (12) 963–965
[10] KozyrevN A, KrukovR E and Kolmogorov D E 2014 *Mechanics and Materials* **682** 495–498
[11] Kozyrev N A, Igushev V F, Kryukov R E and Goldun S V 2012 *Ceramic Flux-additive*patent of the Russian Federation No. 2012104939/02 (007484)
[12] Kozyrev N A, Igushev V F, Kryukov R E and Goldun S V and Kozyreva O E 2012 *Welding FluxIon* patent of the Russian Federation No. 2012108414/02
[13] Kozyrev N A, Titov D A, Kozyreva O E and Starovatskaya S N 2013 Flux *Cored Wire*patent of the Russian Federation No. 2013121468/02
[14] Kozyrev N A, Igushev V F,Titov D A and Kozyreva O 2013 *Flux Cored Wire*patent of the Russian Federation No. 2013128792/02 (042890)
[15] Shurupov V M and Kozyrev N A 2011 *Training Device of Arc Welding patent of the Russian Federation* No. 2011104516/12 (006345)

On the use of slag from silicomanganese production for welding flux manufacturing

N A Kozyrev, R E Kryukov, U I Lipatova and O E Kozyreva

Institute of Metallurgy and Materials Science, Siberian State Industrial University, 42 Kirova Street, Novokuznetsk, 654007, Russia

E-mail: kozyrev_na@mtsp.sibsiu.ru

Abstract. The technologies for manufacturing of welding fluxes with the use of slag from silicomanganese production and dust of gas purification from aluminum production are developed. The new compositions and production technology of welding fluxes are offered. The comparative evaluation of the new compositions and widely used AN-348 flux is provided. It is shown that the quality of submerged arc welding with the use of the developed flux composition is significantly better than the submerged arc welding with AN-348 flux. The effect of fractional composition on high-quality performance of the weld is investigated. The macro- and microstructures, nonmetallic inclusions and the mechanical properties of the weld are examined. It is shown that the introduction of carbon-fluorine containing additive into the flux, based on the dust of gas purification from aluminum production, can significantly improve the whole complex of mechanical properties of the weld, especially characteristics of impact hardness at low temperatures. The conducted research served as a basis for development of submerged arc welding technologies protected by the patents of the Russian Federation.

1. Introduction

Much attention is paid to development of new fluxes and additives to them. A special role is played by the study of the chemical composition influence on the mechanical properties of the weld, on the level of gas saturation and contamination of weld with nonmetallic inclusions of endogenous and exogenous type [1-3]. The existing fluxes for welding of low alloyed steels for weld fused fluxes according to the State Standards (GOST) 9087-81, GOST 52222-2004 and for ceramic fluxes in accordance with GOST 28555-90 provide with the high content of total oxygen and, consequently, an increased amount of non-metallic inclusions in the joints. Today, reduction of non-metallic inclusions in welds during automatic submerged arc welding is achieved by using low oxidizing fluxes. However, such fluxes have poor welding characteristics and are rarely used in welding of low alloyed steels. Moreover, the use of oxidation fluxes leads to weld metal saturation with oxygen due to silica- and manganese-reducing processes.

Currently, the applied alloying systems [4] in welding due to silica- and manganese-reducing processes (1 – low carbon electrode wire and high-manganese flux with high silica content; 2 – low-carbon wire and high-silicon (acid) flux; 3 – medium manganese electrode wire and medium manganese acid flux) have several disadvantages. The use of silicon and manganese as deoxidizers results in the formation of various oxide inclusions, which, due to the transience of the welding process, does not always manage to surface and assimilate by slag, so the concentration of free and bound oxygen in the weld metal is quite high, and the decrease in the values of impact strength of the weld metal especially at low temperatures is observed. To eliminate weld metal contamination and improve the mechanical properties we proposed the use of carbon-fluorine containing additive of grade FD-UFS [5-8], which allows us to carry out deoxidation of the weld with carbon and significantly reduce the contamination level with non-metallic oxide inclusions.

Another significant disadvantage of fused fluxes is their high cost due to the use of expensive natural materials and the costs associated with the preparation of the charge to be melted and flux smelting in special units.

To reduce costs of welding fluxes the waste from metallurgical production, particularly, the slag from silicomanganese production is used. Analysis of the published data shows that in the production of silicomanganese alloys the dumping ladle slags are generated, having the chemical composition suitable for welding fluxes manufacturing. Thus, according to [9] the slag contains: 14–16% MnO; 45–60% SiO2; 7–8% Al2O3; 12–15% CaO; 3–4% MgO with ratio CaO/SiO2=0,52–0,58. From [10]: 47–49% SiO2; 18–20% MnO; 12.2–14% CaO; 7–8% Al2O3; 2.9–3.1% MgO. From [11]: 6.2–8.5% MnO; 45–47% SiO2; 18–23% CaO; 9.2–11.6% Al2O3; 7.6–12.1% MgO; 0.3–0.7% FeO; ≤3% C. In work [12]: 47–49% SiO2; 18–20% MnO; 12.5–14% CaO; 7–8% Al2O3; 2.9–3.1% MgO. In [13]: 3.2–4.5% MnO; 43–47% SiO2; 22–30% CaO; 12–16% Al2O3; 6–10% MgO; 0.3–0.7% FeO; ≈3.5% C. It should also be noted that for carbothermic production of silicomanganese slag may contain up to 20.3% of manganese from charge material [13].

Today in Russia for welding and surfacing of low-alloyed steels the fused fluxes AN-348 produced in Ukraine are widely used [14, 15], therefore, the attention to the issue of import substitution is growing. Based on this fact we considered the possibility of using the slag from silicomanganese production manufacturing flux for welding and surfacing of low-alloyed steels.

2. Experimental research and results

The study was conducted using the flux made of ladle slag from silicon-manganese production smelted in the ore-thermal furnaces by carbothermic method of continuous process, and dust of gas purification from aluminum production (Table 1). Dust of gas purification was mixed with the liquid glass and used for manufacturing of carbon-fluorine containing additive PD-UVC according to the technology protected by the patents of the Russian Federation [16, 17].

Table 1. Chemical composition of the components.

Component	MnO	SiO_2	CaO	MgO	Al_2O_3	FeO	Na_2O	K_2O	F	S	P	C
slag from silicomanganese production	8.01	46.46	22.85	6.48	9.62	0.38	0.36	0.62	0.76	0.17	0.01	
dust of gas purification from aluminum production	0.6	2.33	2.1	0.8	43.27	2.1	10.6	0.8	23.6	0.38	0.10	12.5

Testing of welding modes was performed by ASAW-1250 welding tractor with the use of welding wire Sv-08GA, diameter 4 mm. Butt submerged arc welding of plates 500×75 mm of 16 mm thick was conducted by double-sided welding without bevel. Welding mode: Iw = 700 A; Ua = 30; Vw = 35 m/h, surfacing mode: Iw = 410 A; Ua = 27; Vw = 30 m/h. In the study of the influence of fractional composition on the quality of the weld bead, the following fractions of the ladle slag from silicomanganese production were used: less than 0.45 mm; 0.45-2.5 mm; 2.5-5 mm; 5-10 mm.

The experiments determined that fractions of 2.5-5 mm, 5-10 mm do not provide a qualitative weld (high porosity of weld and slag inclusions), a fraction less than 0.45 mm is associated with obtaining of some "cratering" of the surface, the optimal fraction is 0.45-2.5 mm (Figure 1)

In the future the flux manufacturing was conducted by mixing of fractions 0.45-2.5 mm of ladle slag from silicomanganese production with different percentages of carbon-fluorine containing additive (1st sample – without FD-UFS and from 2nd to 5th samples, 2, 4, 6, 8% FD-UFS respectively,). The chemical analysis of welds is presented in Table 2.

Table 2. Chemical composition of welds.

Sample	C	Si	Mn	Cr	Ni	Cu	Al	S	P
1	0.09	0.61	1.41	0.02	0.06	0.09	0.022	0.020	0.008
2	0.09	0.62	1.40	0.02	0.06	0.09	0.023	0.020	0.008
3	0.10	0.60	1.34	0.02	0.06	0.09	0.013	0.023	0.009
4	0.12	0.66	1.43	0.02	0.06	0.10	0.012	0.027	0.008
5	0.13	0.65	1.36	0.02	0.06	0.09	0.013	0.024	0.008

Metallographic analysis was performed using microscope OLYMPUS GX-51 in bright field at magnifications ranging from ×100 to ×1000 after etching in 4% solution of nitric acid. The grain size was determined in accordance with GOST 5639-82 at magnification ×100. Examination of samples for the presence of nonmetallic inclusions was carried out in accordance with GOST 1778-70. The polished surface was examined at ×100 magnification using microscope OLYMPUS GX 51.

a

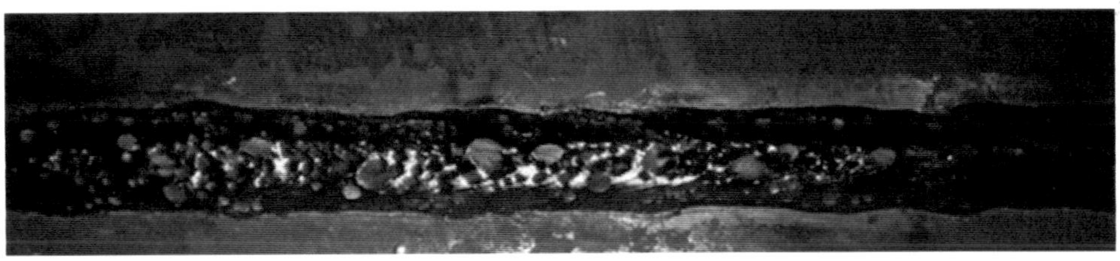

b

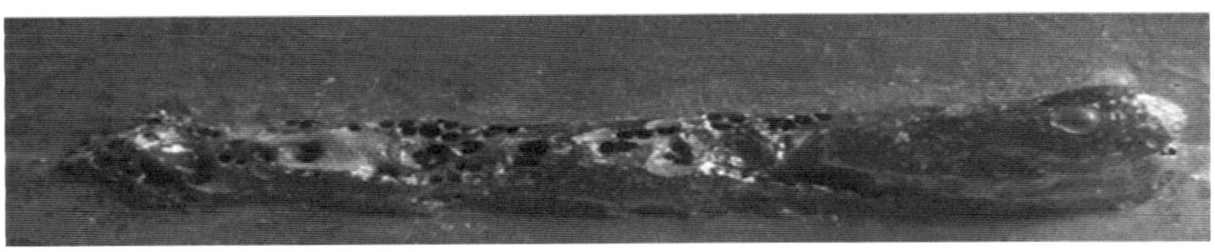

c

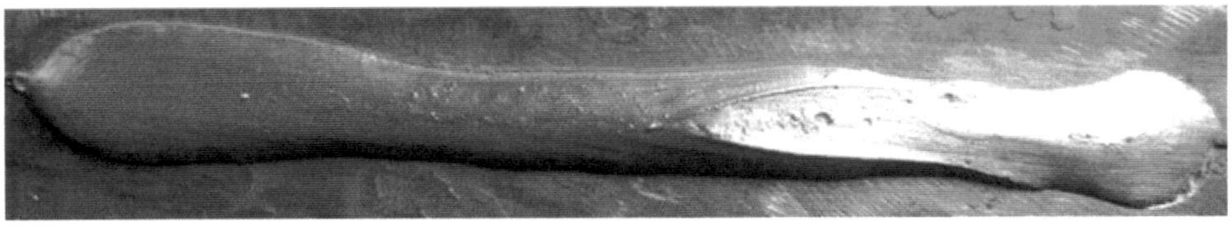

d

Figure 1. The quality of the weld surface (a – fraction 0.45-2.5 mm, b – 2.5-5 mm, c – 5-10 mm, d – less than 0.45 mm).

The resulting welds are shown in Figure 2. Metallographic analysis revealed that the welds of the investigated samples have a ferrite-pearlite structure of widemanstatten character with separate zones of ferrite of acicular structure. In some areas there is striation of ferrite-pearlite structure (Figure 3).

The grain size in the weld structure of sample No.1 according to the grain scale is No 4, 5. The grain size in the weld structure of sample No.2 and No.3 corresponds to 5, 6 and 7. The grain size for samples No.4 and No.5 corresponds to 5 6, 7. Thus, it was found that increased levels of FD-UFS content in the studied fluxes compositions contributes to grain refinement of the weld.

Studies of the nature of non-metallic inclusions showed that in the weld zone of samples there were non-deformable silicates, fragile silicates, point oxide and lineage oxides (Figure 4).

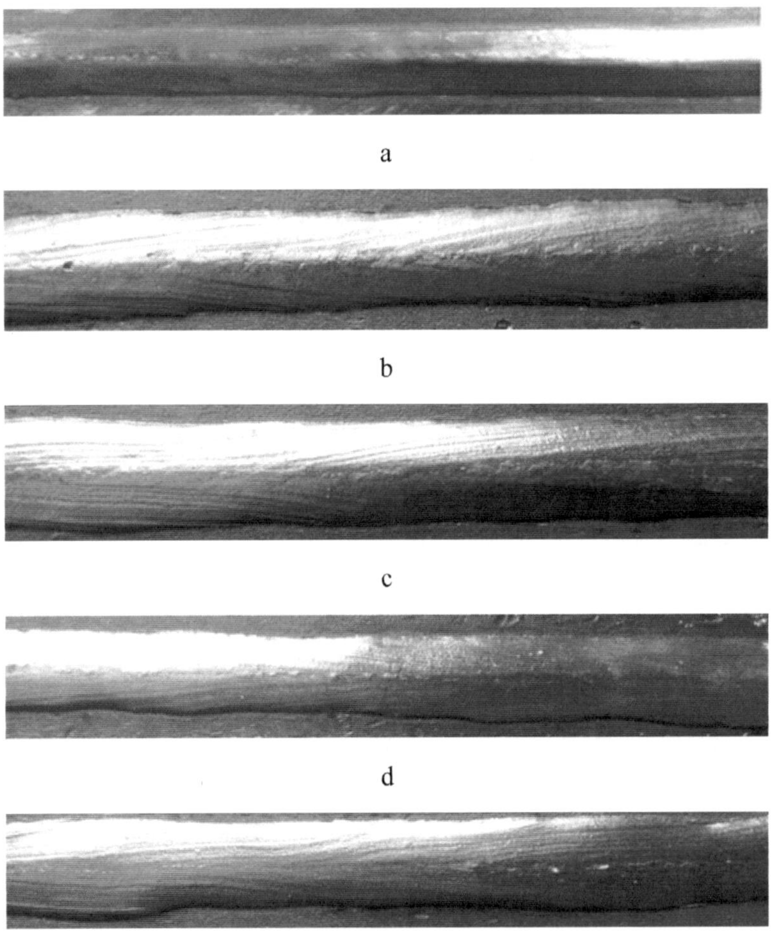

a

b

c

d

e

Figure 2. Beads of samples (a – sample No.1; b – sample No.2; c – sample No.3; d – sample No. 4; e – sample 5).

In the weld zone of sample No.1 the spot oxides were found of 1 a point, the non-deformable silicates, mainly of points 4 b and 3 b, and rarely of point 4 a, fragile silicates, less common, of point 3 b. In the weld zone of samples No.2 and No.3 there are non-deformable silicates of points 2 b, 4 b, and spot oxides of point 1 a.

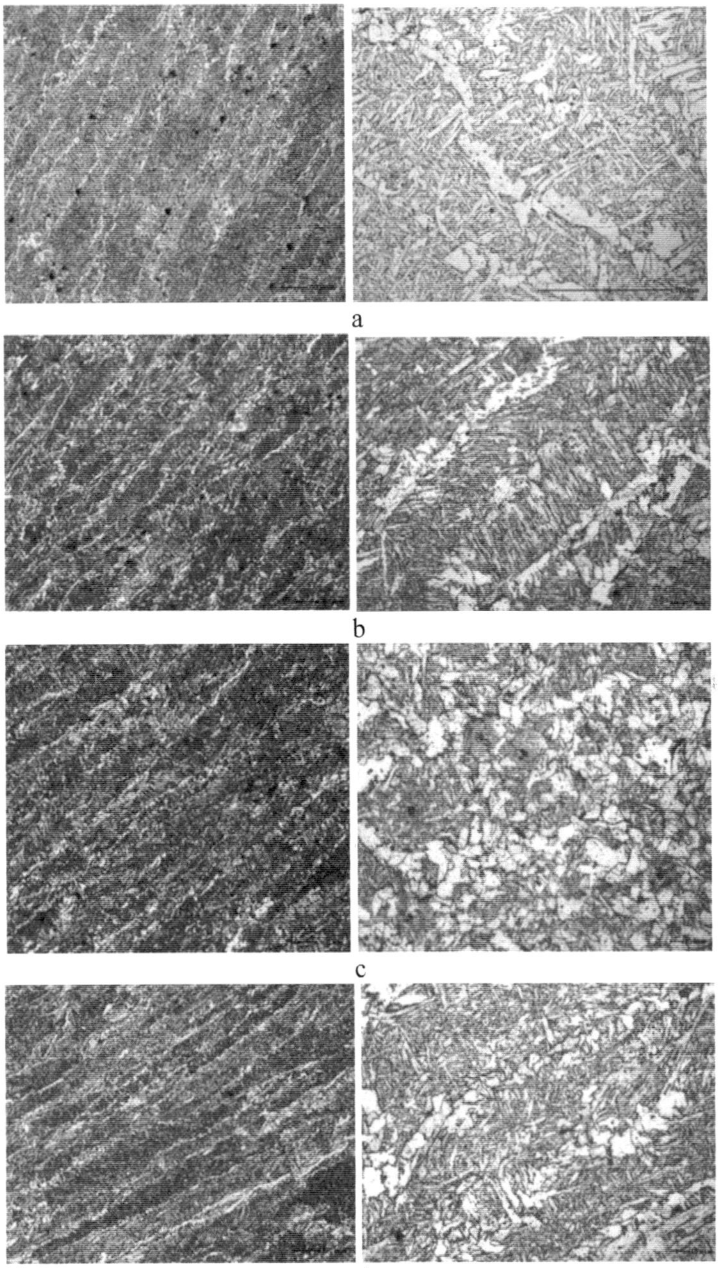

a

b

c

d

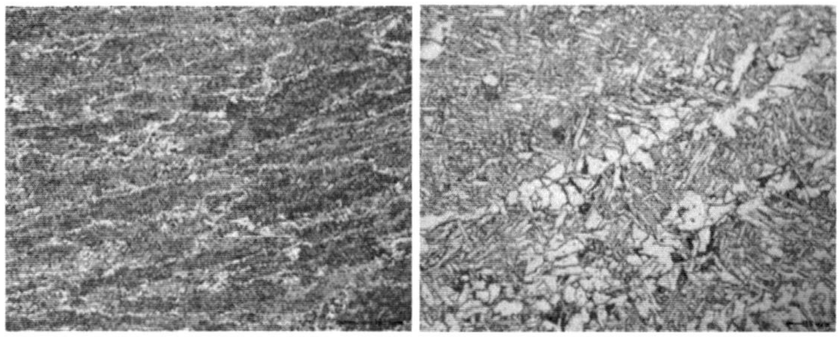

e

× 100 ×500

Figure 3. The weld microstructure of the examined samples (a – sample No. 1; b – sample No. 2; c – sample No. 3; d – sample 4; e – sample No. 5).

In the weld zone of sample No.4 non-deformable silicates of points 2 b, 1 b and spot oxides of point 1 a were found. In the weld zone of sample No.5 there are non-deformable silicates of point 2 b and spot oxides of point 1 a. Introduction of flux carbon-fluorine containing additive (PD UFS) into the flux reduces the contamination level of the weld with non-metallic inclusions and reduces their quantity and sizes (Figure 4). The results of mechanical tests are given in Table 3 and in Figures 5-8. The comparative evaluation of mechanical properties of the flux with the widely used flux AN-348 (GOST 9087-81) showed that the strength properties of the new flux is much higher, however, the value of impact strength at negative temperatures is not satisfactory, which is observed during use of flux AN-348. Introduction of additive FD-UFS into the flux in different ratios (1-5%) significantly increases values of impact strength KCV at a temperature −20 °C.

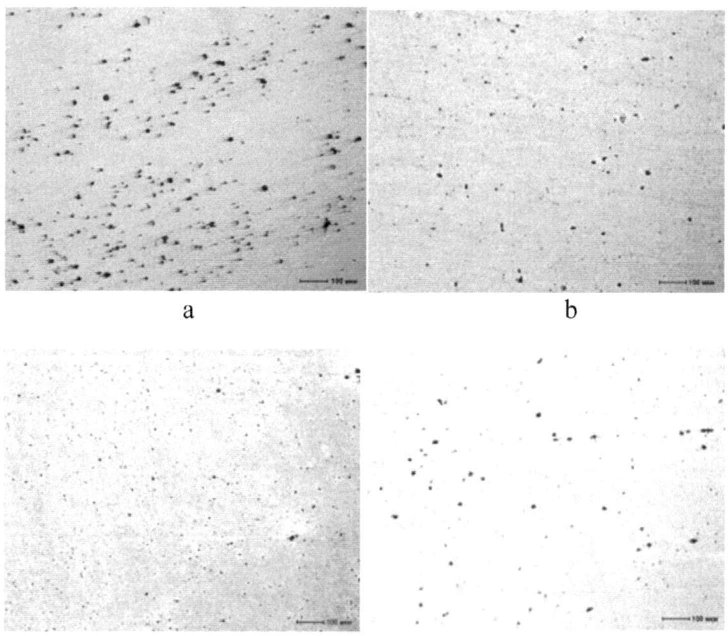

a b

c d

e

Figure 4. The nature of nonmetallic inclusions in the weld zone of samples.

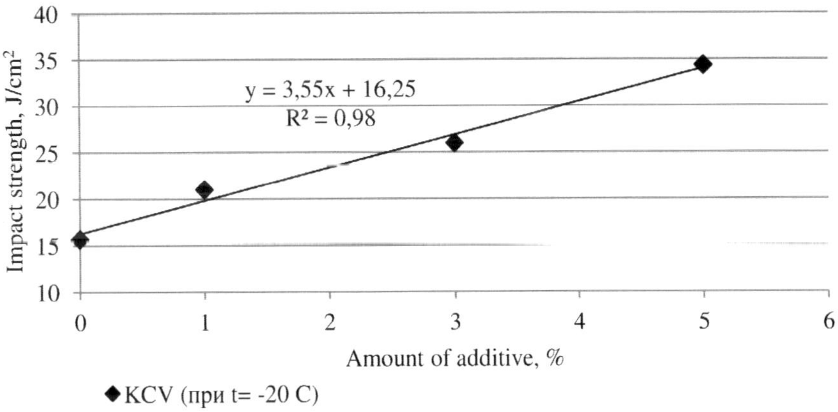

Figure 5. Change of impact strength of metal weld KCV T = −20 °C depending on the content of carbon- fluorine-containing additive in the flux.

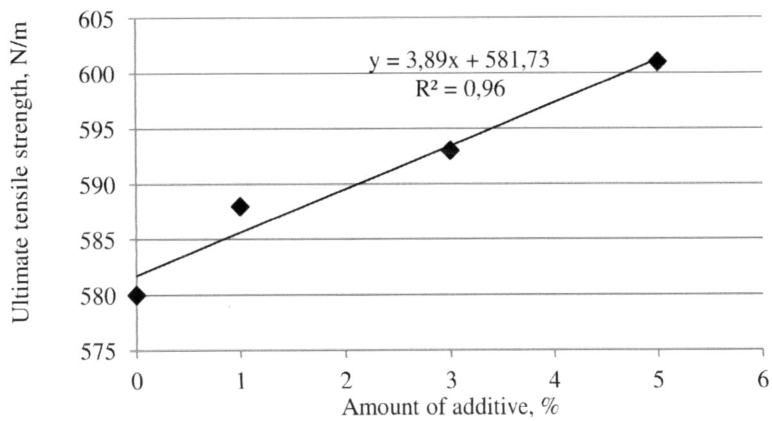

Figure 6. Change in ultimate tensile strength depending on the content of carbon- fluorine-containing additive in the flux.

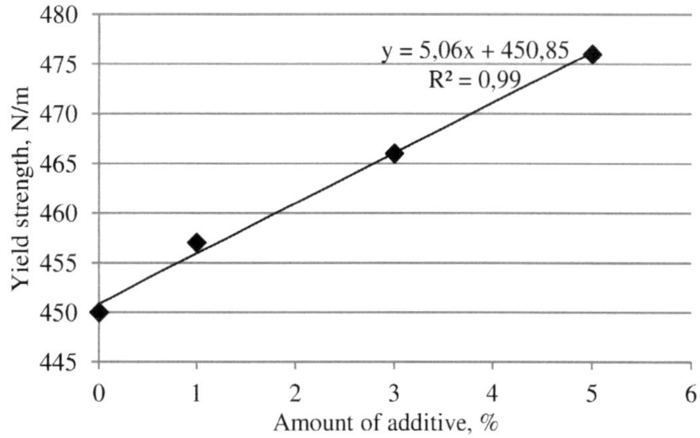

Figure 7. Change in yield strength depending on the content of carbon- fluorine-containing additive in the flux.

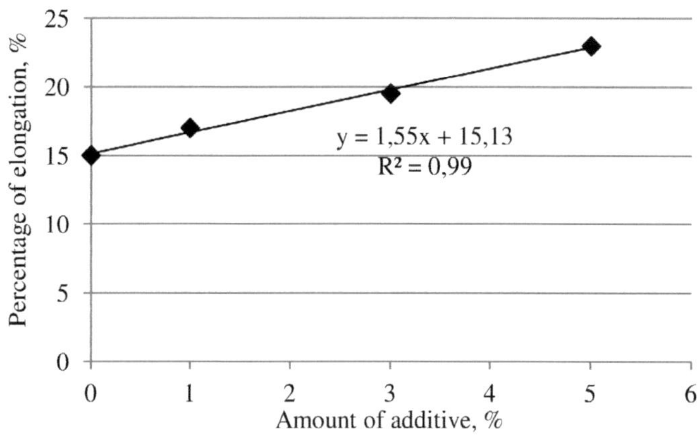

Figure 8. Change in percentage of elongation on the amount of carbon- fluorine-containing additive in the flux.

Table 3. Mechanical properties of welds.

Flux	Ultimate tensile strength, MPa	Yield strength,MPa	Percentage of elongation (δ), %	Impact strength KCV at T= - 20 °C, J/cm² (weld)
AN-348	535 530-543	360 355-368	25 24-26	18 16-21
Slag of silicomanganese	580 576-583	450 447-452	15 14-16	16 15-17
Slag of silicomanganese + 1 % FD-UFS	588 585-592	457 451-463	17 15,5-18	21 18-23
Slag of silicomanganese + 3 % FD-UFS	593 590-596	466 463-469	19,5 19-20	26 20-32
Slag of	601	476	23	35

silicomanganese + 5 % FD-UFS	597-604	472-479	22-24	26-42

*numerator – average values; denominator – minimum and maximum values.

3. Conclusion
As a result of laboratory experiments the conceptual possibility for use of slag from silicomanganese production in manufacturing of welding fluxes was shown. The obtained fluxes are patented in the Russian Federation patents [18, 19].

4. Acknowledgements
The work was performed in Siberian State Industrial University within the project of State Order of the Ministry of Education and Science of the Russian Federation No. 11.1531.2014/k. For tests, investigations, measuring we used equipment in the Collective Centre "Material Science" of Siberian State Industrial University.

5. References
[1] Golovko V V and Potapov N N 2011 *Welding International* **25** (11) 889–893
[2] Amado Cruz Crespoa, Rafael Quintana Puchola et al 2009 *Welding International* **23** (2) 120-131
[3] Volobueva Yu S, Volobueva O S et al 2012 *Welding International* **26** (8) 649–653
[4] Akulov A I, Belchuk G A and Demyantsevich V P 1977 *Technology and Equipment for Fusion Welding* (M.: Mashinostroyeniye) p 432
[5] Kozyrev N A, Igushev V F et al 2014 *Welding International* **28** (5) 403–405
[6] Kozyrev N A, Kryukov R E and Kozyreva O A 2015 *IOP Conference Series: Materials Science and Engineering* 91(2015)012008
[7] Kryukov R E, Kozyrev N A, Galevsky et al 2015 *IOP Conference Series: Materials Science and Engineering* 91(2015)012008
[8] Kozyrev N.A., Kryukov N.E et al 2015 *IOP Conference Series: Materials Science and Egineering* 91(2015)012008
[9] Povolitsky D Ya, Roshchin V E and Malkov N V 1995 *Electrometallurgy of Steel and Ferroalloys* (M.: Metallurgiya) p 592
[10] Gasik M I, Lyakishev N P and Emlin B I 1988 *Theory and technology of ferroalloys production* (M.: Metallurgiya) p 784
[11] Povolotsky D Ya, Roshchin B E et al 1984 *Electrometallurgy of Steel and Ferroalloys* (M.: Metallurgiya) p 586
[12] Gasik M I and Lyakishev N P 1999 *Theory and Electrometallurgy Technology of Ferroalloys* (M.: SP Intermet Inzhiniring) p 764
[13] Ryss M A 1985 *Ferroalloys Production* (M.: Metallurgiya) p 344
[14] Potapov N N, Konishchev B P et al 1989 *Shielding Gases and Welding Fluxes: Handbook* (M.: Mashinostroyeniye) vol 1 p 544
[15] Podgaetsky V V and Rabkin D M 1954 *Fluxes for Automatic and Semi-automatic Welding* (Kiev: AN USSR) 56
[16] Kozyrev N A, Igushev V F et al 2012 *Welding Flux* pat. of the Russian Federation No. 2012108414
[17] Kozyrev N A and Kryukov N A 2013 *Flux Additive* pat. of the Russian Federation No. 2013144914
[18] Krykov N E, Kryukov E N, Kozyrev N A et al 2014 *Welding Flux* pat. of the Russian Federation No. 2014122996
[19] Krykov N E, Kryukov E N, Kozyrev N A et al 2014 Flux *for Mechanized Welding and Surfacing of Steel* pat. of the Russian Federation No. 2014123002

A study on the properties of the deposited metal by flux cored wires 40GMFR and 40H3G2MF

A I Gusev, N V Kibko, N A Kozyrev, M V Popova, and I V Osetkovsky

Institute of Metallurgy and Material Science, Siberian State Industrial University, 42 Kirova Street, Novokuznetsk, 654007, Russia

Email: kozyrev_na@mtsp.sibsiu.ru

Abstract. The effect of chemical composition on the microstructure, hardness and wear resistance of the deposited layer of steels 40GMFR and 40H3G2MF was studied. The influence of the content of alloying elements in the examined steels on the structure of the deposited layer was defined, as well the influence of the structure of this layer on the hardness and abrasive wear. A comparative analysis of two steels was performed. Steel 40H3G2MF showed a greater efficacy for surfacing. It had higher hardness, a lower wear rate and a more fine-grained structure.

1. Introduction

Currently, to protect the hoppers and chutes used for transportation of mined rock and other abrasive materials, the fused wires are used corresponding to the systems C-Si-Mn-V-P and C-Si-Mn-Cr-Mo-V, which are represented by steels 40GMFR and 40H3G2MF [1-11].

2. Experimental research and results

The aim of this study was to investigate the influence of chemical composition on the wear resistance and hardness of the deposited layer. The wire was produced according to the known method [12-15] in the laboratory machine for manufacturing of flux cored wire (FCW). Diameter of the wire was 5 mm, the shell was made of steel strip St3. The respective powder materials were used as fillers, the carbon- and fluorine-containing dust of gas purification from aluminum production with the following chemical composition was used as carbon, mass %: Al_2O_3 = 21-46; F = 18-27; Na_2O = 8-15; K_2O = 0.4 – 6%; CaO = 0.7-2.3; SiO2 = 0.5-2.5; Fe2O3 = 2.1-3.3; C_{tot} = 12.5-30.2; MnO = 0.07-0.9; MgO = 0.06-0.9; S = 0.09-0.19; P = 0.10-0.18.

The surfacing by the produced wire was performed by submerged arc welding with flux AN-26S on the plate of steel St3. The chemical composition of the test samples was defined by X-ray fluorescence method on spectrometer XRF-1800 and atomic emission spectrometer DFS – 71. The chemical composition of the deposited layers is given in Table 1: samples No.1 – 4 are surfaced by wire from steel 40H3G2MF, samples No.5 – 7 by wire from steel 40GMFR. The surfacing was performed using the welding tractor ASAW-1250 at surfacing mode: I = 450 A, U = 30 V, V = 10 m/h. Wear resistance tests were conducted on device 2070 SMT – 1. Tests were carried out at the modes: load 30 mA, frequency 20 rot/ min. The samples hardness was measured with a hardness tester MET-DU.

Table 1. Chemical composition of the deposited layers.

Sample No.	Mass fraction of elements, %								
	C	Mn	Si	Cr	V	Mo	Ni	Co	B
1	0.19	0.52	0.74	2.79	0.14	0.26	0.17	0	0.001
2	0.22	0.62	0.35	2.78	0.02	0.25	0.09	0.04	0.001
3	0.27	0.68	0.49	4.61	0.01	0.42	0.36	0.03	0.001
4	0.43	0.84	0.37	7.04	0.03	0.49	0.42	0.06	0.001
5	0.16	0.32	0.62	0.25	0.19	0.11	0	0	0.003
6	0.10	0.45	0.15	0.23	0.11	0.10	0.08	0	0.003
7	0.23	0.66	0.25	0.95	0.29	0.35	0.34	0	0.003

Metallographic studies of the microsections were performed using an optical microscope OLYMPUS GX-51 in the bright field magnifying in the range of ×100 - 1000 after etching in a 4% alcoholic solution of nitric acid. The fineness of the martensite structure was evaluated by comparing with the standards of the relevant scales and sizes of martensite needles in Table 6 of State Standards (GOST) 8233-56. Longitudinal study of samples of the deposited layer in the presence of non-metallic inclusions was carried out in accordance with GOST 1778-70 with magnification ×100. The grain size was determined in accordance with GOST 5639-82, magnification ×100.

Metallographic analysis showed that the structure of sample No.1 was a fine-needled martensite (point 3, 4) in the primary austenite grains (Figure 1a). Along the austenite grain boundaries the unclosed thin interlayers, presumably consisting of δ-ferrite, are located. The grain size of austenite on the gain scale corresponds to No.5 and No.4 (Table 2).

The examination of the nature of non-metallic inclusions of sample No.1 showed a significant degree of contamination, in particular with non-deformable silicates of points 5b and 4b and spot oxides of points 2a and 3a (Figure 2 a).

As a result of the deposited layer investigation it is shown that its hardness is 44 HRC and the wear rate – $3 \cdot 10^{-5}$ g/min (Table 3).

The structure of sample No.2 is fine-needle martensite (point 4) with zones of troostite inside the former austenite grains. The grain size of austenite corresponds to No.5 and 6 (Table 2).

The contamination degree of sample No.2 with nonmetallic inclusions is much lower compared to sample No.1. In sample No.2 non-deformable silicates are found, mainly of points 2a and 1b and rarely point 3a. Spot oxides are mainly of point 2a and rarely points 1a and 3a (Figure 2b). Thus, in this case a smaller amount of non-metallic inclusions are present, and they have a more dispersed structure.

However, the appearance of troostite in the structure is accompanied by a slight increase in the wear rate of the deposited layer to values $4 \cdot 10^{-5}$ g/min and a decrease in its hardness to 36 HRC (Table 3).

Table 2. Characteristics of nonmetallic inclusions and samples structure.

Sample No.	Contamination with non-metallic inclusions, point			Austenite grain size, point
	Non-deformable silicates	Spot oxides	Line oxides	
1	4 b, 5 b	2 a, 3 a	–	5, 4
2	2 a, 1 b, 3 a	2 a, 1a, 3 a	–	5, 6
3	2 b, 3 b, 4 b	1 a, 2 a	–	6, 7
4	1 a, 1 b	1 a	–	8, 7
5	3 b, 5 b	1 a	–	5, 6
6	1 a, 1 b, 2 b, 3 a	1 a	–	5, 6
7	2 a, 3 b, 4 b	1 a, 2 a	–	6, 7

Table 3. Results of the durability test.

Sample No.	Sample weight at the beginning of test, g	Sample weight at the end of test, g	Loss of weight, (g/%)	Quantity of revolutions	Wear rate g/min	Hardness HRC
1	132.768	132.573	0.195/ 0.26	6214	0.00003	44
2	185.2525	185.036	0.216/0.117	4954	0.00004	36
3	152.968	152.769	0.199/0.13	4790	0.00004	48
4	194.091	193.975	0.116/0.06	4671	0.00002	56
5	130.281	129.398	0.883/ 1.15	5760	0.00015	37
6	155.142	154.58	0.562/0.36	5133	0.00010	28
7	100.89	100.319	0.571/0.56	4590	0.00012	44

It is found that the increased content of carbon from 0.19 to 0.27% and manganese from 0.52 to 0.68% with the simultaneous significant increase of chromium, nickel and molybdenum (sample No.3) provides medium-needle (point 6) and coarse-needle (generally point 7 and rarely 8) martensite in the former austenite grains. The grain size of austenite in the structure of sample No.3 of grain size scale corresponds to No.6 and No.7 (Figure 1c, Table 2). Martensite is of the form of dark needles with regular boundaries; the martensite needle size is 6-19 μm. Along the primary austenite grain boundaries thin layers are located, consisting of δ-ferrite, intermetallides of σ-phase and dispersed carbides (Figure 1c).

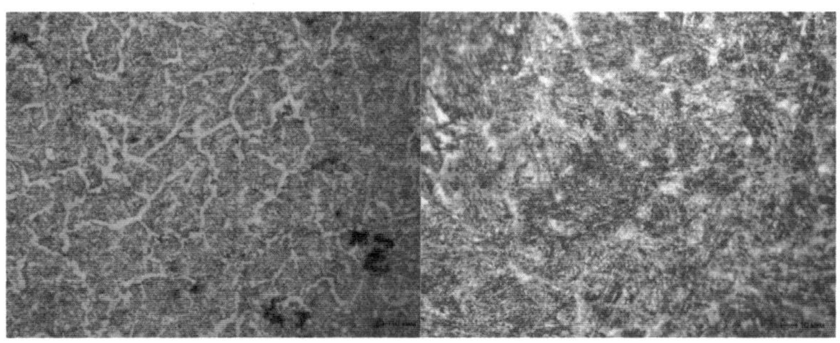

a b

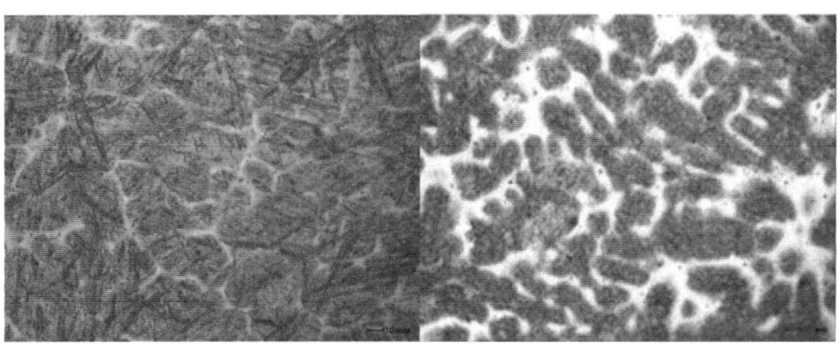

c d

Figure 1. Structure of steel samples 40H3G2MF (a – sample No.1; b – sample No.2. c – sample No.3; d – sample No.4).

The contamination degree with nonmetallic inclusions of sample No.3 is lower than of sample No.1. In this case there are non-deforming silicates of points 2b and 3b and rarely of points 4a and 4b, spot oxides are mainly of point 1a and rarely 2a (Figure 2a).

Increase in the carbon content and alloying elements in the deposited layer provides the austenite grain refinement as compared with sample and No.1 No.2, and also leads to disperse carbides, increasing the hardness values up to 48 HRC, but practically does not influence on abrasive wear. The wear rate of sample No.3 is $4 \cdot 10^{-5}$ g/min (Table 3).

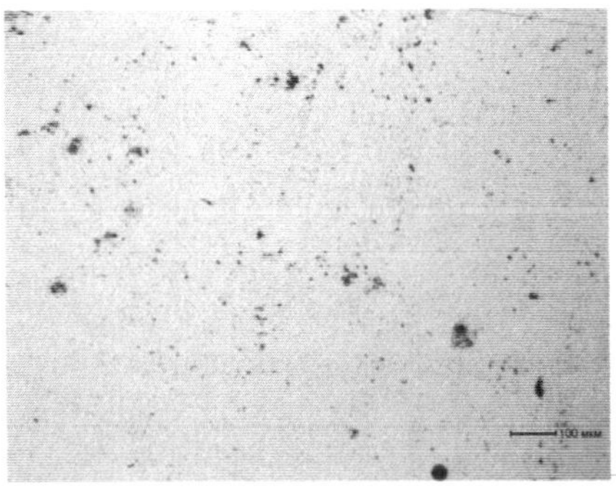

a

b

c

d

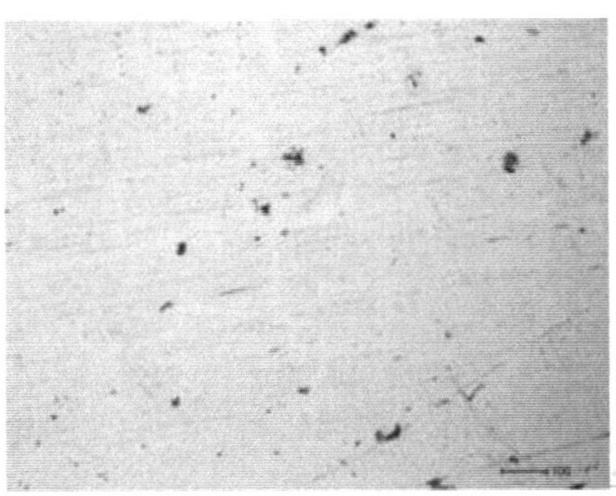

e

f

g

Figure 2. The nature of non-metallic inclusions in the samples (a – sample No.1; b – sample No.2; c – sample No.3; d – sample No.4; e – sample No.5; f – sample No.6. g – sample No.7).

The chemical composition of the deposited layer of sample No.4 has a higher content of carbon and alloying elements in particular 7% of chromium. Metallographic analysis revealed that the primary austenite grain boundaries are clearly marked, fine-needle martensite is formed inside them (point No.3, 4). Between the austenite grains there are extended zones consisting of δ-ferrite, type intermetallids of σ-phase and dispersed carbides. In this case martensite has a more disperse structure (Figure 1 d).

The comparative analysis of the austenite gain size showed that among the rest examined samples welded with FCW made from steel 40H3G2MF sample No.4 has a smaller size of the primary austenite grain. The grain size of austenite in the structure of sample No.4 on a grain size scale corresponds to No.8 and No.7 (Table 2).

The contamination degree with non-metallic inclusions in sample No.4 is the lowest in comparison with other samples: there are non-deformable silicates of points 1a and 1b and spot oxides of point 1a (Figure 2d, Table 2).

Formation of a fine-needle martensite structure and a significant reduction in the degree of contamination with non-metallic inclusions provides higher hardness values up to 56 HRC and reduces wear rate of the deposited layer to $2\cdot10^{-5}$ g/min compared to other samples of steel 40H3G2MF (Table 3).

The chemical composition of sample No.5-7 welded with steel 40GMFR is characterized by a lower carbon content (0.10-0.28%) and alloying elements than of samples No.1-4 welded with steel 40H3G2MF. As a result after welding a low carbon martensite is formed; this causes reduction in hardness and especially in wear resistance (Table 3).

Metallographic analysis showed that the structure of sample No.5 from steel 40GMFR, as well as sample No.1, contains fine-needle (point 4) martensite in the former austenite grains. Along its boundaries thin interlayers of δ-ferrite are located (Figure 3a). The austenite grain size on the graininess scale corresponds to No.5 and No.6 (Table 2).

Non-deformable silicates of points 3b and 5b and spot oxides of point 1a (Figure 2 d) are found in the result of contamination assessment of sample No.5 with nonmetallic inclusions.

Thus, the comparative analysis of samples No.1 and No.5 showed that the latter has a more fine-grained structure and is less contaminated with non-metallic inclusions. However, the structure characteristics of the deposited layer do not provide reduction in abrasive wear. The wear rate of the deposited layer in this case is considerably higher ($15\cdot10^{-5}$ g/min) compared with sample No.1 ($3\cdot10^{-5}$ g/min) (Table 3). Besides, the hardness of the deposited steel 40GMFR (sample No.5) is lower compared to steel 40H3G2MF (sample No.1).

It is found that the structure of sample No.6, with a higher content of manganese and lower quantity of other alloying elements in comparison with sample No.5, is a medium-needle martensite of points No. 5, 6 in the former austenite grains. The size of the martensite needles is 4-12 μm. There are thin interlayers of δ-ferrite (Figure 3b) along the austenite grain boundaries. Austenite grain size corresponds to No.5 and 6 (Table 2).

The contamination degree of sample No.6 with non-metallic inclusions is lower than of sample No.5. In this case non-deformable silicates of points 1a and 1b and rarely 2b, 3a and 3b, spot oxides of 1a point are detected (Figure 2f). Probably such characteristics of non-metallic inclusions influence the reduction of the wear rate of the deposited layer up to $10 \cdot 10^{-5}$ g/min, but also its hardness goes down to 28 HRC (Table 3).

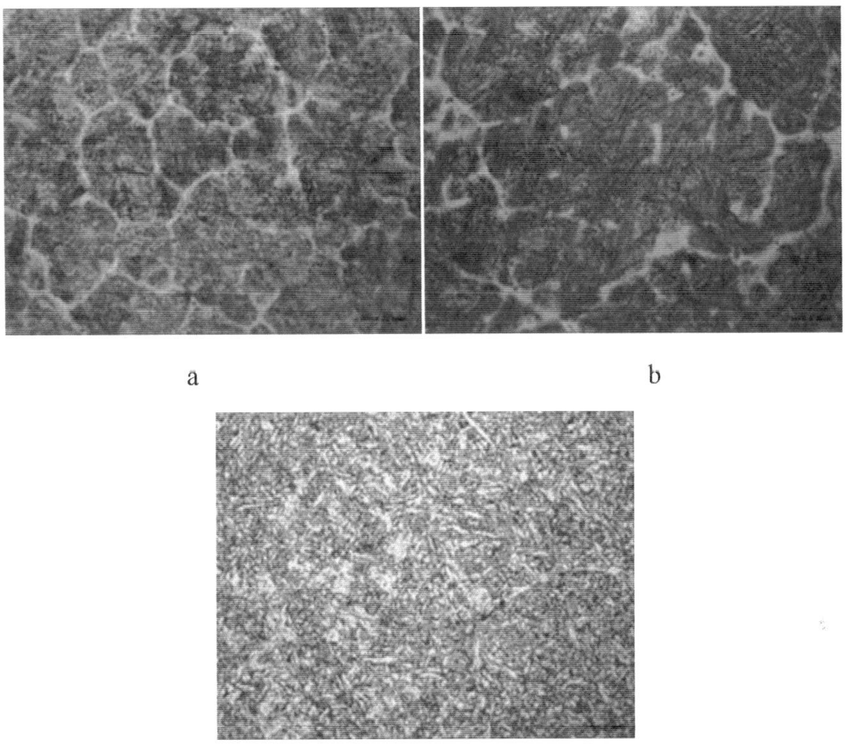

a b

c

Figure 3. Structure of the samples made from steel 40GMFR (a – sample No.5; b – sample No.6, c – sample No.7).

It has been established that a significant increase in the content of carbon, manganese, chromium, vanadium and molybdenum in the composition of the deposited steel provides a structure consisting of fine-needle (point 4) and medium-needle (point 5) martensite. The austenite grain size on the grain size scale corresponds to No. 6, 7 (Figure 3c).

The contamination degree of sample No.7 with nonmetallic inclusions, in particular with non-deformable silicates is lower in comparison with sample No.5: non-deformable silicates are found mainly of points 2b and 3b and rarely of point 4 b; spot oxides mainly of point 1a and rarely 2a (Figure 2 g).

This improvement of the structure decreases wear rate of the deposited layer to values $12 \cdot 10^{-5}$ g/min and increases its hardness values up to 44 HRC (Table 3).

The comparative analysis revealed that the size of the primary austenite grain in the structure of samples welded with FCW made from steel 40H3G2MF ranges from No.4 to No.8, and of samples welded by steel 40GMFR – from No.3 to No.7. The average hardness value of the deposited layer from a more alloyed steel 40H3G2MF and wear rate is 46 HRC and $3\cdot10^{-5}$ g/min. respectively for the samples surfaced by wire from steel 40GMFR – HRC and $10\cdot10^{-5}$ 10 g/min.

Thus, it is shown that the samples surfaced by wire produced from wire of steel 40H3G2MF have a smaller austenite grain size, lower wear rate and higher values of hardness in comparison with the samples from steel 40GMFR.

3. Conclusions

Increase in the content of alloying elements in the surfaced steels 40H3G2MF and 40GMFR provides refinement of austenite grain, which is accompanied by the increase in the hardness of the deposited layer to values of 56 and 44 HRC respectively, and reduction in abrasive wear to a value of 2 and $10\cdot10^{-5}$ g/min. respectively

For the surfacing of bunkers and trenches it is preferable to use FCW from steel 40H3G2MF in comparison to steel 40GMFR. The chosen welding technology provides a specific content and ratio of alloying elements in the steel, which forms a stable dispersed martensitic structure with a low volume fraction of δ-ferrite ensuring high hardness and resistance of the deposited metal to wear.

4. References

[1] Paton B E 1974 *The Technology of Electric Fusion Welding of Metals and Alloys Melting* (M.: Metallurgiya) p 768
[2] Livshits L S et al 1969 *Fundamentals of Alloying of Deposited Metal* (M: Metallurgiya) p 188.
[3] Tolstov I A and Korotkov V A 1990 *Handbook on Surfacing* (Chelyabinsk: Metallurgiya) p 384
[4] Kashchenko G A 1959 *Fundamentals of Materials Science* (Leningrad: Mashinostroyeniye) p 396
[5] Geller Yu A and Rakhshtadt A G 1973 *Materials Science* (M: Metallurgiya) p 448
[6] Gulyaev A P 1977 *Metal Science* (M: Metallurgiya) p 648
[7] Grabin V F 1982 *Metallurgical Fusion Welding* (Kiev: Naukova Dumka) p 416
[8] Shekhter S Ya and Reznitsky A M 1982 *Surfacing of Metals* (M: Mashinostroyeniye) p 71
[9] Barmin L N and Gusev V P 1982 *Modern Methods of Welding and Their Application* (Kiev: IES n.a. E O Paton) pp 40–47
[10] Khasui A 1995 *Surfacing and Spraying* (M: Mashinostroyeniye) p 240
[11] Konyshchev B P et al 1998 *Welding Consumables for Arc Welding* (M.: Mashinostroyeniye) p 544
[12] Patsekin V P and Rakhimov K Z 1979 *Production of Cored Wire* (M.: Metallurgiya) p 80
[13] Esipov V D and Michurin I K 2002 *Welding production* **2** 32–34
[14] Pokhodnya I K 1980 *Production of Cored Wire* (Kiev: Vysshaya Shkola)
[15] Shtokolov S A and Moysov L P 2002 *Welding Production* **7** 39–41

A study on the production of titanium carbide nano-powder in the nanostate and its properties

L S Shiryaeva, S V Rudneva, G V Galevsky and A K Garbuzova

Institute of Metallurgy and Material Science, Siberian State Industrial University, 42 Kirova Street, Novokuznetsk, 654007, Russia

E-mail: kafcmet@sibsiu.ru

Abstract. The plasma synthesis of titanium carbide nano-powder in the conditions close to industrial was studied. Titanium carbide TiC is a wear- and corrosion-resistant, hard, chemically inert material, demanded in various fields for the production of hard alloys, metal-ceramic tools, heat-resistant products, protective metal coatings. New perspectives for application titanium carbide in the nanostate can be found in the field of alloys modification with different composition and destination.

1. Introduction

The developed in the 80s of the 20th century the technology of plasma synthesis of titanium carbide is based on a carbothermic reduction of titanium oxide in the gas phase and is implemented using a three-jet once-through reactor. For plasma flow generation three electric arc plasmatrons EDP-104A are used with the total capacity up to 50 kW. Implementation of the proposed plasma-metallurgical technology provides titanium carbide production in the form nano-powders with nano-particles size 50-70 nm. Along with the advantages there are following disadvantages in this technology: technological and economic inefficiency of using technical propane-butane mixture, requiring for the processing a complex in composition and generation the nitrogen-ammonia-hydrogen plasma, and plasma-metallurgical reactor of the laboratory level.

In this regard, the aim of this work is to study the plasma synthesis of titanium carbide nano-powder in the conditions close to the industrial. In order to achieve this goal the following tasks need to be solved: analysis of the current state of the production and use of titanium carbide; characterization of the three-jet once-through reactor; model-mathematical study on interaction of plasma and raw materials flows; prediction of the basic technological parameters of plasma-metallurgical production of titanium carbide, its physical and chemical attestation and definition of economic indicators.

2. Experimental research and results

To achieve these objectives the three-jet once-through plasma reactor of industrial level was used with power (150 kW), for which the heating engineering, resource and technological characteristics were investigated. The average weight temperature of the plasma flow at the length of the reactor 12 calibres ranges within (5400÷1750) K for a non-lined channel and (5500÷1900) K with its insulation by the cylinder from zirconium dioxide 0.005 m thick; the temperature of the inner surface varies within (942÷490) K and (855÷1549) K, respectively. The specific electrical power reaches 2142

MW/m3, which is much higher than this indicator for a traditional electro-thermal equipment (usually about 0.2 MW/m3). The estimated life of the electrodes is 3125 for a copper anode and 112 hours for a tungsten cathode. The predicted titanium carbide contamination with products of electrode erosion is 0.00012% of copper and 0.000019% of tungsten. On the aggregate characteristics the three-jet plasma reactor of 150 kW can be attributed to a highly efficient, reliable modern electro-thermal equipment.

The plasma synthesis is a metallurgical process difficult to study. This is due to its transience, high temperatures and low volume of carbide formation zone. In such conditions the model-mathematical approach is rather promising involving thermodynamic modeling of synthesis processes, modeling of the interaction of titanium-containing raw materials and gas-coolant, an experimental investigation with use of the planned experiment method. The methodology of this approach is described in detail in [1, 2]. During simulation all kinds of titanium-containing raw materials available were examined: chrome and its oxide TiO2. The natural gas, plasma gas – nitrogen were used as a hydrocarbon raw material.

The thermodynamic modeling of synthesis processes is carried out in order to predict the optimal parameters for producing titanium carbide (ratio of components and temperature), the definition of equilibrium parameters of the synthesis process (conversion degree of raw materials into carbide, composition of gaseous and condensed products), evaluation of the influence of gas-phase reactions on carbide formation, providing an efficient processing of the dispersed raw material in the conditions of plasma metallurgical technologies.

In connection with the use in the synthesis of titanium, oxide TiO_2, hydrocarbon raw material – methane and plasma-forming gas – nitrogen as a titanium-containing raw material the examination objects were the systems C-H-N, Ti-O-C-H-N, Ti-C- H-N. The necessary for the analysis compositions of gaseous and condensed products were calculated by the "constant" method, based on a joint solution of the equation of the mass action, material balance, total number of gas mixture moles, existence of a condensed phase, Dalton's law with the use of software modeling high-complex chemical equilibria "PLASMA" (ISSC SB RAS), which has a built-in data base for the reaction products for oxide-, boride-, carbide- and nitride-forming systems [3]. In the calculations the temperature range 1000-6000 K at a total pressure of 0.1 MPa in the system was considered. Thermodynamic analysis of the selected process variants showed that in the system C-H-N, simulating a high-temperature pyrolysis of methane, in the equilibrium conditions within the temperature range 2800-3800 K the main carbonaceous components of the gas phase are HCN, C_3H and C_2H. In the system Ti-O-C-H-N the titanium carbide formation is possible at a temperature below 3500 K at a stoichiometric and abundant amount of carbon. In the system Ti-C-H-N the formation of titanium carbide is possible in the equilibrium and quasi-equilibrium conditions. In equilibrium conditions when changing the ratio of Ti:C=0.25:(0.0625-0.25) the degree of titanium conversion into carbide varies from 0.25 to 1, and at 1.5-fold abundance of carbonizer carbide comprises 9.1% of free carbon. In the conditions of quasi-equlibrium carbon condensation does not occur. In both cases, titanium carbide formation is thermodynamically possible according to the gas phase reactions involving vapour of titanium and hydrogen cyanide, i.e. according to the scheme "vapour-crystal" that allows us to predict the possibility for reaching its high yield in real conditions of synthesis.

On the basis of model-mathematical research and the experimental data the prediction of technological parameters of titanium carbide production was performed (Table 1). The results of the model-mathematical and experimental studies have allowed us to predict the technological characteristics of the plasma metallurgical production of titanium carbide for various types of titanium-containing raw material and choose the best technology option (Table 2).

Table 1. Technological parameters of titanium carbide production.

Raw material	TiO_2	Ti
Power, kW	150	150
Consumption of plasma-forming gas, kg/s	$9 \cdot 10^{-3}$	$9 \cdot 10^{-3}$

Cooling water consumption, m^3/h	3.7	3.7
The initial temperature of the plasma flow, K	5400	5400
Quenching temperature, K	2800	2800
Mass consumption concentration, kg/kg	0.12	0.10

During calculations the losses of raw materials in the reactor in the amount of 4% and the losses of synthesis chemicals in the capture system 5% were considered.

Table 2. Technological parameters of plasma metallurgical production of titanium carbide.

Technological variant	Variant I (TiO_2 + natural gas)	Variant II (Ti + natural gas)
Conversion degree	0.94	0.94
Product yield, %	92.0	92.0
Productivity, kg/h	2.66	3.7
Intensity of production, $kg/m^3 \cdot h$	2956	4111
Consumption of titanium-containing raw material, kg/kg	1.46	0.87
Gas consumption – coolant gas, kg/kg	9.10	9.10
Natural gas consumption, kg/kg	0.26	0.26
Energy consumption, kW·h/kg	56.4	40.5

On the basis of model-mathematical research and the obtained experimental data the prediction of technological parameters of plasma metallurgical production of titanium carbide was carried out. The optimal technological variant is the process of titanium carbonization by natural gas. The titanium carbide production was carried out in the conditions of NPF "Polimet" on the basis of industrial complex with power 150 kW.

For generation of the plasma flow three electric arc heaters (plasmatrons) EDP-104A were used with power up to 50 kW each, installed in the mixing chamber at the angle of 30° to the reactor axis, were used. The mixing chamber includes a steel water-cooled body with three welded fittings to it with threaded connections for three water-cooled evenly spaced on the plasmatrons circumference and the flange for the camera to be docked with the reactor section and a device for the input of highly dispersed raw material [1]. For injection of finely dispersed raw material a water-cooled tuyere was installed. Its inner diameter was 0.004-0.008 m, so that its lower end was removed from the impact point of the plasma jets at a distance of 0.5-1.0 calibers. The mixing chamber and reactor sections were made of stainless steel. The change in the reactor length and injection of a cold gas (nitrogen) into the plasma flow through the quenching ring, installed at the reactor outlet, makes it possible to quench the synthesis products in different temperature zones. The quenching ring is a hollow metal disk with thickness 0.008 m. The equipment keeping the reactor functioning includes systems of electro-gas and water supply, test instrument, automation, controlling the composition of the plasma-forming and the exhaust gas from the reactor, dosing of the charge and capture of products. Features of a three-jet reactor are given in Table 3.

Table 3. Characteristics of the three-jet reactor.

Reactor characteristics	Values						
Arc power, kW	30	42	55	66	80	110	150
Consumption of plasma-forming gas (nitrogen), 10^4 kg/s	22.20 ÷ 29.60	28.50 ÷ 34.00	32.00 ÷ 40.20	38.10 ÷ 49.60	44.00 ÷ 58.20	57.15 ÷ 80.40	78.30 ÷ 120.50
Service life of plasmatrons*, h	120	110	110	110	110	100	100

*cathodic insertions from thoriated tungsten

Plasmatrons EDP-104A, the design of which is presented in [4, 5], operate on direct current with the following parameters of the electric arc: the arc voltage is up to 250V, current – up to 200 A. The stabilization of the electric arc – gas-swirl due to the tangential entry of the plasma-forming gas and methane through the swirl-insulator. Plasmatrons anodes are copper water-cooled with internal diameter 0.008 m with a practically unlimited service life in the presence of cooling. Plasmatron cathodes consists of copper water-cooled housings and cathode insertions of tungsten, pressed into the body up to its surface (to reduce the electron yield work) with a diameter of 0.003 m and with service life 100-120 hours. Ignition of the electric arc is carried out by breakdown of the gap between the cathode and the anode with high voltage pulse. The arc is drawn into the anode stabilized on the plasmatron axis by the swirled flow of nitrogen. The thermal efficiency of the plasmatrons with power of 50 kW at the minimum consumption of plasma forming gas is 50-52%. This design makes it possible to achieve the fallowing: to reduce the specific erosion of the electrodes and improve their operational life to commercially acceptable levels; apply a more efficient performance of the cathode and the swirl-insulator for the working and the protective gases; install plasmatrons in the mixing chamber of the three-jet once-through reactor without changing its geometric and thermal characteristics.

Power supply of the industrial plasma reactor is carried out by a thyristor of the converter unit series AT4-750/600 having a steeper volt-ampere characteristic and the following operating parameters: power 450 kW; rectified voltage, V-600; rectified current, A-750; efficiency at nominal speed, 96%; supply voltage, kV-6.

Technical nitrogen with an oxygen content of 0.5% is used as a plasma-forming gas. Natural gas with a methane content up to 94% of volume is used as carbonizer. For dosing of the powdered raw material the dosing device of a mixed electromechanical and gas-swirling type with intermittent action and removable cylinder – receiver of the powdered raw material constructed by the Institute of Solid State Chemistry and Mechanochemistry of the Siberian Branch of the Russian Academy of Sciences [6].

Titanium carbide capture system includes a settling chamber where up to 10% of nanopowder is captured and the process gas temperature decreases to 60°C, and two alternately working bag filters (trapping up to 85% of nanopowders) [7, 8]. The bag filter for trapping nanoscale powders, including a housing with a cooling jacket and with pipes for supplying gas powder mixture and the discharge of treated gases, trapping sleeve, regeneration system with the installed coaxially outside the sleeve chamber with radial gas-tight partitions and pipes for supplying the regenerating gas, collector for the nanodispersed powder with s slide gate device. With the nanopowder accumulation the gate device is periodically opens and nanopowder falls out into the collector. The filter cloth is a metal gauze from chromium-nickel steel of twill weave. The use of the bag filter of this design allows to separate titanium carbide from gases, enhance the degree of its trapping, prevent particle coarsening and their surface oxidation, reduce the volume of regenerating and cleaning gas, reduce the necessary filter surface. Plasma synthesis products were investigated by X-ray, chemical, electron microscopy, derivatographic analyses, as well as by BET to determine by their specific surface. Also methods of transmission (TEM) and scanning electron microscopy (SEM) were used.

The plasma synthesis product is the titanium carbide TiC. Titanium carbide content in the products of synthesis is 92.88-93.42%. The micrographs of titanium carbide are given in Figure 1. The nanopowder titanium carbide is represented by spherical aggregates ranging in size from 150 to 600 nm, formed by the particles of the cubic form within a wide size range – from 10 to 60 nm. The faceted form of titanium carbide particles indicates their formation according to the mechanism of "vapour – crystal", presumably during the interaction of titanium vapour and hydrogen cyanide.

The presence of aggregates of various volumes in the synthesis products indicates a high possibility for the further nanoparticles coarsening by their coagulation when the temperatures are lowered.

Carbide is accompanied by free carbon represented by spherical particles of sizes preferably 10-30 nm.

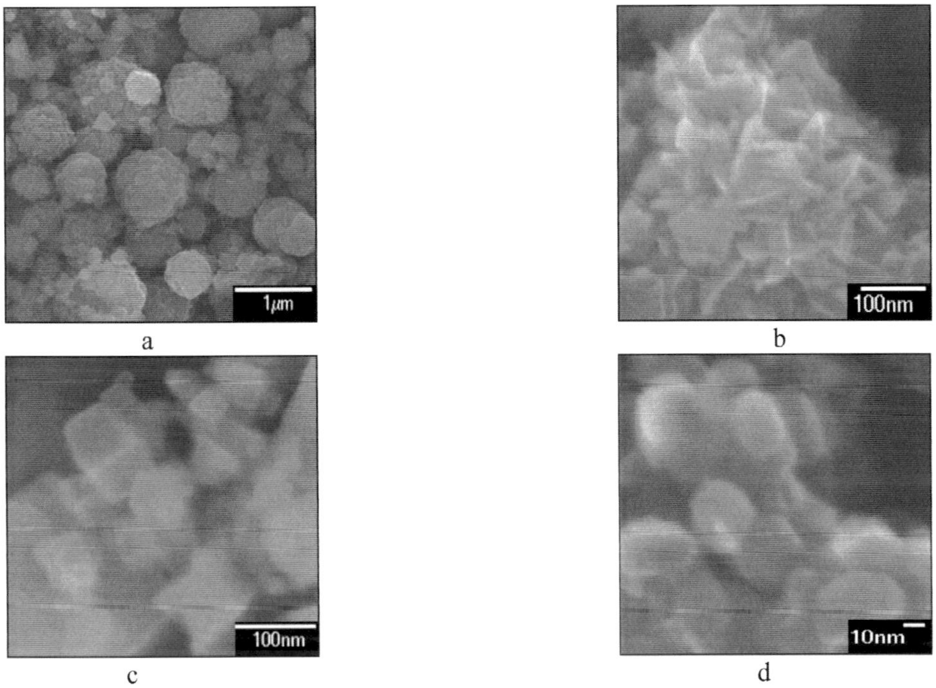

Figure 1. Micrographs of titanium carbide nanopowder – SEM: a – in the condition after synthesis; b – morphological pattern of the aggregate; c – ensemble of nanoparticles; d – a cube-shaped nanoparticle.

The optimal values of technological factors, and the main characteristics of titanium carbide correspond to the following: titanium powder fineness – -5 μm; quantity of number carbonizer, % from the stoichiometric – 120-140; the initial temperature of the plasma flow, K – not less than 5400; quenching temperature, K –

2600÷2800; titanium carbide yield, % mass – 92; productivity, kg/hr – 3.7; rate, kg/h·m^3 – 4111 (which is 30-50 times higher than the rate of traditional metallurgical productions); specific surface, m^2/kg – 33000-35000; particle size, nm 34-36; shape of the particles – spherical

The synthesis products contain titanium monocarbide with a face-centered cubic lattice with the parameter a=0.4323 nm, which is 0.0004 nm less than of the bulk crystals. This may be due to a non-equilibrium state in the particles smaller than 100 nm of near-surface layers, resulting in deformation (compression) of crystal lattices, atoms shift from ideal positions, the appearance of micro stresses [9, 10]. Microstress of the lattice, evaluated by the ration value Δa / a is $(0.92 \pm 0.10) \cdot 10\text{-}3$. Free pyrolyticcarbon is formed, presumably, during methane decomposition in the amorphous state and does not appear in the diffraction patterns.

Estimation of economic efficiency of the plasma metallurgical production of titanium carbide shows that at the sale price of 305 USD/kg it can be competitive on the world market of nanomaterials, where today the leading foreign manufacturers of carbide nano-powders established the price ranging within 500-2000 UDS/kg.

3. Conclusions

On the basis of the model-mathematical and technological research the optimal values of parameters for nano-titanium carbide synthesis and its physical and chemical characteristics are found. It is found that during carbidization of titanium powder by methane in the conditions of plasma nitrogen flow within the temperature range 5400-2600 K the product of plasma synthesis is titanium carbide TiC.

IOP Conf. Series: Materials Science and Engineering **150** (2016) 012034 doi:10.1088/1757-899X/150/1/012034

The phase and chemical compositions, dispersion, morphology and oxidation of synthesis products were investigated. The following main technical and economic parameters of the proposed technology as productivity, intensity, cost, and a sale price were defined.

4. References

[1] Nozdrin I V et al 2012 *Proc. of Higher Educ. Inst. Ferrous Metallurgy* **2** 13–18
[2] Nozdrin I V, Galevsky et al 2011 *Proc. 3rd All-Russian Conf.* (SibSIU) 60–64
[3] Nozdrin I V, Galevsky G V et al 2011 *Steel in Translation* **41** (10) 799–804
[4] Nozdrin I V, Shiryaeva L S et al 2011 *Steel in Translation* **41** (8) 644–648
[5] Galevsky G V, Rudneva V V et al 2011 *Electric Arc Heater of Gas Nitrogen-oxygen Mixture for a Three-jet Once-through Chemical Metallurgical Reactor* pat. of the Russian Federation No. 2011112115/07
[6] Rechkin V N, Gusev A A 1986 *Dosing Device for Powder-gas Mixture* pat. of the Russian Federation No. 775795/28–13

IOP Conf. Series: Materials Science and Engineering **150** (2016) 012035 doi:10.1088/1757-899X/150/1/012035

Evolution of dispersion of high temperature chromium compounds

L S Shiryaeva, I V Nozdrin, V V Rudneva and G V Galevsky

Institute of Metallurgy and Material Science, Siberian State Industrial University, 42 Kirova Street, Novokuznetsk, 654007, Russia

E-mail: kafcmet@sibsiu.ru

Abstract. Chromium compound – boride Cr_3B_2 and carbide Cr_3C_2 are hard, wear-resistant, chemically inert materials, demanded for production of protective coatings of metals and cermets as components and alloying additives of tungsten free solid alloys. Future prospects for expansion of boride and chromium carbide usage are related to their production in the form of nanopowders. The researches into change of the particles shape and size of high-temperature chromium compounds in the conditions of plasma flow were carried out. There is coarsening of boride and carbonitride particles of nanoscale level at the reduction in the linear velocity of their growth.

1. Introduction

The priority directions of modern machine building require materials that can be used in the conditions of simultaneous exposure to high temperatures, high loads and corrosive environments. In this regard, the role and significance of high temperatire synthetic superhard compounds – carbides, borides, nitrides, oxides and their combinations is constantly rising. Among them are chromium compounds – borideCr_3B_2 and carbide Cr_3C_2which are hard, wear-resistant, chemically inert materials, demanded for production of protective coatings of metals and cermets as components and alloying additives of tungsten free solid alloys. [1-3]. Future prospects for expansion of boride and chromium carbide usage are related to their production in the form of nanopowders. Boride and chromium carbide in the nanostate opens up new directions for their application, including for electroplating, surface and volumetric modification of metal alloys and polymers. All this demonstrates the need for further development of the technological base of boron – carbon-containing chromium compounds. Comparison of different technologies for producing powders of refractory compounds indicates that to achieve the nanoscale level, first of all, the technology based on the use of highly concentrated energy flows for the raw material gasification and formation of the desired product at a volume condensation from the gas phase is required. Among such techniques the plasma method is characterized by relative simplicity, and it is the most studied and competitive method [4].

Currently, the empirical approach to the study of nano-products enlargement in the conditions of plasma synthesis is the most informative, establishing the connection between the characteristics of dispersion with the temperature as the main factor controlling the dispersion, as another important factor – the mass of condensate concentration in the flow is maintained close to the maximum during the plasma processing of the powdered raw material.

The aim of this work is to study changes in the particles shape and size of high temperature compounds of chromium – boride Cr_3B_2 and carbide Cr_3C_2 in the conditions of plasma flow.

2. Experimental research and results

Considering plasma syntheses of diboride and chromium carbonitride as three-stage processes comprising transition of the initial raw material into the vapour state, the synthesis itself and formation of nano-disperse product (condensation, coalescence, crystallization, coagulation), it should be noted that the least studied of these stages is the latter. This is due to the short duration of the processes of nano-disperse product formation (up to 20-25 ms) and the absence of reliable diagnostic and monitoring tools operating in these conditions. In such circumstances, a model-mathematical approach is very promising.

In [5] the authors offered a generalized mathematical model of carbide formation during plasma synthesis of silicon carbide, which includes sub-models "Evaporation of raw materials" and "carbonization of raw materials" providing, in combination with the software complex, the execution of multiple research and engineering design calculations of the reactor parameters and the efficient plasma processing of various types of silicon-containing raw material into carbide. However, the model does not contain sub-models "Condensation", responsible for the formation of nano-disperse product, which generally reduces the significance of the results achieved by the authors [7].

In [5, 6] a theoretical analysis of the possibility of building a condensing block of the model within the technological process of volume condensation was performed, providing quantitative predictions of dispersion for the obtained nano-products. The analysis is based on the scheme of formation of the desired nano-disperse product by converting the supersaturated single-component vapour into a the disperse condensation followed by the evolution of its dispersed composition first by liquid-drop coalescence, and then – aggregate coagulation.

According to the authors [5, 6], the volume condensation can be described by the equation of Szilard-Farkas, but much simplified due to the theoretical and experimental indeterminacy of a number of factors included in it and adapted to the system without chemical interactions with spatially homogeneous conditions, i.e., far from reality. In [5, 6] there is no assessment of the results reliability for prediction of dispersion of solid particles of condensation origin, specific examples and simulation results, which limits the technological feasibility of this approach.

Taking into account the estimation nature of the results obtained in the result of theoretical modeling of condensation processes, the authors in [7 - 9] conducted for a number of substances an experimental study of the temperature dependence of the average sized particles as they rise in the plasma flow. The statistical processing of the results for the temperature conditions of coagulation was is made in the form of dependence

$$\bar{d} = A \cdot T^{-m} \tag{1}$$

where $\bar{d}$ – particle average size;

A – coefficient depending on physical properties of the condensate;

T – temperature.

The results, described in [6, 7], are shown in Table 1.

The temperature dependence of the composition and dispersion of the condensed products of diboride and chromium carbonitride synthesis were based on probe selection of condensate from different areas of the reactor. Sampling was carried out in the temperature range 2600-2000 K.

Table 1. Temperature dependence of the particles size of various substances during their growth in the plasma flow.

Substance	Temperature range, K	Variation $d \cdot 10^9$, m	Coefficients in equation $d = A \cdot T^{-m}$		Information sources
			A	m	
Ni	2200–800	70–121	$(2.413\pm0.559)\cdot10^{-6}$	0.451±0.090	[6]
W	3000–2200	30–46	$(1.023\pm0.246)\cdot10^{-4}$	1.423±0.297	[6]
W	2300–1800	43–85	40.231±5.112	2.401±0.554	[6]
$Si+Si_3N_4$	2200–1650	28–59	$(0.924\pm0.193)\cdot10^{-4}$	1.304±0.251	[6]

The temperature of synthesis products in different zones of the reactor was assumed to be equal to the average weight temperature of the flow. At each temperature level, the samples were taken from the central zone of the plasma flow, i.e., along the reactor axis, three times at intervals of 10 minutes, and the results of quantitative determinations were averaged.

The phase and chemical composition of the synthesized condensed products diboride and chromium carbonitride were studied by the methods of X-ray and chemical analysis, and their morphology and dispersibility were determined by transmission and scanning electron microscopy. In the first case the studies were conducted using a transmission electron microscope EF/4-M/P "Karl Zeis" with an accelerating voltage 65 kV.

While preparing the material the weight 0.1 g was mixed with 500 ml of 50% ethanol solution and treated by ultrasound in the disperser 3 DH-1 for 15-30 min. A drop of the resulting slurry was applied on the coal substrate with thickness 0.10-0.25 μm and dried at a temperature of 323 K. During work with the nanopowders the magnification was 40000. In the second case a scanning electron microscope JSM-6700F with an accelerating voltage 0.5-30 kV and resolution 0.1 nm at 15 kV and 0.22 nm at 1 kV was used. For preparation of the material on the preliminary rolled indium plate of size 10x8x1 mm a nanopowder was sprayed; the sprayed layer was pressed in, and the remains were removed by surface blowing. Further with the magnification up to 300000 the particles were examined impregnated into the metal matrix. The specific surface area of nanopowders was determined by the thermal desorption of argon in the installation IPM NAN made in Ukraine in accordance with the State Standards (GOST) 23401-78 (with some changes introduces after January 01, 1985). The essence of the method consists in determining the argon amount adsorbed on the surface of nanopowders from the flow of an argon-helium mixture with the predetermined concentration (usually argon 5-7%, helium 93-95%) at a liquid nitrogen temperature, followed by its desorption into the same mixture at a temperature increase up to 293±5 K. Sample weight is 0.05-0.10 g. The samples treatment is performed by argon blow at a rate $(0.5\text{-}0.8)\cdot10^{-6} m^3/sec$ at a temperature 673 K for 40-50 min. The error of the surface area measurement does not exceed 5%. The average particle size was calculated by the formula

$$d = \frac{6}{0{,}8 \cdot S_{spec} \cdot \rho} \quad (2)$$

S_{spec} – the specific surface of the sample, m^2/kg; ρ – material density, kg/m^3.

The experimental studies were carried out for three technology variants of chromium boride synthesis CrB_2 in the nitrogen-hydrogen plasma flow of powder mix containing chromium – boron, chromium chloride – boron, chromium oxide – boron, and one the most optimal technological variant of chromium carbonitride $Cr_3(C_{0.80}N_{0.20})_2$ – carbonization of chromium powder with natural gas in the plasma flow of nitrogen [9].

IOP Conf. Series: Materials Science and Engineering **150** (2016) 012035 doi:10.1088/1757-899X/150/1/012035

Micrographs of chromium diboride nanopowder obtained by TEM with the use of the prepared material (Figure 1) provides visualization of individual particles, definition of their size and statement of nanoscale, confirmation particles shape close to spherical.

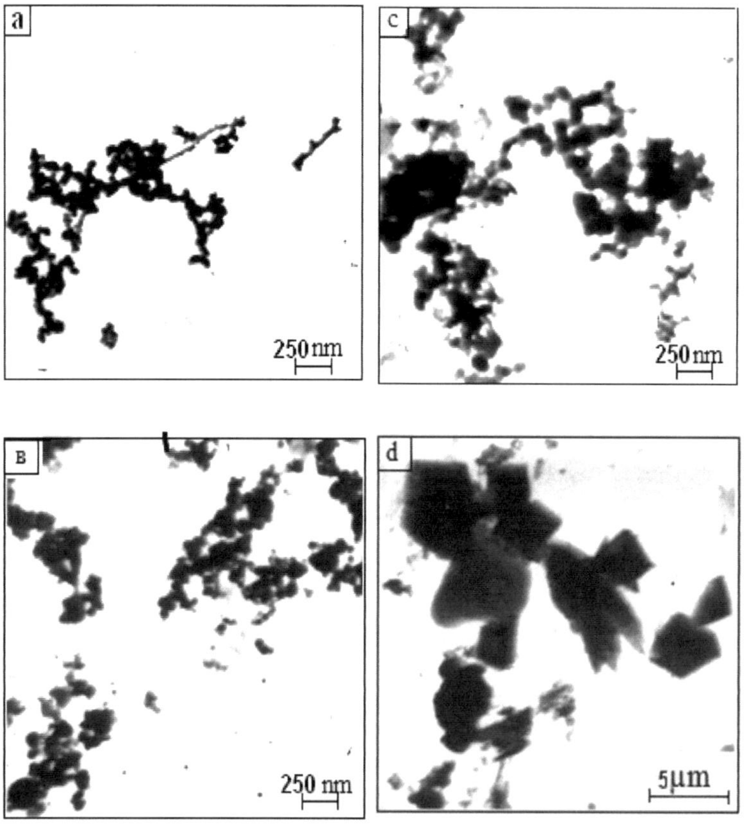

Figure 1. Micrographs (TEM) of chromium diboride of plasma synthesis from chromium (a), its chloride (b), oxide (c) and the magnesium-thermal synthesis from chromium (d).

Analysis of the micrograph also helps to establish the basic nano-particle size range corresponding to 20-50 nm. The average and maximum size of nano-particles of chromium diboride is 42.0 and 90.0 nm. The spherical shape of nanoparticles suggests that the boride formation proceeds according to the mechanism "vapour- melt-crystal".

Micrographs of raster electron microscopy of chromium diboridenano-powder are shown in Figure 2, from which it follows that the nanopowder in the as-received condition is presented by aggregates of spherical shape with various sizes – from 150 to 500 nm, formed by spherical particles of a wide sizes range – from 20 to 80 nm, the number of which in the aggregate depends on its size.

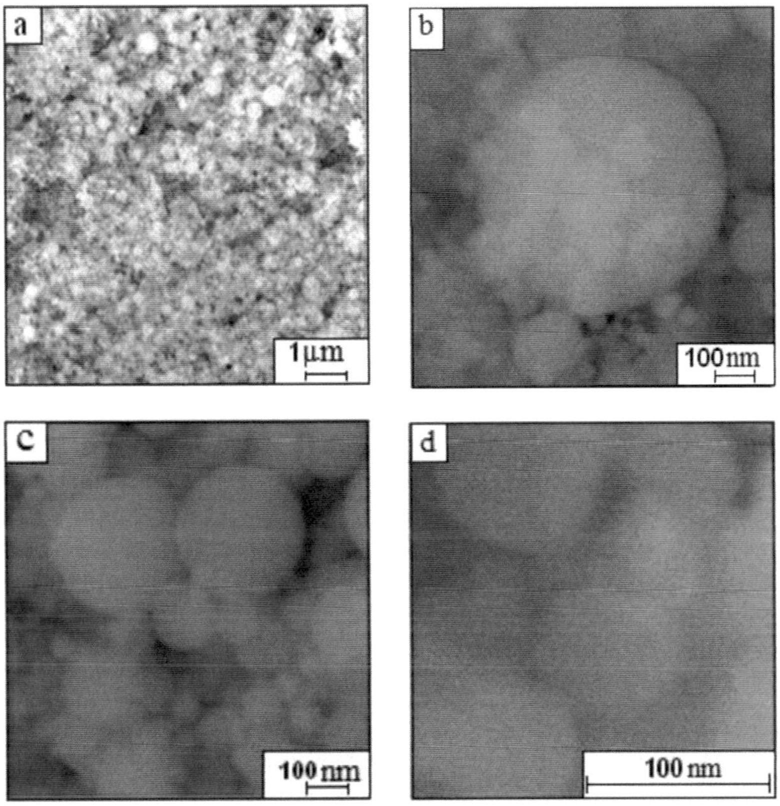

Figure 2. The micrographs (raster electron microscopy) of chromium diboride nanopowder of plasma synthesis from chromium boron-containing mix: a – in the as-received condition; b – morphological pattern of the aggregate; c – the ensemble of particles and aggregates; d – the individual particles.

Nano-level and morphology of the particles can be considered as products of chromium microdroplets boriding, produced during the volume condensation of its vapour, liquid-drop coalescence and crystallization, and the presence in the examined samples of aggregates of various sizes indicates a high likelihood of further nanoparticles coarsening at the lowered temperatures by their coagulation [10].

Indeed, in the studied temperature range (2600-2000) K we can see coarsening of boride particles with the reduction of linear velocity of their growth (Figure 3a). Linear velocity of the nanoparticles growth is $(0.86 \div 2.65)\cdot 10^{-6}$ m·sec^{-1}.

The average size of the diboride nanoparticles increases from 46 to 52 nm and varies with the temperature as follows:

$$d = (2.30 \pm 0.04)\cdot 10^{-6} T^{-(0.471 \pm 0.105)}. \quad (3)$$

Consequently, the diboride nanopowder coarsens in 1.15 times.

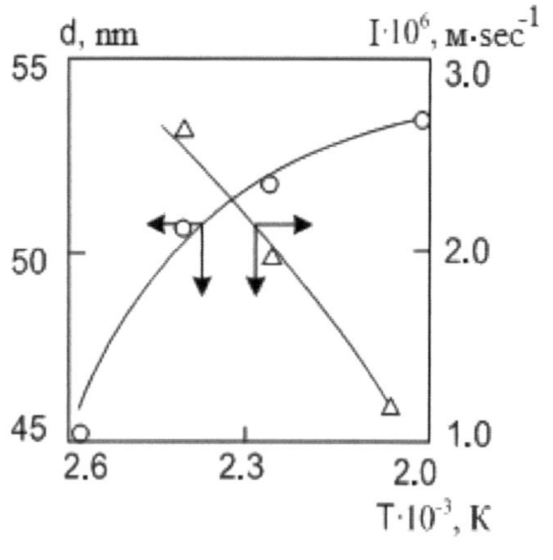

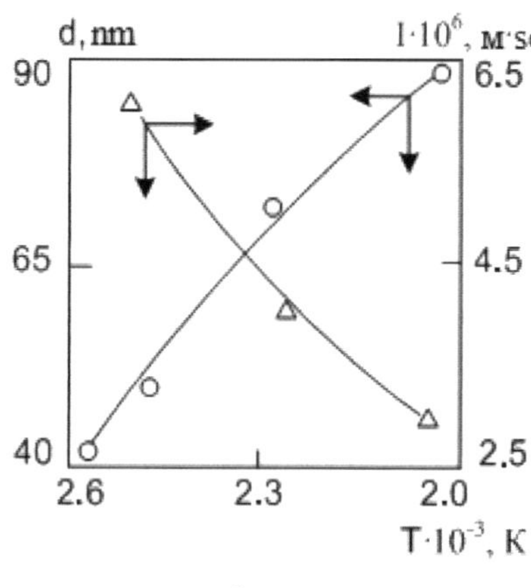

Figure 3. Dependence of particle size (d) of chromium boride (a), chromium carbonitride (b) and the linear velocity of their growth (I) on the flow temperature (T).

The micrographs of chromium carbonitride, obtained by transmission electron microscopy (TEM) and scanning electron microscopy (SEM) using the prepared materials, are shown in Figures 4 and 5, respectively.

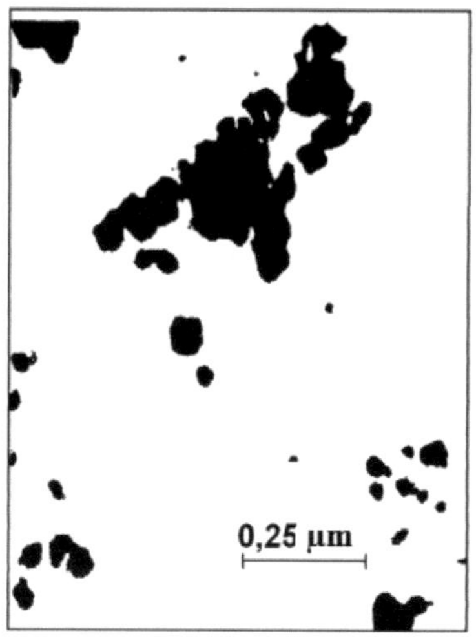

Figure 4. Micrograph (TEM) of nano chromium carbonitride.

Nanopowder of chromium carbonitride is represented by aggregates of spherical shapes with a size from 600 to 150 nm, formed by globular particles with a rather wide size range – from 20 to 80 nm, the number of which in the aggregate depends on its size. Nano-level and morphology of particles allow us to consider them as products of chromium microdroplets carburizing, generated during volumetric condensation of its vapor, liquid-drop coalescence and crystallization, and the presence in the examined samples of aggregates of various sizes indicates a high likelihood of further coarsening of the nanoparticles at the lower temperature by their coagulation. Indeed, in the studied temperature range (2600-2000) K we can see a steady carbonitride particles coarsening during natural decrease of linear velocity of their growth (Figure 2b). The linear velocity of the nanoparticles growth is $(2.96\div6.04)\cdot10^{-6}$m·sec^{-1}.

The average size of the carbonitride nanoparticles increases from 41 to 90 nm and varies alongside with the temperature:

$$d = (726 \pm 35.1) \cdot T - (2.96 \pm 0.44). \quad (4)$$

Consequently, carbonitride nanopowder coarsens in 2.2 times.

The results analysis of the dispersion evolution of high-temperature chromium compounds indicates the increased influence of the temperature on the processes of coarsening during transition from the liquid-drop coalescence to aggregate coagulation. The linear velocity of the particle growth was $(0.2–1.2) \cdot 10^{-6} m \cdot sec^{-1}$, and the achieved level of dispersion did not exceed 80-120 nm. Such behaviour of the dispersion change with the temperature for high temperature chromium compounds may be due to a special energy state of nano-dispersed systems.

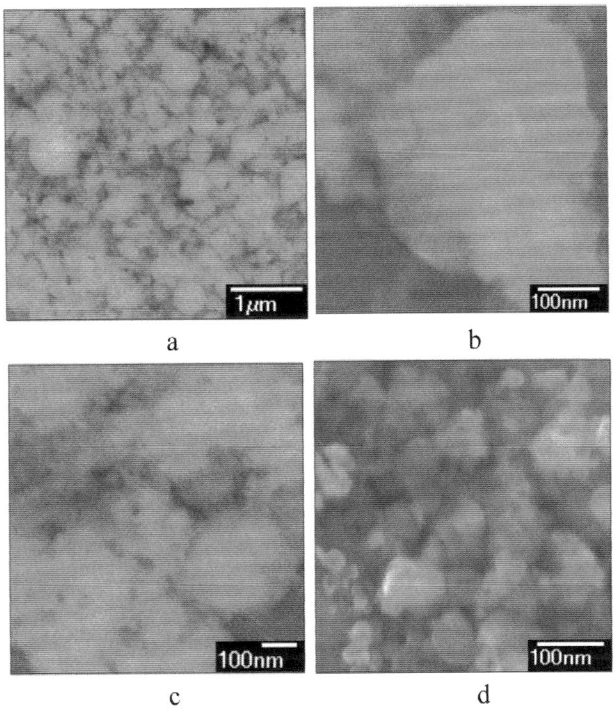

Figure 5. The micrographs (SEM) of nano- chromium carbonitride: a – exterior view; b – morphological pattern of the aggregate; c – ensemble of particles and aggregates; d – the individual particles.

Nano-sized level of the powder provides increased stresses in the crystal lattice, accumulation of line and point defects, and defects of a grain boundaries type. The boundary is the major defect in the nano-dispersed systems. In such systems, in contrast to conventional powder systems and polycrystals the ratio of the total area of boundaries to their volume is by 4-5 orders of magnitude greater than, for example, in the polycrystal with a grain size ≈ 100 μm.

3. Conclusions

Researches of the change in the particles shape and size of high temperature chromium compounds in the conditions of plasma flow were conducted. There is a coarsening of boride and carbonitride particles of a nanoscale level at the reduction of the linear velocity of their growth. Such behaviour of the dispersion variation of carbide and chromium boride nano-powders might be due to the capacity of the initial nanoparticles of metallic compounds to the solid phase coalescence (primary recrystallization) inside the aggregates.

The results showed that the achieved level of dispersion of high temperature chromium compounds does not exceed the nano-scale level (80-120 nm), which opens up new areas of their application, including electroplating, surface and volumetric modification of metal alloys and polymers.

4. References

[1] Shiryaeva L S et al 2014 *Electroplalting and Surface Treatment* **2** 51–57
[2] Oskolkova T N and Budovskikh E A 2013 *Metal Science and Heat Treatment* **55** (1–2) 96–99
[3] Oskolkova T N and Budovskikh E A 2012 *Metal Science and Heat Treatment* **53** (11–12) 608–610
[4] Nozdrin I V, Shiryaeva L S et al 2011 *Steel in Translation* **41** (8) 644–648
[5] Rudneva V V, Galevsky G V and Yurkova E K 2007 *Proc. Conf. on Metallurgy: Technologies, Control, Innovations, Quality* (Novokuznetsk, SibSIU) pp 221–226
[6] Rudneva V V, Galevsky G V et al 2007 *A Study and Improvement of Reactor for Plasma Metallurgical Production of Nono-Silicon Carbide* (Novokuznetsk: SibSIU) p 15
[7] Samsonova G V 1976 *Physical Properties* (M.: Metallurgiya) vol 2 p 600
[8] Alekseev N V 1987 *Proc. on Plasma Processes in Chemical Industry* (Chernogolovka: INKhP AN SSSR) pp 54–72
[9] Nozdrin I V, Galevskii G V et al 2011 *Steel in Translation* **41** (10) 799–804
[10] Pegov V S and Evgenov I V 1987 *Proc. on Physics and Chemistry of Ultradisperse Systems* (M.: Nauka) pp 180–186

Strain-stress simulation and comparison of different welding sequences during manufacturing of packing vacuum cover

Yanhu Wang[1], Xizhang Chen[1,2] and Sergey Konovalov[2,3]

[1] School of Mechanical and Electrical Engineering, Wenzhou University, 276 Xueyuan Middle Rd, Lucheng, Wenzhou 325035, China
[2] Hualian Machinery Group CO.LTD, Area A China Import and Export Fair Complex, Wenzhou, 325000, China
[3] Department of Physics n.a. Prof. V.M. Finkel, Siberian State Industrial University, 42 Kirova Street, Novokuznetsk, 654007, Russia

E-mail: kernel.chen@gmail.com

Abstract. Double ellipsoid heat source model was used to simulate the welding strain-stress evolution during the welding of a vacuum cover for packing machine. Results showed that each welding process has influence about the deformation and residual stress distribution over product. Stress concentration area was found in an area of the cover and the improved structure, manufacturing technologies were proposed to minimize residual stress after simulation and comparison of different manufacturing process, it is an available method to improve the product quality and reduce residual stresses.

1. Introduction

Vacuum cover is one of the most important components for packing machine, its fabrication quality is the key for the working performance. Welding is the main technology used to fabricate the vacuum cover, one of the most challenges is the rupture of the whole cover result from external air pressure. The welding technologies of austenitic stainless steel are well established, but the mechanism of strain-stress formation and distribution are not completely understood yet, especially for those components with complex welding structure and sequences of manufacturing. Austenitic stainless steels have a high thermal expansion coefficient and a lower heat transfer coefficient, which would result in tensile residual stress due thermal gradient from welding process [1]. Residual stress of stainless steel welded joints contributes for stress corrosion cracking. Deformation will also lead to some problems in assembly process, for example, it reduces the mechanical resistance of the structure and seriously affects the quality of welded parts. So, understanding about welding distortion and residual stress for this mechanical component are particularly important [2].

In this paper, SYSWELD software was used to simulate strain-stress evolution during welding fabrication of a vacuum cover. The temperature field distribution, welding deformation and stress were calculated and analyzed, which provides a theoretical basis for the reduction of welding defects and the improvement of product quality.

2. Simulation

2.1. Heat source

There are three common heat source models for hot-working process, i.e. two-dimensional (2D) Gaussian model, double ellipsoid model and three-dimensional (3D) conical Gaussian model. 2D Gaussian model is mainly used for surface heating, such as surface heat treatment. Double ellipsoid model is mainly used for fusion welding. 3D conical Gaussian model is mainly used in some high-energy beam such as laser, plasma and electron beam. It is convenient to calculate by double ellipsoid heat source model, because the TIG welding technology is used. Double ellipsoid heat source model was shown in Figure 1 [3-5].

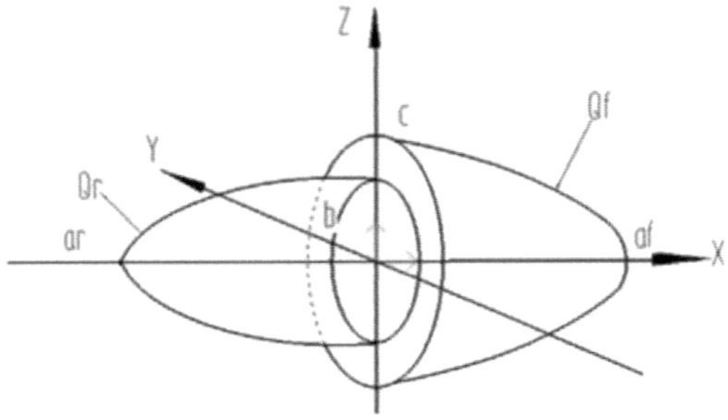

Figure 1. Double ellipsoid heat source model.

Where a_f, a_r, b and c in Figure 1 are the main Gaussian parameters, a_f is the length of the front half ellipsoid, a_r is the length of the rear half ellipsoid, b is the depth of penetration of the ½ and c is the weld width of 1/2.The welding parameters used are: welding current 380 A, welding voltage 20V, welding speed 5mm/s, melting depth 3 mm, penetration 3 mm, melting width 4 mm and the efficiency 0.7. When the heat source is checked, only a part of the 2D and 3D grids are checked. After repeated correction, the heat source function is determined, save it to a function library for using when the simulation is calculated.

2.2. Model building

A 3D model is built according to the product dimension with the solid works software. Dimensions are shown in Figure 2. The 3D model is shown in Figure 3.Tungsten Inert Gas Welding (TIG) was used in each diagonal seam welding. Since the entire piece is not a ramp, they need two welds on each corner. The number 1 to 8 is weld representative in Figure 3(a) and (b).Figure 3 (a) is the current welding sequence and Figure 3 (b) is the welding sequence after weld adjustment. Welding sequence is in accordance with numbers in the figure.

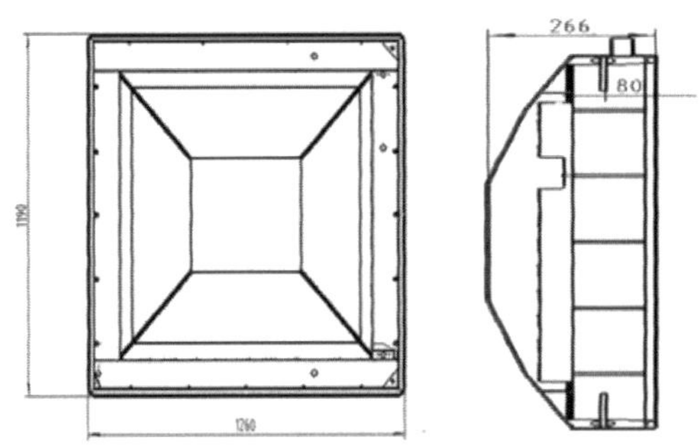

Figure 2. Dimensions of vacuum cover.

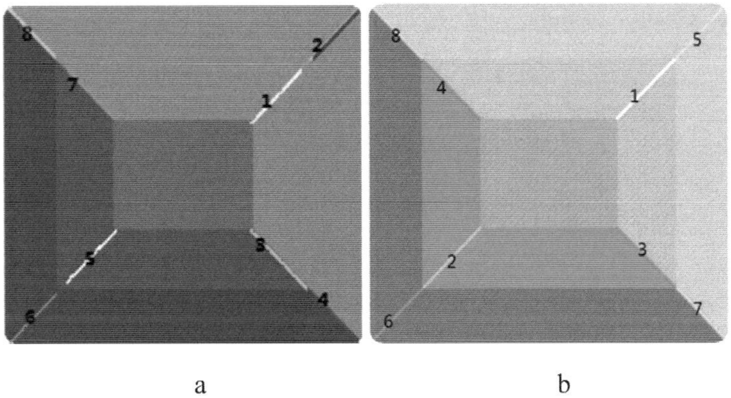

Figure 3. Three-dimensional solid model of vacuum cover (a – current welding sequence; b – welding sequence after weld adjustment).

2.3. Meshing

The complete 3D model is imported into Visual Mesh to mesh. The 3D meshing process is shown in Figure 4. The quality of the grid directly determines the accuracy and efficiency of the calculation. To ensure the accuracy of the premise, the main parts of the grid are denser, the results are not significantly affected by the use of sparse partition. For example, the area away from the weld can choose the larger size of the grid [6]. The mesh elements in the weld seam and the surrounding area are refined. Because of irregular shape, we also use a single area division to ensure the weld area near the surface is totally hexahedral mesh. Area away from the weld, sparse partitioning and a small amount of tetrahedral mesh are allowed. The completed grid is shown in Figure 5. The mesh model contains 54627 units, 38160 nodes, and a total of 77 groups.

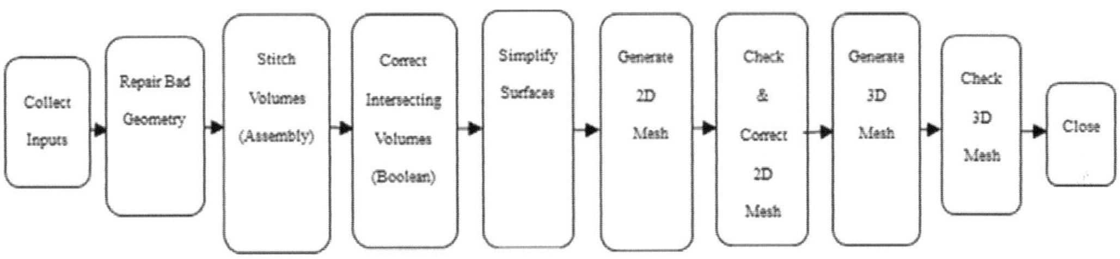

Figure 4. 3D mesh process.

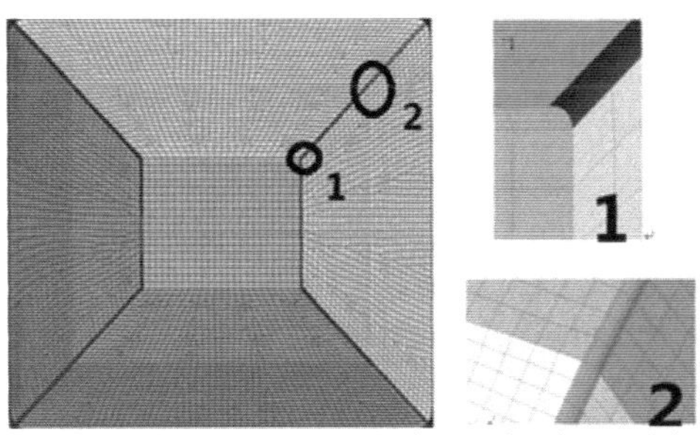

Figure 5. Meshing model (number 1 is the enlarged view of a the weld combined with the parts top, number 2 is the enlarged view of a combination of the two welds at the slope).

2.4. Pretreatment and simulation

SYSWELD software was used for simulation. Boundary conditions are required for the calculation of the welding stress and strain field. The main purpose of applying boundary conditions is to prevent the rigid displacement during calculation and when the results cannot converge. However, the applied boundary conditions can not seriously prevent the free relief of stress and the free deformation during welding. Austenite stainless steel 0Cr18Ni9 is used as base material in the welding of vacuum cover.

There are 8 weld seams as shown in Figure 3 (a) and (b). The simulated welding sequence is the same as the actual works. According to the results of the first weld simulation, the average temperature of the weld heat section is exported as curve Figure 6, the most dramatic change period of temperature is choose as all the other seams' welding heat source.

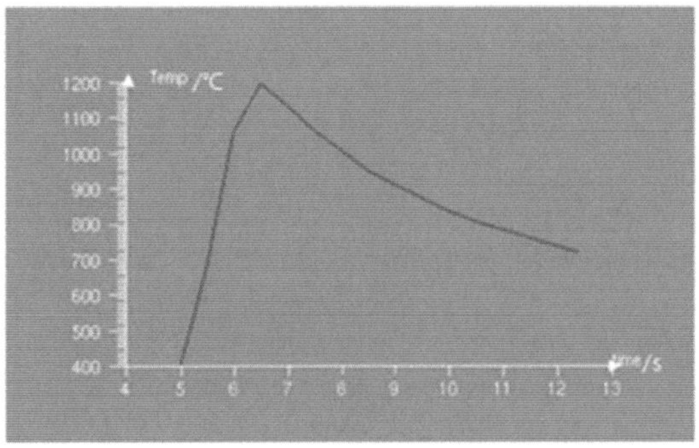

Figure 6. Heat source.

3. Result and analysis

The deformation and stress of each welding stage are shown in Figure 7 and Figure 8, separately. Number 1 to 8 indicates each weld seam sequence. Welding deformation is frequently occurred in production of welding structures. It will not only affect the accuracy and appearance of the welded structure, but also reduce mechanical properties and then result in fracture [7]. According to the size and of the shape of this component, variable simulation results shown in Figure7 (a) and (b),the maximum strain occurred at the fourth corners, as well as at the beginning of the welding seam. Because of the influence of the first weld seam to the latter one, it caused the amount superposition of the deformation. The deformation increases linearly with the increase of the number of the welding seams. It may be caused by the increasing open groove angle from top to bottom. The groove gaps of the four corners are the largest, where the deformation is greater than other parts. Comparing Figure 7 (a) with (b), we find that at the first welding process, the weld distortion is significantly more than the second welding process corresponding to the deformed position. In the current welding sequence the maximum deformation is 1.699 mm. The maximum amount of deformation of welding sequence after weld adjustment is 1.593 mm. It results in 7% less deformation.

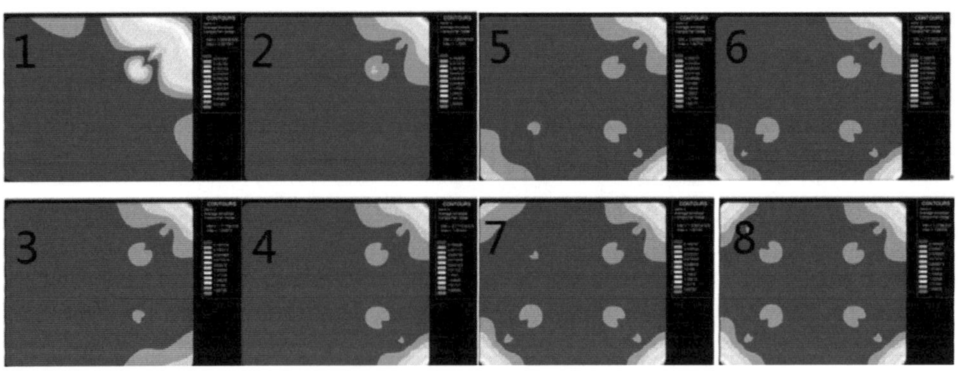

a

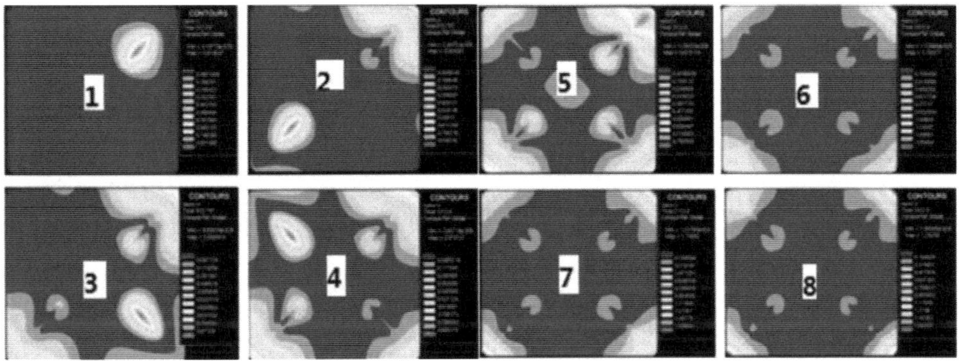

b

Figure 7. Strain cloud image (a – strain cloud image of the current welding sequence; b – strain cloud image of the welding sequence after weld adjustment).

Thermal elastic-plastic analysis, inherent strain method, viscoelastic plastic analysis and coupling effect theory considering the phase transformation and thermal stress are commonly used to study the welding stress and strain. The thermal elastic-plastic analysis method is to calculate the thermal stress and strain in the process of welding thermal cycle [8]. Analysis of thermal elastic-plastic behavior of metallic materials is very complicated. The finite element method can give the deformation history and temperature change of each part of the deformable body in the numerical simulation of the metal welding process. Through stress cloud in Figure 8 (a) and (b), we know the stress at the center of the weld is the largest and therefore each seam center is the stress concentration area. Each weld seam has an effect of welding stress superposition on the next weld. The stress is also correspondingly become larger with the increasing number of the welds. The location of the maximum stress will also change. Comparing stress clouds in Figure 8 (a) and (b), we find weld stress in each seam area is very small, but with the next welding seam the stress increased rapidly, stress amplitude change significantly under the current welding sequence. And, the stress value increases rapidly at each of the weld areas, which is obviously different from the stress distribution obtained after sequence adjustment. The strains after welding sequence adjustment produce less deformation, which would improve the service life of vacuum cover and reduce the possibility of welding defects formation. Generally, the larger weld area stress is, more possible is the defects appear in the weld seam, such as the cracks.

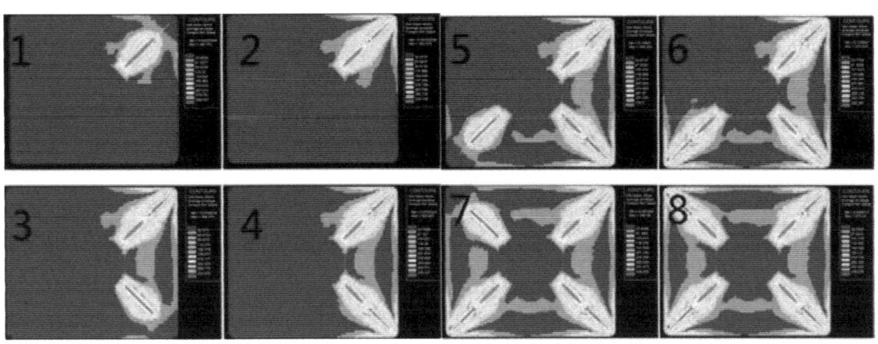

a

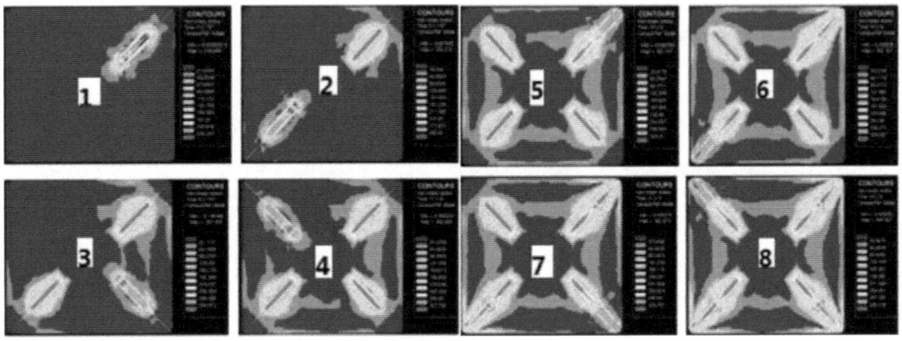

b

Figure 8. Von Mises stress cloud image (a – Von mises stress cloud image of current welding sequence; b – Von Mises stress cloud image of the welding sequence after weld adjustment).

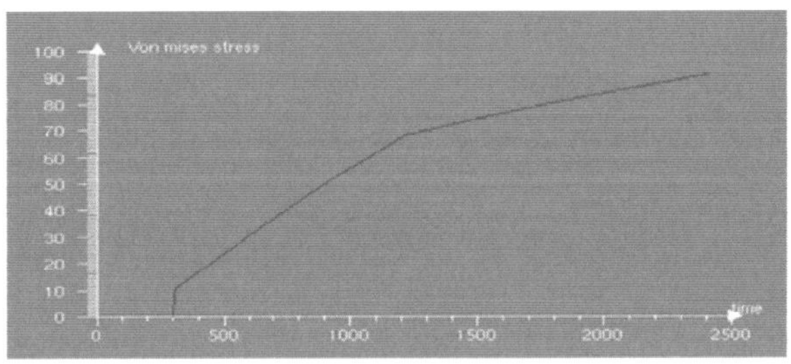

a

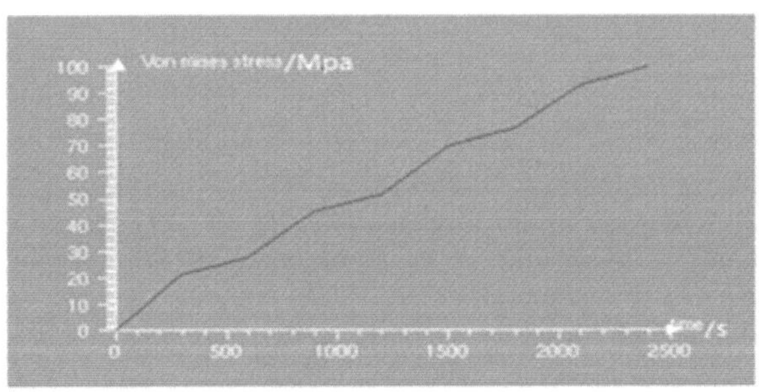

b

Figure 9. The average stress curve (a – the average stress curve of current welding sequence; b – the average stress curve of the welding sequence after weld adjustment).

From Figure 9 (a) and (b), we know the average stress of the welding parts is increasing with the increase of the number of welding. But by comparison Figure 9 (a) and (b), for the current welding sequence, the average stress of the work piece at the beginning of the second welds (the first weld end time is 300s) stepwise increases, this change will have a greatly effect on the work-piece, while the welding sequence after weld adjustment, the average stress of the work-piece with the first weld gradually increased. The influence of the welding seam on the welding parts includes the stress and strain distribution, as shown in Table1.

Table1. Simulation result data (current welding sequence; b:welding sequence after weld adjustment)

Number (Weld process types)	Maximum strain (mm)	Maximum stress value (Mpa)	The position of the maximum strain	The distribution position of the maximum stress
1 (a)	0.510	309.767	The 2th corner weld	The 1st weld seam
2 (a)	1.608	329.706	The 2th corner weld	The 1st weld seam and in 2nd half before the beginning of the weld
3 (a)	1.681	333.078	The 2th corner weld	The 1st and 3rd weld seam and in 2nd half before the beginning of the weld
4 (a)	1.685	336.337	The 2th corner weld	The 1st and 3rd weld seam and in 2nd and 4th half before the beginning of the weld
5 (a)	1.698	334.800	The 2th and 4th corner weld	The 1st, 3rd and 5th weld seam and in 2nd and 4th half before the beginning of the weld
6 (a)	1.699	336.381	The 2th, 4th and 6th corner weld	The 1st, 3rd and 5th weld seam and in 2nd, 4th and 6th half before the beginning of the weld
7 (a)	1.698	334.916	The 2th, 4th and 6th corner weld	The 1st, 3rd, 5th and 7th weld seam and in 2nd, 4th and 6th half before the beginning of the weld
8 (a)	1.700	336.578	The 2th, 4th ,6th and 8th corner weld	The 1st, 3rd, 5th and 7th weld seam and in 2nd, 4th, 6th and 8th half before the beginning of the weld
1 (b)	0.831	248.888	The 1st weld seam	The 1st weld seam
2 (b)	0.840	332.216	The 2nd weld seam and the corners of the 5th seam weld	The 1st weld seam but the value increases
3 (b)	0.871	352.439	The 3rdweld seam and the corners of 5th seam weld	The 1st and 2nd weld seam but the value increases
4 (b)	0.885	360.493	The 4thweld seam and corners of the 5thand 6th seam weld	The 1st, 2nd and 3rd weld seam but the value increases
5 (b)	0.796	362.187	The 5th weld seam, the	The 1st, 2nd, 3rd and 4th weld

			corners of the 5th, 6th and 7th seam weld	seam but the value increases
6 (b)	1.554	362.161	The correspond corners of the fifth seam weld	The 1st,2nd,3rd, 4th, 5th weld seam
7 (b)	1.590	362.673	The corners of the fifth and sixth seam weld	The 1st, 2nd ,3rd ,4th ,5th and 6th weld seam but the value increases
8 (b)	1.593	363.302	The corners of the fifth , sixth and seventh seam weld	The 1st,2nd,3rd, 4th, 5th,6th and 7th weld seam but the value increases

4. Conclusions

The deformation and residual stress of the welded parts are successfully simulated by using SYSWELD software under two welding process. With the increasing of weld seam numbers, the stress of the welding increases and the maximum stresses are at the intersection of the two welds. Computational simulation results guided us to establish the best welding sequence for structure of the vacuum cover. By comparison with simulated results, we can see that the welding sequence after weld adjustments have reduced deformation and as consequence reduced residual stress in the structure The results have been used to improve the welding fabrication of vacuum cover of packing machine.

5. Acknowledgements

This work was partly sponsored by National Natural Science Foundation of China (No.51575401), Zhejiang Provincial Natural Science Foundation (No. LY16E050007). We appreciate Mr. Cristiano da Silva Varzim to polish the language and review the paper before submit to publication.

6. References

[1] Xizhang Chen, Shuyan Zhang, Jingjun Wang and Joe Kelleher F 2015 *Materials and Design* **76** 26–31
[2] Xizhang Chen, Xing Chen, Huili Xu and Bruce Madigan 2015 *Int. J. Adv. Manuf. Tech.* **80** (5–8) 1197–1211
[3] Peilin Li and Hao Lu 2011 *Transactions of the China Welding Institution* **11** 66–70
[4] Jinzhou Zhang 2010 *Journal of Yangtze University* **3** 50–53
[5] Yu Kang,Yuejin Ma, Wenlin Zhang, Jianguo Zhao and Wei Me 2008 *J. Agric. Univ. Hebei* **2** 36–38
[6] Amaraeh, Mebanin and Allaloun 2005 *Proc. CAOL 2005 and Int. Conf. on Advanced Optoelectronics and Lasers* (USA: IEEE) 146–154
[7] Ruiyun Ming 2010 *Welding Technology* **12** 44–46
[8] Wei Li, Zefeng Wen, Lei Wu and Xuesong Jin 2010 *J. of Mech. Eng.* **46** (10) 95–101

Effect of electron beam treatment on structural change in titanium alloy VT-0 at high-cycle fatigue

S V Konovalov[1], I A Komissarova[1], D A Kosinov[2], Yu F Ivanov[3], O V Ivanova[4] and V E Gromov[1]

[1] Department of Physics n.a. Prof. V.M. Finkel, Siberian State Industrial University, 42 Kirova Street, Novokuznetsk, 654007, Russia
[2] Department of Research Projects, Siberian State Industrial University, 42 Kirova Street, Novokuznetsk, 654007, Russia
[3] Institute of High Current Electronics, Siberian Brunch of the Russian Academy of Sciences, 2/3 Akademichesky Ave, Tomsk, 634055, Russia
[4] Department of Higher Mathematics, Tomsk State Architectural and Construction University, 2 Solyanaya Sq., Tomsk, 634000, Russia

E-mail: konovserg@gmail.com

Abstract. Changes in the surface of the fractured structure of commercially pure titanium VT1-0 under treatment by low-energy high-current electron beams and the subsequent cycle fatigue to the failure were analyzed by transmission scanning and transmission electron diffraction microscopy. The increase in the fatigue life of samples in 2.2 times after treatment by electron beams was established. An assumption was made that the increase in the fatigue life of titanium, grade VT1-0, was due to the formation of a lamellar substructure conditioned by high-velocity crystallization of the titanium surface layer.

1. Introduction

With the development and improvement of technology the materials requirements are constantly tightening. They are defined, in particular, by the desire to improve the components operating temperatures, the realized velocities of movement and transmitted pressures in the movable matings of machines and simultaneously to reduce the weight of parts per unit of power. In some cases, the use of conventional structural steels and alloys cannot meet these complex requirements. In addition, the creation of new volumetric alloyed materials, which is still the main method of increasing their reliability and durability, is becoming increasingly problematic due to the scarcity and high cost of alloying elements [1-7].

In this regard, it is economically and technically feasible to provide structural strength due to surface hardening by creating microlayers with the modified structural-phase state in accordance with the operational requirements. For this purpose, in recent years, the concentrated energy flows are widely used. They can significantly reduce the processing time, which in some cases makes it possible to embed the appropriate equipment into a unified manufacturing technology, as well as reduce the energy costs for treatment due to the fact that only those parts which require hardening and protection are exposed to the concentrated energy flows [8].

In our previous works [9-11] we conducted a study to determine the influence of treatment by electron beams on the fatigue life of steels of different structural classes, and silicon-based alloys. The regularities of the formation and evolution of the phase composition and defect metals substructure were identified. The present paper is a continuation of the previous studies. Its purpose is to examine the fracture surface, phase composition and state of dislocation substructure of commercially pure titanium of grade VT1-0 destroyed in the result of fatigue tests.

2. Materials and methods

As a research material samples of commercially pure titanium VT1-0 were used (up to 0.18 Fe; up to 0.07 C; up to 0.04 N; up to 0.1 Si; up to 0.12 O; up to 0.004 H; 0.3 other impurities; remainder Ti, mass %) [12]. The fatigue tests were carried out in a special unit according to the scheme. The samples with dimensions 120*12*4 mm^3 were produced in accordance with the State Standards (GOST) 25.502-79 and had a symmetrical stress concentrator with 20 mm in radius. The minimal thickness of the sample in the area of stress concentrator was 5 mm. The testing temperature – 300 K, the frequency of bending loading was 10 Hz at a load of 20 MPa. Irradiation of the sample surface, prepared for fatigue testing, was carried out on the unit "SOLO" [13] with the following parameters: the electron energy 16 keV; pulse repetition rate 0.3 s^{-1}; duration of the beam electron pulse 150 μm; density of electron beam energy 25 J/cm^2; quantity of impact pulses – 3.

The investigations of the fracture surface were carried out by scanning electron microscopy (Tesla BS-301). The analysis of the phase composition and condition of the material defect substructure was performed by transmission electron diffraction microscopy (JEM-2100). Foils for research were prepared by ion thinning of blanks, cut by electric spark method from the sample parallel to the fracture surface as close to it as possible.

3. Results and discussion

High-cycle fatigue testing of samples from commercially pure titanium VT1-0 revealed a significant increase in the fatigue life of the material irradiated by intense pulse electron beam of submillisecond exposure duration (16 keV, 25 J/cm^2, 150 μsec, 3 pulses, 0.3 sec^{-1}). Namely, we detected an excess (relative to the non-irradiated samples) of fatigue life of the material in average by ≈2.2 times. Obviously, this effect is associated with the transformation of the structure of titanium surface layer in the conditions of high-speed heating and cooling caused by electron beam irradiation.

Simulation of thermal processes occurring in the surface layer of metals and alloys under the influence of electron beam showed that exposure of the material surface to the electron beam of micro- and submillisecond duration makes it possible to realize the conditions of ultra-high heating rates (up to 10^8deg/sec) of the surface layer to the melting temperature and evaporation, and to form in the surface layer a temperature gradient 10^7-10^8deg/m, providing cooling of the surface layer due to heat removal into the cold volume of the material at a rate up to 10^7deg/sec [14]. As a result of this impact in the surface layer within the time interval of a single irradiation pulse the conditions for the formation of non-equilibrium state are created that allows submicro- nanostructured multiphase structures to be registered, which are characterized by properties, unrealizable via the conventional methods of treatment. [13]

The numerical solution of the problem of finding the temperature field [15], which is formed in the surface layer of titanium irradiated with an intense electron beam allowed the thickness of the modified layer to be estimated, as well as the maximal temperature which can be achieved on the exposed surface, the temperature gradient, the heating and cooling rates, time period of existence of the material in different aggregate states. The thermal conductivity equation is solved using the finite difference method [16]. Thermal characteristics of titanium were taken from [17]. The thickness of the surface layer for thermal calculations $d = 0.5 \cdot 10^{-3}$ m, the observation period – 600 μsec. The calculations were performed for the intense electron beam with energy density 25 J/cm^2 and pulse duration 150 μsec. It was shown that titanium irradiation at a single-pulse mode by the electron beam at the parameters described above leads to the melting of the surface layer with thickness 18.1 μm

(Figure 1). This layer is divided into two sublayers: the surface layer, thickness 10.4 μm, which is a single phase layer (liquid) (Figure 1, layer 1.) and the intermediate layer, thickness 7.7 μm, which is a two-phase layer (liquid + solid state) (Figure 1. layer 2). The lifetime of the first layer is 80 μsec; the time interval from the start of titanium melting to its complete crystallization is 147.2 μsec.

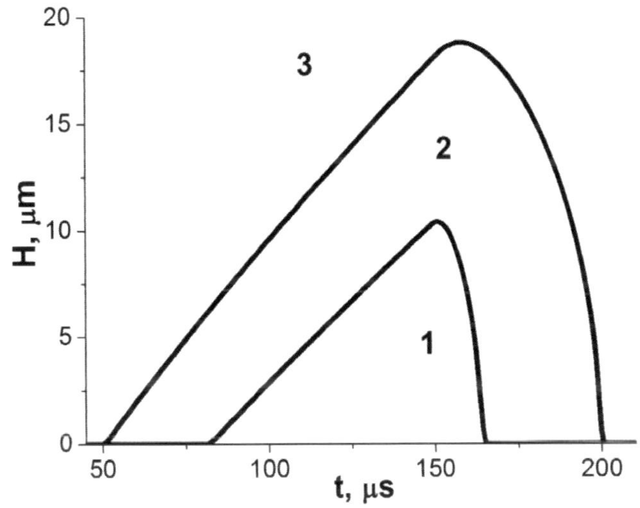

Figure 1. Thickness dependence of layer 1 (liquid) and layer 2 (solid + liquid phase) on the observation period during irradiation of the titanium surface by electron beam having an energy density 25 J/cm^2 (pulse duration 150 μs); zone 3 – zone of the solid phase (the layer of heat influence).

Study on the surface morphology, formed as a result of fatigue tests to fracture of commercially pure titanium, irradiated by electron beam, allowed the thermal calculations to be validated. It was found that part of the sample adjacent to the surface of the electron beam irradiation had a multilayered structure: surface layer (thickness: 9.5 μm) (layer 1 in Figure 2) and the intermediate layer (thickness 10 μm) (layer 2 in Figure 2). The surface layer has a columnar structure, which apparently indicates a greater crystallization rate of titanium. From the analysis of Figure 2 it follows that layer 1 appeared in connection with the melting and high-speed crystallization of single-phase state, and layer 2 is a layer formed during the crystallization of the two-phase state of titanium.

Phase composition and the defective substructure of the surface layer destroyed during the fatigue tests of VT1-0 titanium samples and previously irradiated by an intense electron beam were analyzed by transmission electron diffraction microscopy of thin foils. In parallel with the irradiated samples the fractured titanium samples in the initial state were subjected to the electron microscopy studies. As a result of the research, it was found that the commercially pure titanium in the initial state is a polycrystalline aggregate. After the fatigue tests we observed the dislocation substructure represented by randomly distributed dislocations or dislocations, forming an irregular reticular substructure. The scalar dislocation density was $2.8 * 10^{10}$ cm^{-2}. The same dislocation substructure was detected in the grain volume of commercially pure titanium in the surface layer and in the layer located at a distance of 150-200 μm from the treated surface.

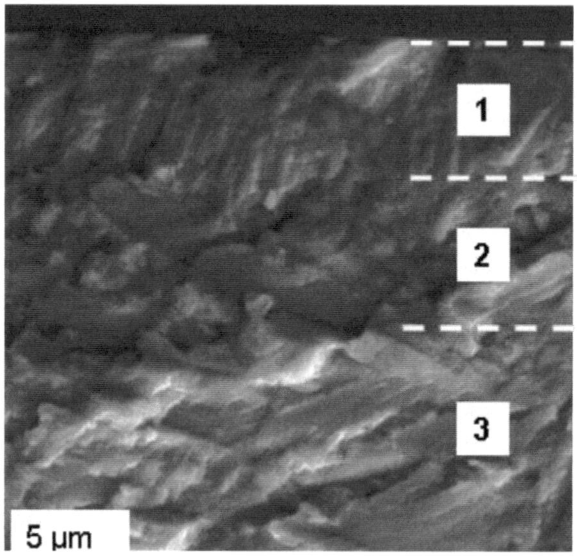

Figure 2. The surface of titanium fracture after electron-beam treatment.

The structure of the surface layer of commercially pure titanium VT1-0 after electron beam treatment and high-cycle fatigue is significantly different from the structure of the material in the initial state. The surface layer with thickness of 10 µm has a lamellar structure (Figure 3a). The plates α-Ti, are generally located at the angle of 70-90° to the irradiated surface. As the distance from the irradiated surface increases the laminas are randomly located in relation to the exposed surface. At a depth of 12-15 µm with lamellar structure we found a subgrain structure (Figure 3b), which is the main type of structure of the material in the layer located at a distance of 30-40 µm from the irradiated surface (Figure 3c). The grain structure close to the structure of the non-irradiated samples after their destruction was detected at a depth of 80 m (Figure 3d).

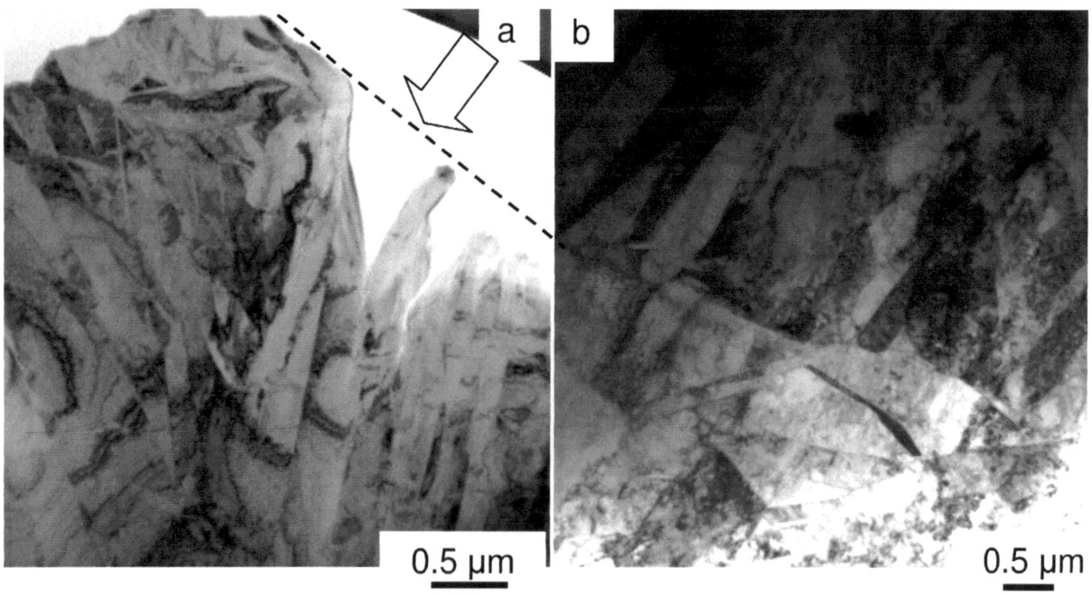

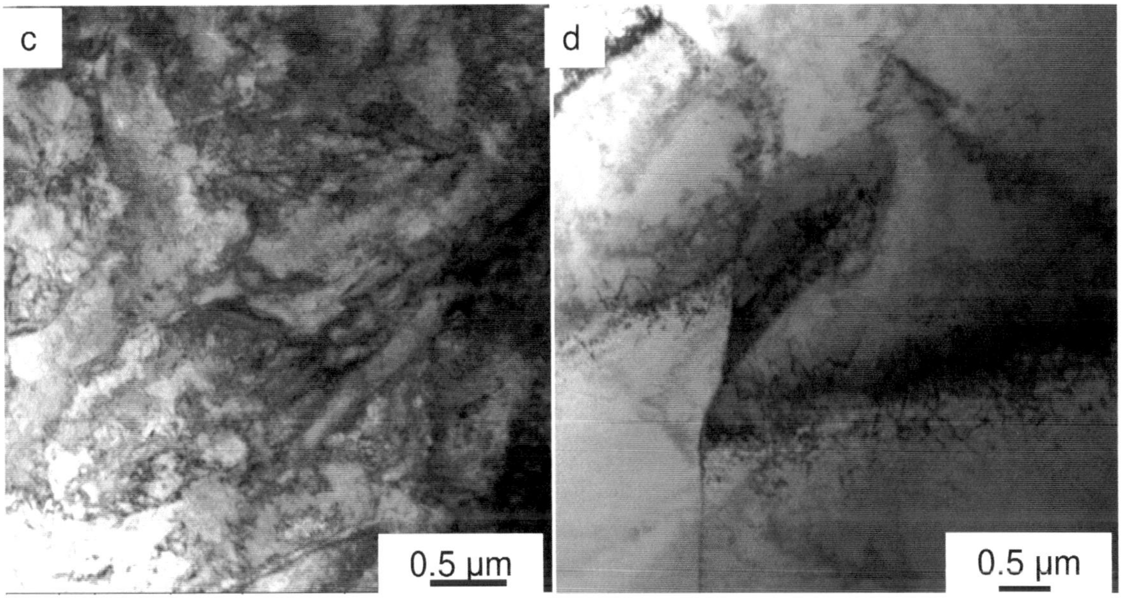

Figure 3. Electron microscope image of titanium structure, grade VT1-0, subjected to electron-beam treatment and fatigue tests up to sample destruction.

Comparing the results of thermal calculations and information obtained by means of electron diffraction microscopy of thin foils, we can conclude that the surface layer of a lamellar structure was formed as a result of the titanium crystallization from a single-phase state; the sub-layer located below with a mixed structure was formed during the crystallization of the two-phase state. Deeper layers should be attributed to the heat affected zone.

4. Conclusions

The treatment of commercially pure titanium VT1-0 by intense pulsed electron beam of submillisecond duration (16 keV, 30 J/cm2, 150 µsec, 3 pulses, 0.3 s-1) was performed. Fatigue life tests of samples from commercially pure titanium VT1-0, irradiated by intense pulsed electron beam, and non-irradiated samples were carried out. After fatigue tests of samples subjected to electron beam treatment, an increase in fatigue life was established which is on average in 2.2 times higher in comparison to the samples without treatment. The studies of the fracture surface and defect substructure destroyed during fatigue tests of samples from commercially pure titanium were performed. It showed that electron-beam processing of titanium leads to the formation of multilayer structure.

The layer adjacent to the irradiated surface has only a lamellar structure; the following layer has a mixed structure represented by laminas and subgrains. The layer located at a distance (30-40) µm from the irradiated surface has, mainly, a subgrain structure. Numerical calculations of the temperature field, which is formed in the titanium surface layer irradiated by the intense electron beam, were made. It is shown that the layer with the structure of lamellar morphology, adjacent to the irradiated surface, was formed by a high-speed single-phase crystallization (liquid phase) of titanium; the layer lying beneath with lamellar subgrain structure was formed as a result of a high-speed two-phase crystallization (liquid phase + solid phase); the layer of subgrain structure is a layer of thermal influence. The physical cause for the multi-fold increase of the fatigue life of commercially pure titanium VT1-0 is the formation of a lamellar substructure due to high-speed crystallization of the titanium surface layer.

5. Acknowledgements

The work was supported by the grant of the President of the Russian Federation for state support of young Russian scientists – doctors of sciences (Project MD-2920.2015.8), the state task No. 3.1496.2014/K and the scholarship of the President of the Russian Federation for young scientists and graduate students engaged in advanced scientific research and developments in priority areas of modernization of the Russian economy in the years 2016-2018 (Project SP-3590.2016.1). Thermal calculations and radiation of commercially pure titanium VT1-0 by intense electron beam on the installation "SOLO" was financed by the Russian Scientific Foundation (Project No. 14-29-00091).

6. References

[1] Schijve J 2009 *Fatigue of Structures and Materials* (Berlin: Springer)
[2] Terentyev V F 2002 *Fatigue of Metallic Materials* (M.: Nauka) p 248
[3] Zhang C, Gao M and Zeng X 2016 *J. of Materials Processing Technology* **231** 479–487
[4] Nogami S, Guan W et al 2015 *Hasegawa: Fusion Science and Technology* **68** 607–611
[5] Chen Xizhang, Yuan Qibing, Madigan Bruce and Xue Wei 2015 *Corrosion Science* **96** 178–185
[6] Chen Xizhang, Shen Zheng, Chen Xing, Yucheng and Huang Qunying 2011 *Fusion Engineering and Design* **86** 2943–48
[7] Chen Xizhang, Fang Yuanyuan, Li Peng, Yu Zhenzhen, Wu Xiaodong and Li Dongsheng 2015 *Materials and Design* **65** 1214–21
[8] Gribkov V A, Grigoryev F I et al 2001 *Perspective Radiation-beam Technologies of Materials Treatment* (M.: Krugly God) p 528
[9] Gromov V E, Ivanov Yu F, Vorobiev S V and Konovalov S V 2015 *Fatigue of steels modified by high intensity electron beams* (Cambridge International Science Publishing Ltd, Cambridge)
[10] Gromov VE, Vorobiev S V, Konovalov S V, Romanov D A, Semina O A and Kazimirov S A 2016 J. *Surf. Investigation* **10** 83–87
[11] Ivanov Y, Alsaraeva K et al 2015 *Materials Science and Technology* **31** 1523–29.
[12] Borisova E A, Bochvar G A et al 1980 *Titanium Alloys. Metallography of Titanium Alloys* (M.: Metallurgiya) p 464
[13] Rotshtein V, Ivanov Yu and Markov A 2006 *Materials Surface Processing by Directed Energy Techniques* (Elsevier) ed Y Pauleau chapter 6
[14] Ivanov Yu F, Koval N N et al 2013 *High Temperature Material Processes* **17** (4) p 241–256
[15] Samarsky A A 1997 *Introduction to Numerical Methods* (M.: Nauka) p 271
[16] Samarsky A A 1989 *Theory of Difference Schemes* (M.: Nauka) p 616
[17] Babichev A P, Babushkina N A et al 1991 *Handbook on Physical quantities* (M.: Energoatomizdat) p 1232

Change of deformation characteristics and dislocation substructure of nonferrous metals under influence of magnetic field

D V Zagulyaev, S V Konovalov, V V Shlyarov, E A Anuchina, I A Komissarova and V E Gromov

Department of Physics n.a. Prof. V.M. Finkel, Siberian State Industrial University, 42 Kirova Street, Novokuznetsk, 654007, Russia

E-mail: konovserg@gmail.com

Abstract. The objects of the study were polycrystalline copper of M00b grade and commercially pure titanium BT1-0. Microindentation was carried out on the samples of titanium BT1-0 in the initial state, immediately after magnetic field exposure of 0.4 T and after certain time intervals. The defect substructure of cooper samples M00b, subjected to loading to failure in the creep mode under the influence of magnetic field of 0.35 T and without it, was investigated by the methods of electron diffraction microscopy. It was revealed that the effect of magnetic field exposure on commercially pure titanium BT1-0 leads to the decrease in microhardness with the subsequent stabilization during the time that depends on the processing parameters. And the effect of the magnetic field exposure on copper during the process of creeping results in the redistribution of dislocation substructure types. Also, there are changes in quantitative characteristics of dislocation substructures.

1. Introduction

Development of production technologies at the present stage of technological progress forces to search for materials with high performance properties, thus, contributing to the development of new methods of energy impact on materials. Since the 1980s the question of influence of weak magnetic fields on the real structure and plasticity of "non-magnetic" materials [1-5], including the metal ones (Al, Cu, Zn, etc.), has been actively studied [6-11].

One of the factors responsible for the physical and mechanical properties of materials is their structure. Changes in the structure characteristics occur during the plastic deformation. In this connection it is necessary to have comprehensive information about the changes in the material structure close to the source that caused its complete fracture [12-15]. In that context the purpose of work was to investigate the magnetic field effect on the change in properties of non-ferrous metals undergoing plastic deformation. Commercially pure titanium was subjected to microindentation, and commercially pure copper was tested for creep.

2. Materials and methods

We used samples of titanium BT1-0 with dimensions of $0.4\times1\times1$ cm^3. Samples were prepared by annealing at 800 K during 2 hours followed by furnace cooling, mechanical polishing, chemical etching in the solution composed of hydrofluoric and nitric acids with the addition of water in the ratio of 1:1:6. The etching time was empirically adjusted. After etching the samples were thoroughly washed with running water and dried. The average grain size in the initial state was 79.1 μm.

Part of the research is carried out on polycrystalline copper of M00b grade. Flat samples with the dimensions of the working part 150×5×0.46 mm^3 were used. Before creep testing the material structure was equilibrated by recrystallization annealing during 2 hours at a temperature of 700 °C with the subsequent cooling in water. Tests in the creep mode under constant tensile stress $\sigma = 130$ MPa and a temperature of 25 °C were performed up to fracture. The first half of the samples was deformed under the conditions of external magnetic field with induction 0.35 T, the second half – under the normal conditions [16].

Microhardness measurements were performed with the help of microhardness tester HVS-1000A by the method of micro-Vickers for three variants: without the magnetic field influence, immediately after exposure to magnetic field, and after certain time intervals; magnetic field exposure time was constantly varied.

The microhardness value was expressed as an average of at least 30 measurements. Quantitatively the effect of the magnetic field was characterized by the relative change in microhardness:

$$Q = (HV - HV_0)/HV_0 \times 100\%,$$

where HV is the microhardness value of the sample exposed to magnetic field, HV_0 – the initial value of the microhardness.

For the creep tests like in [17, 18] the upgraded testing machine for studying processes of plastic deformation in metals was used. The permanent magnet was applied as a source of magnetic field. The field density was controlled by the milliteslameter TPU with a precision up to 1×10^{-3} T.

The research of the dislocation substructure of samples (DSS) was performed by the transmission electron microscopy of foils with the help of electron microscope EM-125. The foils were prepared from the material adjacent to the fracture zone and at some distance from it. Images of the material fine structure were used for the classification of morphological features of the structure.

3. Results and discussion

3.1.1. Influence of magnetic field B=0.4 T on titanium microhardness

Dependences of the relative change in microhardness on the magnetic field exposure time 0.4 T in the interval from 0.75 hour to 2 hours, obtained from the experimental data, are presented in Figure 1.

It can be seen that at these processing parameters the decrease in microhardness with its subsequent stabilization for all processing times (except for 2 hours) takes place. Stabilization of microhardness occurs according to the same exponential law, regardless of the processing time. The processing time that equals to 2 hours is characterized by the residual influence, even after 24 hours, which is about 2 ± 1.3%. The dependence of the relative change in microhardness on the processing time, given in Figure 2, shows the initial effect of the magnetic field.

The analysis of this dependence shows that a linear decrease in microhardness takes place as the exposure time in the field increases. The presented dependence obeys a linear law $Q = - 4t + 2$. The magnetic field effect is not observed for the exposure times less than 0.5 hour, which indicates the existence of a certain time threshold below which the effect is not observed.

Since the initial effect of the magnetic field depends on the exposure time in it, it becomes relevant to analyze the stabilization dependencies of microhardness value on the magnetic field exposure time. This dependence is shown in Figure 3. It can be seen that it is of exponential character.

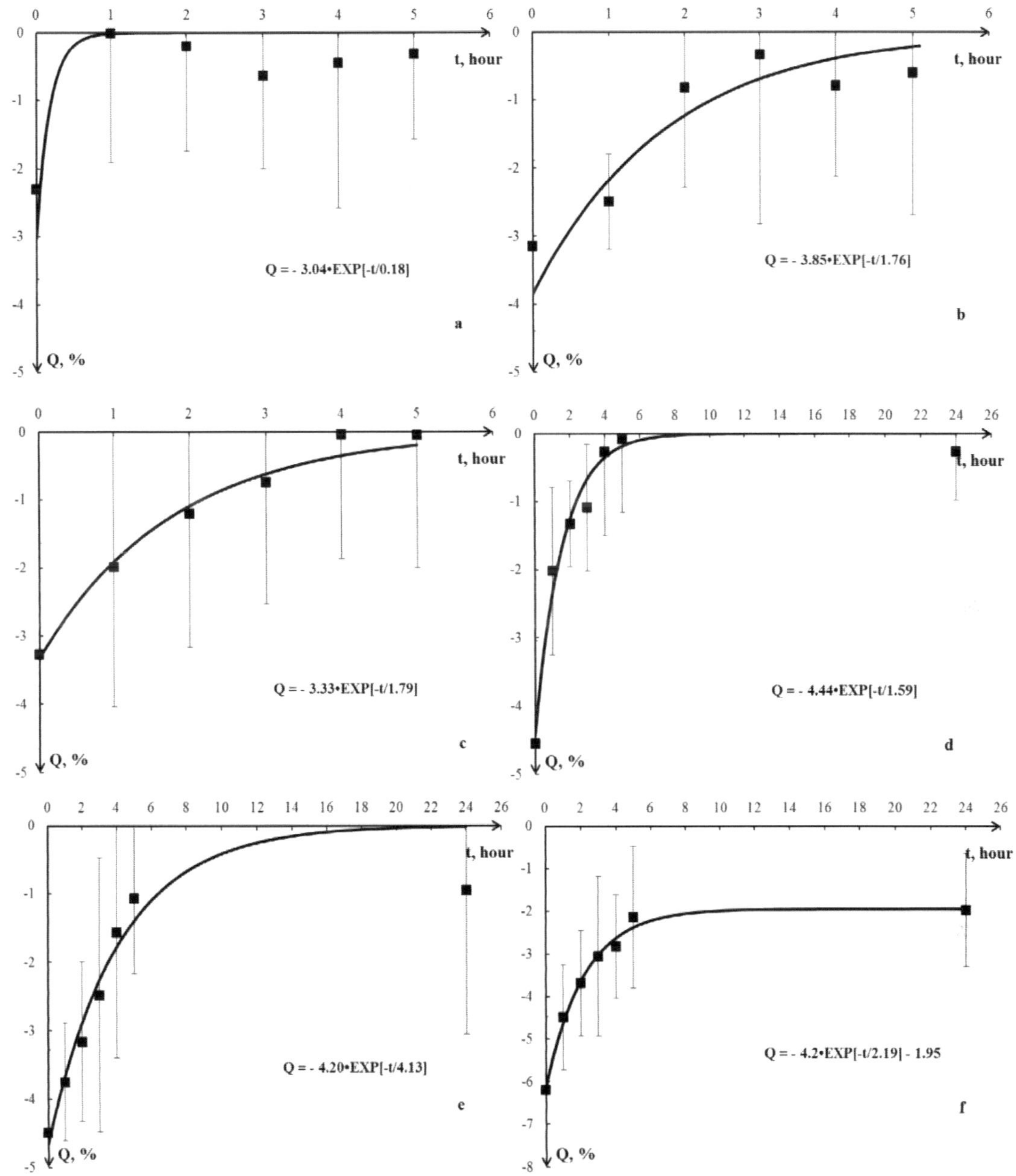

Figure 1. Dependence of the relative change in microhardness (Q) on the time elapsed after the treatment by magnetic field of 0.4 T during 0.75 h (a), 1 h (b), 1.25 h (c), 1.5 h (d), 1.75 h (e), 2 h (f).

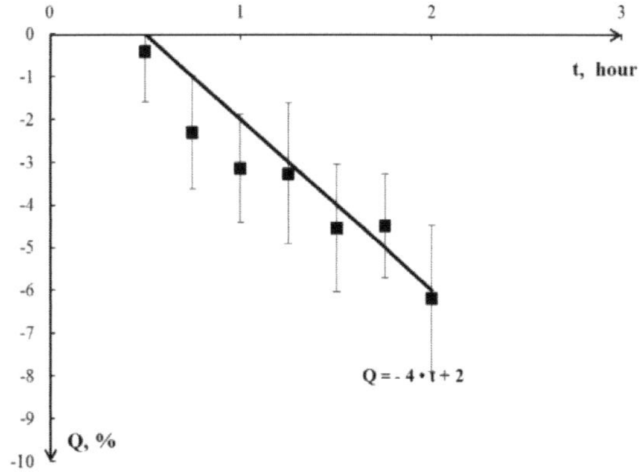

Figure 2. The initial effect of magnetic field of 0.4 T on the titanium micro-hardness.

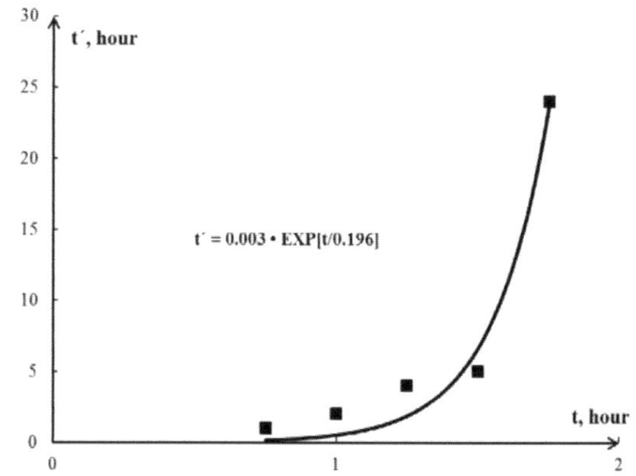

Figure 3. The dependence of the stabilization time of microhardness (t′) on the magnetic field exposure time (t) 0.4 T.

3.2. Magnetic field effect B=0.35 T on the change in dislocation sub-structure of commercially pure copper at creep

In the fractured areas of copper samples the structure in the form of fragments and dislocation chaos is found. Flexural extinction contours indicate the internal stresses arising during the creep (Fig. 4). The main dislocation sub-structure is a cellular one. Also the strip dislocation substructure is present in the material.

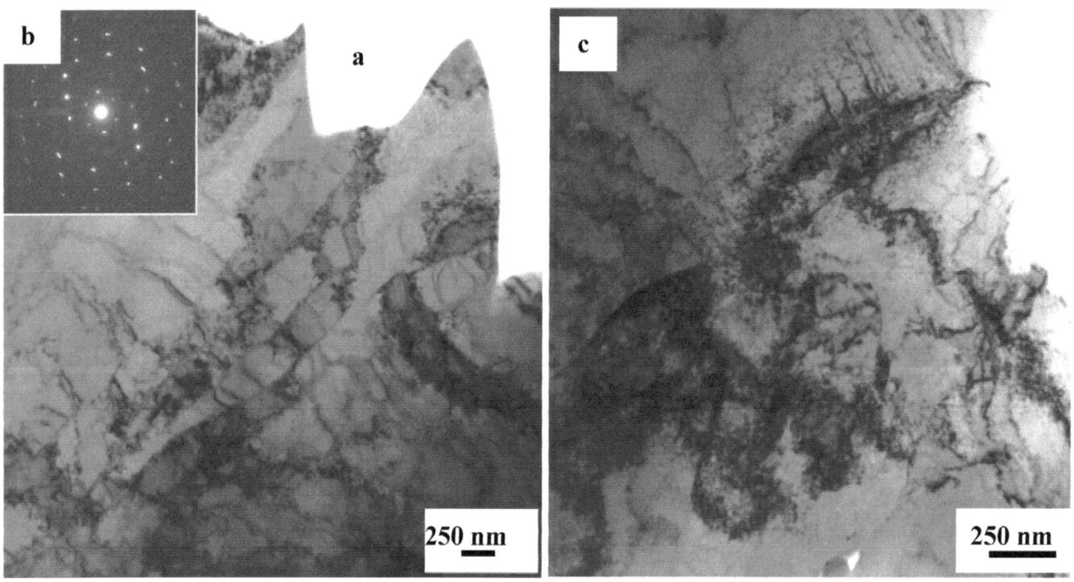

Figure 4. The electron microscopic image of the layer structure adjacent to the fracture surface of copper.

As the distance from the fracture surface increases (the foils prepared at a distance of 85, 325 and 635 μm are investigated), the strip and cellular dislocation substructures are observed in copper. In the strip substructure and in the cells the dislocation substructure in the form of randomly distributed dislocations is found.

The analysis of the density of flexural extinction contours from the distance to the fracture surface of copper is carried out. This characteristic is linearly related to the density of stress concentrators. It is found that the density of flexural extinction contours decreases exponentially as the distance from the copper fracture surface increases, ranging from 1200 cm^{-2} to 195 cm^{-2}.

The magnetic field effect on copper leads to the redistribution of dislocation substructure types. In addition to the identified types of dislocation substructures the sub-grain structure is formed with the sub-grain sizes of a nanoscale interval.

The analysis of changes in the scalar dislocation density from the distance to the fracture surface of copper, tested for creep under the conditions of magnetic field exposure and without it, is carried out. It is found that the difference in the values of scalar density of dislocations in the copper, fractured without a magnetic field and exposed to the magnetic field, is most clearly revealed in the material adjacent to the fracture surface. The analysis of internal stress fields, formed during the creep of copper under the conditions of magnetic field exposure and without it, is performed. They are predominantly located at the intraphase interfaces: grain boundaries and grain boundary intersections; sub-grain boundaries and sub-grain boundary intersections; boundaries of a strip substructure. Besides, these contours are detected at the fragmented sub-boundaries. As a parameter characterizing the bending-torsion of the copper crystal lattice the density of flexural extinction contours is used.

The similarity in the number of stress concentrators in cooper, deformed by the magnetic field and without it, is established. There is a pronounced maximum of stress concentrator density in the material, located at a distance of ~4 mm from the fracture surface. At this depth the maximum of scalar density of dislocations is identified and the maximum of relative content of a strip sub-structure in copper is located [16]. Consequently, the main source of bending-torsion of the crystal lattice of copper during creep are the interfaces formed in the stripe dislocation substructure.

4. Conclusions

The magnetic field effect on commercially pure titanium BT1-0 leads to the decrease in microhardness with its subsequent stabilization over the time that depends on the processing parameters. The typical feature in determining the relative microhardness value immediately after the magnetic treatment is the increase of the exposure time resulting in the inevitable build-up effect. The initial effect is characterized by the linear dependence on the processing time for the value of field density of 0.4 Tesla. Stabilization time of microhardness is of exponential nature for the induction of 0.4 Tesla.

It was found that during the destruction of commercially pure copper without magnetic exposure on the creep process, regardless of the distance to the fracture surface, the main type of dislocation substructure is cellular. Along the grain boundaries the strip substructure is detected.

The differences in quantitative characteristics of dislocation substructures are revealed and the gradient nature of the change in the amount of stress concentrators as the distance from the fracture surface increases, regardless of the magnetic effect, is identified.

In the case of a magnetic field effect on the process of creeping the main type of dislocation substructure, regardless of the distance to the fracture surface, is a strip dislocation substructure. In some instances, the grains with the structure of dislocation chaos, cellular and mesh substructure are identified.

The reduction in the density of flexural extinction contours is established as the distance from the fracture surface increases during fracture in the magnetic field.

5. Acknowledgements

The work is financed by the grant of the RFBR (project No. 16-32-50135_mol_nr), Russian President for state support of the young Russian scientists – doctors of sciences (project MD-2920.2015.8), state task No. 3.1496.2014/K and the scholarship of the President of the Russian Federation to young scientists and postgraduate students (project SP-3590.2016.1).

6. References

[1] Golovin Yu I and Morgunov R B 2001 *Solid State Physics* **5** 827–832
[2] Osipyan Yu A, Golovin Yu I, Lopatin D V, Morgunov R B, Nikolaev R K and, Shmurak S Z 1999 *JETP Letters* **2** 110-113
[3] Golovin Yu I, Morgunov R B, Lopatin D V, Baskakov A A, Evgenyev Ya E 1998 *Solid State Physics* **11** 2065–68.
[4] Alshits V I, Darinskaya E V, Koldaeva M V and Petrzhik E A 2012 *Solid State Physics* **2** 305–312
[5] Alshits V I, Darinskaya E V, Koldaeva M V and Petrzhik E A 2013 *Solid State Physics* **2** 318–325
[6] Urusovskaya A A, Alshits V I, Smirnov A E 1997 *JETP Letters* **6** 470–474
[7] Alshits V I, Urusovskaya A A, Smirnov A E and Bekkauer N N 2000 *Solid State Physics* **2** 270–272
[8] Peschanskaya N N, Smirnov B I and Shpeizman V V 2008 *Solid State Physics* **6** 997–1001
[9] Smirnov B I, Shpeizman V V, Peschanskaya N N and Nikolaev R K 2002 *Solid State Physics* **10** 1915–18
[10] Klypin A A 1973 *Metal Science and Heat Treatmen* **8** 2–6
[11] Soyka A K, Sologub O I, Shepelevich V G and Sivtsova P A 2015 *Solid State Physics* **10** 1947–49
[12] Zagulyaev D V, Konovalov S V, Litvinenko N G, Komissarova I A and Gromov V E 2013 *Tsvetnye Metally* **4** 74–77
[13] Golovin Yu I, Morgunov R B, Lopatin D V, Baskakov A A and Evgenyev Ya E 1998 *Solid State Physics* **11** 2065–68
[14] Alshits V I, Darinskaya E V, Koldaeva M V and Petrzhik E A 2003 *Crystallography* **5** 826–854
[15] Zagulyaev D V, Konovalov S V, Yaropolova N G, Ivanov Y F, Komissarova I A and Gromov V E 20015 *J. Surf. Invest.* **9** (2) 410–414
[16] Konovalov S V, Zagulyaev D V, Yaropolova N G, Komissarova I A, Ivanov Yu F and Gromov V E 2015 *Russian Journal of Non-Ferrous Metals* **56** (4) 441–448
[17] Gromov V E, Ivanov Yu F, Stolboushkina O A and Konovalov S V 2010 *Materials Science and Engineering: A.* **527** (3) 858–861
[18] Konovalov S V, Danilov V I, Zuev L B, Filip'ev R A and Gromov V E 2007 *Physics of the Solid State* **49** (8) 1457–59

IOP Conf. Series: Materials Science and Engineering **150** (2016) 012039 doi:10.1088/1757-899X/150/1/012039

Automated system of control and diagnostics of cast-steel defects in the mass production

S V Knyazev[1], A A Usoltsev[1], D V Skopich[2], E A Fatyanova[2] and A E Dolgopolov[1]

[1] Department of Materials Science, Foundry and Welding Engineering, Siberian State Industrial University, 42 Kirova Street, Novokuznetsk, 654007, Russia
[2] JSC "IndasKholding", 2 Suvorova Street, Novokuznetsk, 654000, Russia

E-mail: krookia@mail.ru

Abstract. To solve the problem of efficient control over technological process in the foundry the mathematical model is required, that would connect the the input and output parameters of the object. To collect the data on cast-steels defects an automated system was developed that greatly simplifies the process of technical control and allows the problem of process parameters optimization to be solved using the criterion of defects minimization in the cast-steels

1. Introduction

The technological process of mass production of cast-steels is a complex multi-parameter system, the state of which may be to assess by the quality of finished products. Examination of reasons leading to appearance of rejects and defects in the cast-steels associated with deviations in smelting and casting technology, properties of molding and core sand, forming conditions, as well as the design features of pattern equipment, is usually carried out based on the experience and intuition of process engineer and does not always allow the issue of stabilization and optimization of technological processes (TP) to be resolved quickly and in a timely manner. The role of this issue is particularly important in serial and mass production as material losses from reject, corrections, and mechanical processing of defective cast-steels are the heaviest.

Reject of cast products can be divided into two components – a reject of metal (steel) due to mechanical properties and chemical composition, and the presence of uncorrectable casting-process defects exceeding the critical characteristics in accordance with the State Standards (GOST), specification, customer requirements. The problems causing the reject due to steel mechanical properties are solved by the predictive models. It is possible to predict and control the mechanical properties of steel during its smelting. Having a reject prediction by one of the parameters, specifying the current values of the elements of the chemical composition and final conditions, it is possible to calculate the steel chemical composition, which is still possible to achieve for this melting and which will ensure the absence of reject. It was found that for the analysis of tendencies in the quality of cast-steels two key indicators from two groups of mechanical properties can be used – yield strength and impact viscosity on the V-shaped sample – KCV [1].

In the foundry practice the range of technological parameters variation is defined by GOST and the enterprise standards. It should be noted that these ranges are generally quite broad and comprise 10% of the nominal or more. Often during manual TP operation even these very broad ranges are violated. In addition, not all monitored parameters are sufficiently informative and their effective control is difficult.

Studies have shown that non-optimal sampling frequency, insufficient volume of samples and sample sizes are also the cause of deviations of technology parameters. Thus, maintaining a "hard" technological regime does not guarantee that it is optimal to obtain a minimal level of defects in cast-steels. Elimination of disadvantage of such regime is possible when using a "flexible" technology, which allows the production process (PP) to be adapted to changing external disturbances on the criterion of a minimal cast-steels defects [2].

Reliable results of assessment of the PP state in terms of identifying the reasons for casting-process defects and development of measures aimed at their elimination can be obtained by combined system analysis of cast-steels and a complex of technological testing devices [3].

2. Results and discussion

The most common defects of cast-steels of a complex configuration with a wall thickness 15-30 mm and weight up to 500 kg, manufactured in the single sand-clay molds with a lot of bars, are: hot cracks, blow-holes and porosity, clogging, scabs and burning-in. These types of defects are detected by external examination of cast-steel surface; their size can be estimated with a minimal use of measuring equipment.

The nature and size of surface defects provides information on the presence and size of internal defects of cast-steels. In order to identify correlations between the surface and internal defects of type "discontinuity" a selective quantitative visual and non-destructive testing of cast-steels before and after their machining was conducted by a special technique.

It was found that the correlation coefficients for the considered types of defects are $r = 0.71 \div 0.86$ at the critical value $r_{cr} = 0.56$ and 95% of confidence probability.

The presence of significant correlations allows us to conclude that the results of monitoring of surface defects for this class of cast-steels and manufacturing methods can serve as an objective assessment of level of their defects.

To solve the problem of effective TP control in the foundry we need to have a mathematical model connecting the input and output parameters of the object. Information and diagnostic model (**IDM**) of the TP control subsystem on the level of cast-steels defects is built using mathematical and statistical methods of pattern recognition theory for two groups of parameters: parameters of technology and the level of casting defects [4]. The first task, solved during diagnosis, is to get information about the average level of defects (**ALD**) in the control batch of cast-steels for all considered types of defects.

For the collection of data on defects of cast-steels a formalized procedure was developed, which greatly simplifies the process of technical control and helps to organize the input of results into a computer in a dialogue mode. In this process the problem is solved, which almost eliminates the influence of a subjective perception of the operator-inspector on the assessment of signs of casting and process defects, both on the qualitative and quantitative levels.

The calculated data **ALD** of the control batch of cast-steels are used to diagnose the state of the TP and to make recommendations for its optimization.

In accordance with the terminology used in the pattern recognition, the diagnostic information consists of two components: system of features- assessments **ALD** for all considered types of defects and system of images-assessments of the TP state.

By the method of statistical classification [5, 6] from these components was formed a training diagnostic sample (**TDS**), for which the experimental sample data from the cast-steels connection ALD with the values of process parameters were used.

TDS is a table, each row of which contains the components of the feature and the corresponding image (Table 1).

Table 1. The form of **TDS** presentation.

No.	Feature components Values of TP parameter					Feature components Normalized ALD value				
	1	2	3	…	L	1	2	3	…	K
1	$X_{1,1}$	$X_{1,2}$	$X_{1,3}$	…	$X_{1,L}$	$Y_{1,1}$	$Y_{1,2}$	$Y_{1,3}$	…	$Y_{1,K}$
2	$X_{2,1}$	$X_{2,2}$	$X_{2,3}$	…	$X_{2,L}$	$Y_{2,1}$	$Y_{2,2}$	$Y_{2,3}$	…	$Y_{2,K}$
3	$X_{3,1}$	$X_{3,2}$	$X_{3,3}$	…	$X_{3,L}$	$Y_{3,1}$	$Y_{3,2}$	$Y_{3,3}$	…	$Y_{3,K}$
…	…	…	…	…	…	…	…	…	…	…
N	$X_{N,1}$	$X_{N,2}$	$X_{N,3}$	…	$X_{N,L}$	$Y_{N,1}$	$Y_{N,2}$	$Y_{N,3}$	…	$Y_{N,K}$

$X_{i,L}$ – value of the TP parameter;
$Y_{i,K}$ – normalized value of ALD;
$i=1 \div N$ – number of TDS line;
L, K – number of process parameters and types of foundry and process defects in TDS.

To provide optimal conditions for diagnosis in terms of recognizability during the TDS preparation the value of confidence interval for all considered parameters was taken into account.

TDS is used for pattern recognition – the probable technological situation, conditioning the appearance of a set of defects in the control sample. The control sample is a group of cast-steels for the aggregate defects of which the line with characteristic components is formed, for which the TP diagnosis was performed.

At this stage of technological situation recognition the Hamming distance is taken as a measure for the assessment of the control line proximity to TDS line [7]:

$$d_i = \sum_{k=1}^{K} \left| y_{i,k} - y_{j,k} \right| \tag{1}$$

where $y_{i,k}$ – normalized values of ALD for j control batch.

During the preliminary recognition usually the subgroup of image compounds is detected, i.e. several technological situations, in which appearance of a set of casting-process defect corresponding to the control sample, might occur.

For clarification of the content of the situation and its quantitative assessment the identified subgroup of attributes is subjected to the repeated recognition. In this case, Euclidean distance is taken as a proximity measure of lines [7]:

$$d_i^2 = \sum_{k=1}^{K} \left| y_{i,k} - y_{j,k} \right|^2 \tag{2}$$

The line of images in the diagnostic table, which has the lowest value d^2 is a probable technological situation conditioned the emergence of defects in the control sample.

To use the diagnostics results recommendations should be provided helping to develop optimal control actions on TP.

To achieve a higher level of technological situation recognizability the change in the nature and content of the **TDS** is required along with the accumulation of information. In order to adapt **IDM** to changing production conditions the subprogram of its self-education is organized. Implementation of the self-educating subprogram consists in the update of data and elimination of redundant information. Synthesis of new diagnostic line image-feature takes place, based on the analysis of a series of control samples.

In cases when the occurrence of defects depends on the violation of the technological process and can not be associated with control actions, the parameters, which do not correspond to standards, are identified, and percentage of reject due to violation of the technology is calculated.

The TP control subsystem on the level of casting defects is mainly used for PP stabilization, identification of tendency for TP rejection and generation of control actions aimed at their elimination.

A significant delay of information due to the interval duration from the beginning of its formation until its inspection after the preliminary cleaning and large volume to be controlled is not always possible to effectively use the subsystem for PP optimization.

A promising method to diagnose PP is to use the complex of technological probe-sensors having a greater sensitivity to the formation of defects than the industrial castings. In this case, the information about the TP state, deviation of process parameters or their unfavorable combination can be obtained much earlier, before the increase in reject of cast-steels or their defects. This information will help to optimize the parameters of TP through preventive measures of technological and organizational character aimed at PP correction.

In the analysis of defects and cast-steels reject in the foundry shops it was established that during casings into of expendable sand-clay molds 80% of all types of defects are blow-holes and porosity, hot cracks and clogging.

Qualitative and quantitative diagnosis of PP according to the listed types of defects is carried out by a complex consisting of four technological samples [3]. These probes help to reproduce specific types of defects and obtain their quantitative evaluation based on the combination of PP variables factors.

The first two samples in the complex, designed to detect propensity for hot cracking in castings, depending on the steel chemical composition, temperature and casting speed, the deceleration degree of linear shrinkage by the mold.

The first technological sample represents five bars of different lengths with elements of deceleration of shrinkage by the mold common thermal node. The critical rod length corresponding to its destruction at different PP factors is determined by calculation [8, 9] and experimentally. The Assessment of PP trends to hot cracking is the number of bars with hot cracks in the conjugation with the thermal sample node.

The second technological sample is a traditional grille with a thermal node and a massive middle core, where the formation of hot crack takes place, by the degree of crack opening the alloy hot brittleness is evaluated.

The third sample provides a quantitative differentiated assessment of the causes for gas defects. It consists of two annular elements with a common gating system. The total area of the gas defects in each element, determined by X-ray diffraction method, is proportional to gas generating capacity of the rods made of the same sand and sodium silicate mixture, as well as the degree of steel deoxidation.

The last sample serves for the probabilistic assessment of the causes of clogging, depending on the physical and mechanical properties of the molding sand and conditions of forming [10]. A quantitative measure of clogging formation probability is the number of the destroyed sections of various heights of the sample print after its removal from the half-mold.

All technological samples are mounted on the molding plate and produced simultaneously with the production cast-steels. The first three samples are poured with metal and analyzed for the position of castings shake-out, the result of the fourth sample is obtained right after the completion of the forming process.

The diagnostic information characterizing PP includes two components: quantitative estimates of the level of defects in samples and values of technological parameters. From these components based on the experimental data a training sample (**TS**) is formed, further used for evaluation of probable PP parameters PP, conditioned the appearance of a certain level of samples defects (Table 1). TS has N lines, the number of which is specified in the course of data storage. Each line is characterized by assessments of the level of sample defects and PP parameters. Parameters L contain the most significant ones influencing the defects appearance: the temperature and the casting speed; content of carbon, sulfur and residual aluminum in the steel; humidity, strength, gas permeability, compressibility, and the content of organic impurities, friability – for the molding sand; strength and gas permeability and friability – for core sand. The listed parameters are selected according to the results of statistical investigations of the reasons of defects in steel cast-steels, they are characterized by high information content, absence of meaningful correlations in the field of their determination and the possibility for their effective change.

Information and diagnostic model of the TP control subsystem as for information on technological samples defects is similar to the control subsystem model for the level of defects in cast-steels. The information characterizing the PP state, is composed of two components – quantitative estimates of the samples defectiveness and process parameters values.

Analysis of the current state of PP is conducted the following way: the information on the set and quantitative characteristics of samples technological defects obtained at some point in time, is entered into the computer, running the diagnostic algorithm. The algorithm provides a comparison of the presented levels of defects in samples with the possible situations of the training sample. Any situation from the recognized ones is characterized by a well-defined set of PP parameters, purposeful change of which reduces defects in the samples and correlation associated with it the level of casting defects [3].

Usage of samples depends on the value and nature of external disturbances. During the stable low level of casting defects and small external disturbances the complex of samples is used one – two times for the flow in a shift.

3. Conclusions

Experimental and industrial tests of the quality control system of casting showed recognition rate up to 80%, which is acceptable for solution of technical problems.

For the development of the system and production of more informative data on the casting defects it is recommended the introduce innovative non-destructive control methods (NCM) – X-ray tomography on the basis of the betatron to control internal technological casting defects in the cast-steels of type "discontinuity" (now they are not controlled) and thermal unit based on the infrared thermal imager for detection of surface cracks (the control is difficult has a selective character). These NCMs allow the complete additive control to be performed in the flow of internal and surface defects of cast-steels and store these data in electronic form for each manufactured casting. This information will complement the product passport with new data, raise control of castings and products manufacturing to a new level, with analysis of various components failure during their operation, mathematical modeling of components cast structures, calculation of their reliability and lifetime prediction.

Development and implementation of models into production and parallel automation of production control processes (decision-making) will minimize the impact of human factors on the quality of products, stabilize and optimize the production process of cast products.

4. References

[1] Knyazev S V, Skopich D V et al 2015 *Proc. SibSIU* **2** (12) 31–33.
[2] Knyazev S V, Antipenko V I et al 1992 *Foundry Production* **4** 26–27
[3] Antipenko V I and Knyazev S V 1987 *Soviet Castings Technology (Liteinoe Proizvodstvo)* 7 34
[4] Vasilyev V I 1983 *Handbook on System Recognition* (Kiev: Naukova Dumka) p 421
[5] Igolkin V N and Borisovich V N 1978 *Statistical Classification Based on the Sampling Distribution* (L.: LSU) p 104
[6] For A 1989 *Perception and Image Recognition* (M.: Mashinostroyeniye) p 272
[7] Eliseev I I and Rukovishnikov V O 1977 *Grouping, Correlation, Images Recognition* (M.: Statistika) p 144
[8] Balandin G F 1979 *Fundamentals of the Theory of Casting Formation* (M.: Mashinostroyeniye) part 1 p 328
[9] Balandin G F 1979 *Fundamentals of the Theory of Casting Formation* (M.: Mashinostroyeniye) part 2 p 336
[10] Zhukovsky S S 1989 *Strength of the Mold* (M.: Mashinostroyeniye) p 288

A study on structure and tribological properties of the electroerosion coating Mo-Ni-Cu, formed by the mixed method on copper

D A Romanov[1], E N Goncharova[1], V E Gromov[1] and Yu F Ivanov[2]

[1] Department of Physics n.a. Prof. V.M. Finkel, Siberian State Industrial University, 42 Kirova Street, Novokuznetsk, 654007, Russia

[2] Institute of High Current Electronics, Siberian Brunch of the Russian Academy of Sciences, 2/3 Akademichesky Ave, Tomsk, 634055, Russia

E-mail: romanov_da@physics.sibsiu.ru

Abstract. Multi-layered coating from immiscible components based on the system Mo-Ni-Cu was formed by the combined method of electro-explosive sputtering and subsequent irradiation by high-intensity pulse electron beam of submillisecond duration of influence on the surface of electrical copper contact (M00 grade of copper). The structure and phase composition studies of the applied coating as well as its mechanical and tribological properties are carried out.

1. Introduction

About a quarter of the world's silver production is consumed by electrical technologies industry for manufacture of interrupting electrical contacts of low-voltage equipment. Because of the relatively high cost of such contacts, the developments of non-silver interrupting contacts, which can provide reliable operation of the commutation devices at low prices, are carried out. Copper, as a substitute of silver in the electrical contacts, has properties set which allow alloys, based on copper, with increased resistance to arc wear and welding, to be created. Thus, composite alloys from cooper are widely applied in the electrical technology as a contact material for low and high-voltage circuit breakers, inserts of plasmatrons, electrodes of welding machines, etc. [2]. Refractory metals, carbides, oxides and nitrides are applied as strengthening phases in such materials. Composites are obtained by methods of powder metallurgy [2, 3] and advanced technique of electro-explosive sputtering (EES) [4-6].

The aim of this work is to research into the structure, elemental and phase composition, mechanical and tribological properties of the coating based on Mo-Ni-Cu, formed by combined method on the copper substrate.

2. Materials and methods

The coating based on Mo-Ni-Cu, formed by the combined method of electro-explosive sputtering and electron-beam treatment on the copper substrate (electrical copper of grade M00), served as a material for examination [2-5]. EES was carried out on the electro-explosive unit EVU 60/10M [7].

It includes capacitive energy storage and pulsed plasma accelerator, consisting of coaxially-butt system of electrodes with a conductor placed on them, flash chamber localizing explosion products and passing into the nozzle, through which it flows into the evacuated process chamber with residual pressure 100 Pa. The electro-explosion takes place due to the high-density current passing through the conductor during the capacitor discharge.

Coatings were deposited on the samples of annealed copper grade M00 with sizes 15×15×5mm. The mode of thermal and power influence on the irradiated surface was set by charge voltages election of capacitive energy storage, according to which the absorbed power density was calculated [7]. The electro-explosive sputtering was conducted with the use of a composite electro-explosive conductor [8] for deposition of coatings. The conductor consisted of a copper double layer foil with a powder sample of molybdenum and nickel. Their weights were: molybdenum – 0.217 g, nickel – 0.120 g and copper – 0.277 g. The power density, absorbed by the irradiated surface during sputtering, was 4.1GW/m^2, copper nozzle diameter of the flash chamber of the plasma accelerator – 20 mm, the distance of the sample from the nozzle section was 20 mm.

The subsequent treatment by high-intensity pulse electron beam was carried out on the unit SOLO (Institute of High Current Electronics Siberian Brunch of the Russian Academy of Sciences) [9] at the following parameters of the electron beam: 18keV, 55 J/cm^2, 100 µs, 0/3s^{-1}, 10 pul.

The studies of the phase composition and defect substructure of the coating were carried out by the methods of scanning and transmission electronic diffraction microscopy.

The wear resistance tests were carried out at a dry friction according to the scheme disc-ball on a high-temperature tribometer TNT-S-AH0000, CSEM, Switzerland. The indenter (counterbody) was a ball with a diameter 3 mm from hard alloy VK8. Tests were carried out at room temperature and relative humidity of 50% under the following conditions: normal load on the indenter 8 H, sliding velocity 3.5 cm/s, the track diameter 4 mm, the number of revolutions (cycles) 5000. The wear rate of surface layer was evaluated after profilometry of the formed track with the help of the laser optical profilometer Micro Measure 3D Station (Stil, France). Deterioration parameter was calculated by the formula:

$$V = \frac{2 * \pi * R * A}{F * L} \quad [\text{mm}^3/\text{N}\cdot\text{m}],$$

where R – the track radius [mm], A – cross-sectional area of the wear groove [mm^2], F – the value of the applied load [H], L – the distance covered by the ball [m].

3. Results and discussion

The coatings formed by the method of electric explosion of the conductive material, is typically characterized by a highly developed relief (lappings, drops, micropores) [7]. The subsequent irradiation of such coatings by high-intensity pulsed electron beam at a mode of surface layer melting under the influence of surface tension forces leads to smoothing of the irradiated surface (Figure 1 a, d).

At the optimal mode of irradiation the surface layer is formed with an island structure (Figure 1, b, d); in the islands volume the submicro-crystalline structure is found, the characteristic image of which is given in Figure 1 c, f. It should be noted that the optimization of the irradiation mode was performed by carrying out thermal calculations that the thermal field in the surface layer of the materials to be modeled.

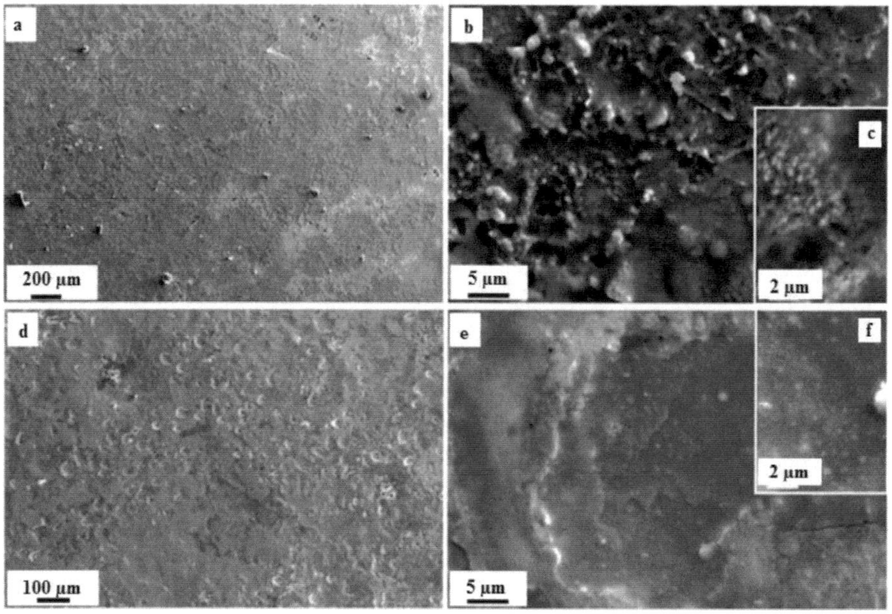

Figure 1. Structure of the coating surface Mo-Ni-Cu, formed on the cupper sample and irradiated by high intense pulse electron beam.

The formed coating is multi-layered (Figure 2). Analyzing the results presented in this figure the surface layer can be distinguished (Figure 2 b, c, layer 1), firmed in the result of coatings melting by high intensity electron beam; transitional layer characterized by a large amount of micropores (Figure 2 b, layer 2), and coating layer adjoining to the substrate (Figure 2, b layer 3).

At the same time, in the near-surface volume of the substrate, adjoining to the coating, the structure with a submicron size of crystallites is identified (Figure 2 b, layers 4 and 5). The copper substrate is a polycrystalline aggregate, the grain sizes of which are units-tens of micrometers (Figure 2 a, layer 6).

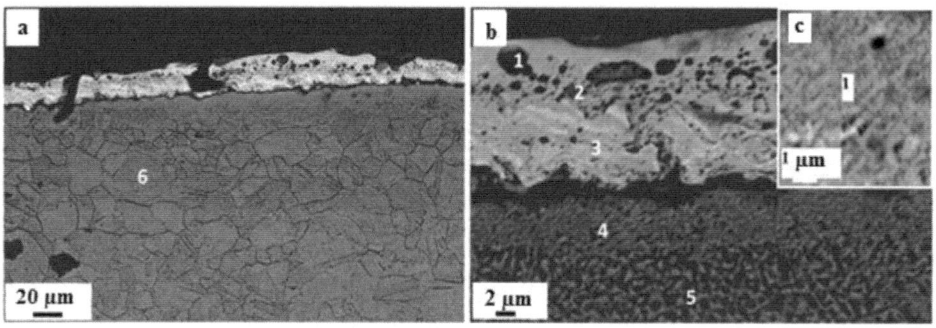

Figure 2. Structure of the transversal section of the coating Mo-Ni-Cu, formed on the copper sample and irradiated by high intensity pulse electron beam.

The performed investigations show that the identified layers differ not only it their structure but in the elemental composition, which is proved by the results of micro X-ray spectral analysis of the coating elemental composition given in Figure 3 and 4 in Table 1.

Table 1. Elemental composition of the modified layer presented in Figure 3a.

Zone No.	Element, at. %

	Mo	Ni	Cu	C	O
1	2.3	6.2	16.4	67.6	7.1
2	27.0	14.3	11.6	36.6	10
3	49.5	17.3	15.1	0.0	16.8

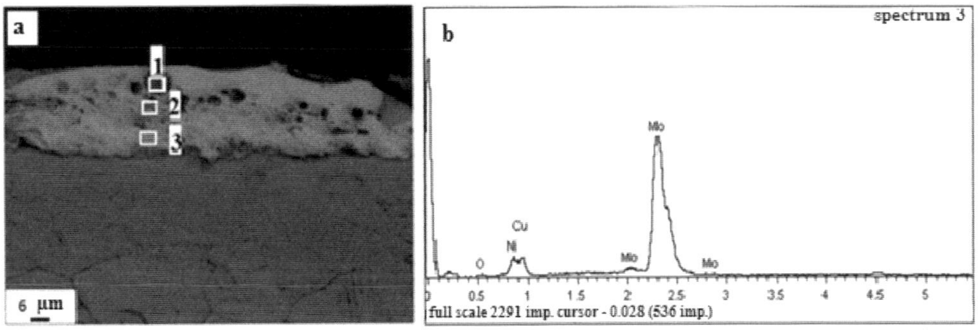

Figure 3. Structure of the transversal section of the coating Mo-Ni-Cu, irradiated by high intensity pulse electron beam (a); b – energy spectra obtained from the zone 3 (a). Scanning electron microscopy.

Analyzing the results in Figure 3 and in Table 1, it can be noted that the elements of the sputtered coating along the thickness of the modified layer are distributed unevenly. And namely, the concentration of the alloying elements Mo and Ni increases as the distance from the coating surface rises. Additionally to the sputtered elements in the coating the atoms of carbon and oxygen are found. The presence of oxygen and carbon atoms indicates the possibility of carbides and metal oxides formation in the coating.

The phase composition and the defect substructure of coatings were analyzed by the methods of transmitting electron diffraction microscopy of thin foils. The coating layer located at a depth of ≈15 μm was examined.

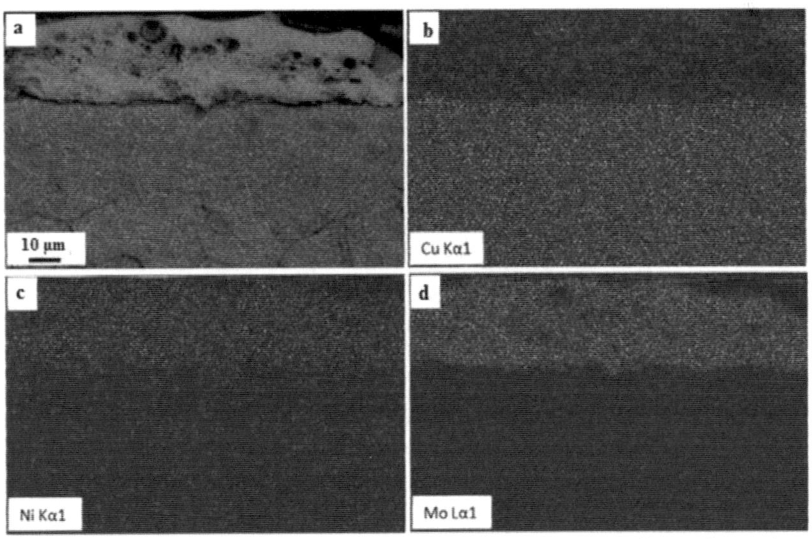

Figure 4. Structure of the transversal coating section of the system Mo-Ni-Cu, formed on the copper sample and irradiated by high intensity pulse electron beam, revealed during mapping of the elemental composition of the material [13]; a – image, obtained by imposition of the X-Ray radiation of the basic elements in the material; b – image obtained in the X-Ray radiation of copper atoms; c –nickel atoms; d – molybdenum atoms.

Electron-microscopic image of the defect coating substructure is presented in Figure 5. The analysis of micro electron-diffraction patterns obtained from such zones of the material makes it possible to conclude that the copper has this substructure. Thus, in the copper matrix the reticular (Figure 5 a), cellular-reticular, band and fragmented dislocation substructures are identified.

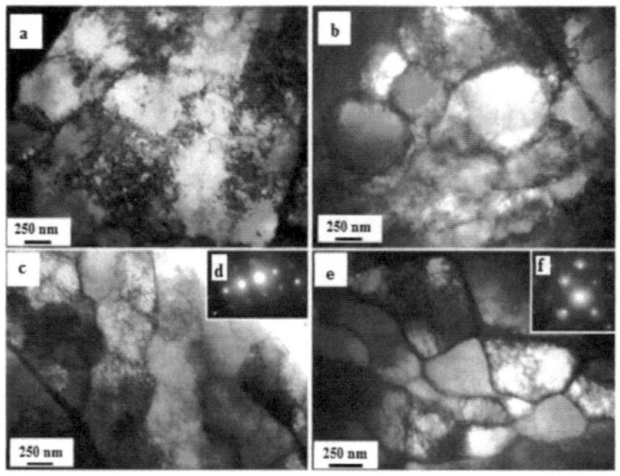

Figure 5. Electron microscopic image of defective coating substructure of Mo-Ni-Cu composition, formed on the copper; a, b, c, e – bright fields; d, f – micro electron-diffraction patterns for (c) and (e).

Alongside with a single-phase there are two-phase zones of the foil, and namely, the zones, containing copper and molybdenum. The typical images of such structures are presented in Figure 6. It can be clearly seen that in the process of high speed crystallization the two-phase structure of lamellar (Figure 6 a-c) or cellar type (Figure 6, d) is formed. In the first case molybdenum in the form laminas is located in the copper matrix (Figure 6 a-c); in the second – molybdenum in the form of interlayer contours the copper cellars (Figure 6 d-e).

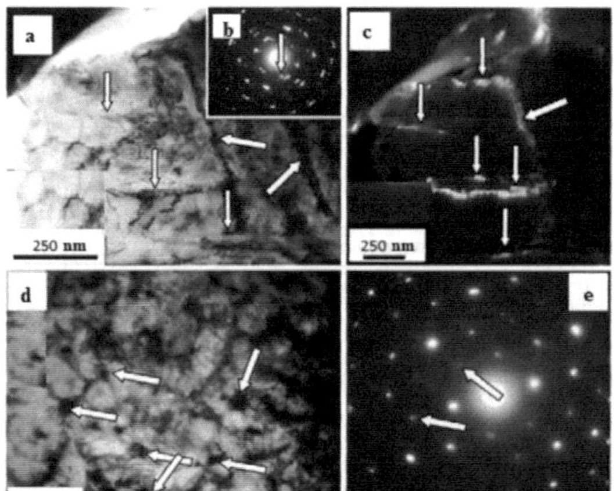

Figure 6. Electron-microscopic image of the coating structure of Mo-Ni-Cu composition, formed on the cooper; a, d – bright fields; b, e – micro electron-diffraction patterns for (a) and (d); c – dark field obtained in the reflections [111]Cu + [110]Mo; the arrows in (a) and (d) indicate the molybdenum particles; in (b) – reflections, in which a dark field was obtained (c); in (e) – reflections of type [002]Mo and [220]Mo.

The strength characteristics of the coating were analyzed to determine the hardness value by the methods of nano-hardness measurement. Alongside with the nano-hardness the value of Young modulus was determined. The results are given in Figure 7.

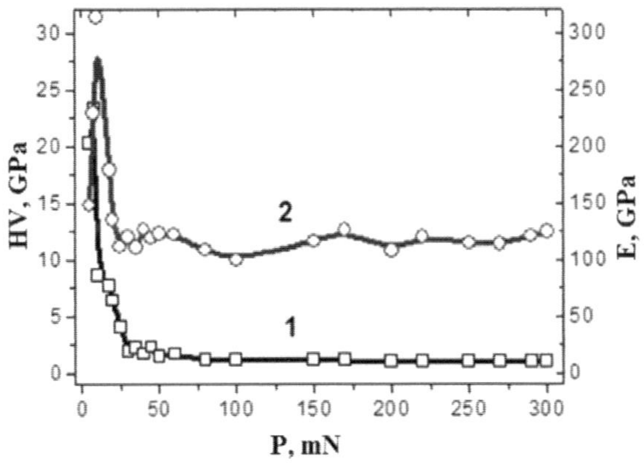

Figure 7. Dependence of hardness value (curve 1) and Young modulus (curve 2) of the coating with composition Mo-Ni-Cu, treated by electron beam from the load on indentor.

Analyzing the presented in Figure 7 dependencies, it can be noted that at low loads on indentor (5-25 mN) the hardness and Young modulus achieve rather significant values, that points on the strong influence of the oxide film formed on the material surface. At load on the indentor not higher 30-40 mN, hardness and Young modulus of the examined coating practically do not change, and consequently, correspond to the hardness of material coating.

From the reference data it follows that microhardness of copper is 0.6 GPa [3], microhardness of system coating Mo-Ni-Cu is ≈1.0 GPa. Thus, coating hardness excesses the hardness of copper substrate in ≈1.7 times. Young modulus of commercially pure copper changes within range (110-120) GPa [7], which is close to coating Young modulus of the system Mo-Ni-Cu.

One of the important characteristics of the electrical contacts is their durability. Results of durability tests for the coating with composition Mo-Ni-Cu are shown in Table. 2. Analysis of the presented in Table 2 measurement results of specific wear volume (V) and the average value of the friction coefficient (f) shows the coating system based on the system Mo-Ni-Cu formed on the surface of copper is characterized by high values of wear resistance and close values of the friction coefficient (f).

Table 2. Tribological test results of coating with composition Mo-Ni-Cu formed on copper.

Mode	V, 10^{-6}, $mm^3/N\cdot m$	V(initial) V(coating)	<f>	f(min)	f(max)
Cu initial	76.7		0.332	0.116	0.412
Mo-Ni-Cu	33.4	2.3	0.408	0.2	0.442

4. Conclusions

- By the electron plasma combined method including electro-explosive spattering and subsequent irradiation by high-intensity pulse electron beam on the surface of copper a multi-phase multi-layer coating on the basis of the system Mo-Ni-Cu was formed, characterized by high values of hardness and wear resistance.
- The specific wear volume of the coating of the system Mo-Ni-Cu exceeds the similar parameter of electrical copper by more than 2.3 times.
- The coefficient of friction values of the coating system Mo-Ni-Cu and electrical copper have similar values.

5. Acknowledgements
The authors are grateful for the financial support:
1. Russian Foundation for Basic Research in the framework of a research project No. 16-32-60032 mol_a_dk;
2. Grant of the President of the Russian Federation for state support of young Russian scientists – candidates of sciences MK-4166.2015.2;
3. State Task of the Ministry of Education and Science No. 2708 and 3.1496.2014/K to perform the scientific research;
4. Russian Science Foundation (project № 14-29-00091) for supporting the choice of the optimal exposure modes of surface coating by high-intensity pulse electron beam on the unit SOLO;
5. As well as the centre of "Materials Science" SibSIU for partial execution of research.

6. References
[1] Stockel D 1983 *Entwicklungsrichtungen den Werkstoffer fur Elektrische Kontakte. Metall* (W. Berlin) 1 30–36
[2] Berezin V B 1983 *Reference Book on Electrical Materials* (M.: Energoatomizdat) p 399
[3] Zuhailawati H 2008 *J. of Alloys and Compounds* **464** 185–189
[4] Romanov D A, Budovskikh E A and Gromov V E 2011 *J. of Surface Investigation. X-ray, Synchrotron and Neutron Techniques* **5** (6) 1112–17.
[5] Budovskikh E A, Gromov V E and Romanov D A 2013 *Doklady Physics* **58** 82–84
[6] Romanov D A, Goncharova E N, Budovskikh E A et al 2915 *Inorganic Materials: Applied Research* **6** (5) 536–541
[7] Romanov D A et al 2014 *Electro-explosive Sputtering of Wear and Electroerosive Resistant Coatings* (Novokuznetsk: Poligrafist) p 203
[8] Romanov D A et al 2013 *Composite Electrically Exploded Conductor for Electro-explosive Alloying of Metals and Alloys Surfaces* pat. of the Russian Federation No. 2478732
[9] Koval N N and Ivanov Yu F 2008 *Izvestiya Vuzov. Fizika* **5** 60–70
[10] Ivanov Yu F, Koval N N et al 2013 *High Temperature Material Processes* **17** (4), 241–256
[11] Ivanov Yu F et al 2015 *Izvestiya Vuzov.* Fizika **58** (4) 46–51
[12] Ivanov Yu F et al 2015 *Izvestiya Vuzov. Fizika* **58** (6/2) 91–95
[13] Brandon D and Kaplan U 2006 *Microstructure of Materials. Methods of Research and Control* (M.: Tekhnosfera) p 384

FORMATION OF THE INCREASED WEAR-RESISTANT PROPERTIES OF HARDOX 450 STEEL BY DEPOSITED COATINGS

S V Konovalov[1], V E Kormyshev[1], E V Kapralov[1], Y F Ivanov[2], O V Zavatski[1] and V E Gromov[1]

[1] Department of Physics n.a. Prof. V.M. Finkel, Siberian State Industrial University, 42 Kirova Street, Novokuznetsk, 654007, Russia
[2] Institute of High Current Electronics, Siberian Brunch of the Russian Academy of Sciences, 2/3 Akademichesky Ave, Tomsk, 634055, Russia

E-mail: konovserg@gmail.com

Abstract. The structure-phase conditions formed during the deposition of surface coatings on Hardox 450 steel by the wire comprising C, V, Cr, Nb, W were examined by the methods of X-ray phase analysis and transmission electron diffraction microscopy. It was established that after the hardened layer was deposited, the wear-resistance increased 153 times and the coefficient of friction in the material decreased 2.5 times. It was concluded that the increased properties of the surface coating were due to the formation of martensitic structure and the occurrence of high volume fraction of carbide phase inclusions.

1. Introduction

Problems of surface restoration and hardening of machine and mechanism elements appeared when the mechanical engineering became one of the basic branches of industrial production and continue to be relevant nowadays.

Among the technologies of restoration and deposition of hardening coatings the leading position belongs to the arc surfacing method due to the deposition of coatings on the surfaces of large scale equipment operating under conditions of abrasive wear. For example, on the components of groundcare, construction and mining equipment. In these cases the choice of the arc surfacing method is conditioned not only by the possibility of deposition volume coatings, up to several millimeters of thickness, but also the possibility of formation of high-strength and wear-resistant structures in the surface coating by adjusting the composition of the deposited electrode. Thus, there is a specific interest in application of powder electrodes (wires) for deposition of coatings with the desired physical and strength properties, which allow the phase composition of the deposited material to be varied greatly by combining the powder components during the electrode production [1-9].

In our earlier works [10, 11] we studied the features of structure formation and properties of wear-resistant coatings deposited on Hardox 400 steel under conditions of strengthening particles synthesis from the components of flux-cored wires, including the coatings deposited by the flux-cored wires containing carbide- and boride-forming metals, carbon and boron in the structure-free state that ensure the formation of a coating structure with high operational properties.

The aim of this work is to expand the research and to conduct the metallographic analysis of the impact of deposition by the flux-cored wire containing C, V, Cr, Nb, W on the martensitic steel Hardox 450 and on the change in structure and properties of the formed coating.

2. Materials and methods

Hardox 450 steel ((weight %): 0.19-0.26% C; 0.70% Si; 1.6% Mn; 0.025% P; 0.010% S; 0.25% Cr; 0.25% Ni; 0.25% Mo, 0,004% B; the rest – Fe) was used as a base material. As well as Hardox 400 steel, Hardox 450 steel is intended for conditions under which special requirements for wear-resistance alongside with good cold bending properties and weldability are imposed. Field of application for Hardox 450: dump truck bodies, containers, crushers, screeners, feeders, skip winders, knife cutting edges, conveyors, buckets, knives, gears, sprockets, etc. [12].

The coatings were deposited by the flux-cored wire, having the following chemical composition: (weight %): 1.4% C; 1.0% V; 7.0% Cr; 8.0% Nb; 1.2% W; the rest – Fe. Single- and double-layer coatings were applied. Double-layer coating was carried out perpendicularly to the single-layer coating. The coating was deposited in the shielding gas, having the chemical composition: Ar 82%, CO_2 18% with arc current 250 – 300 A and arc volts 30 – 35 V.

The studies of the phase composition and defect substructure of steel and deposited metal were carried out by the methods of X-ray structure analysis and transmission electron diffraction microscopy. The foils were produced by the method of electrolytic thinning of plates cut from the deposited metal (layer located at half the thickness of the deposited metal), contact zone of the deposited metal and the steel at a distance of 15 mm from the contact zone in the volume of the steel sample.

3. Results and discussion

Microhardness analysis along the cross-section of the formed coating after single-layer deposition revealed the following. The maximum value of microhardness 11 GPa is observed in the layer with 5 mm of thickness near the surface. Then it non-monotonically decreases and constitutes 6.5 GPa in the base material. Thus, the deposition of coating leads to an almost two-fold increase in microhardness.

Formation of the second deposited layer results in the increase of microhardness up to 12.5 GPa. This value is observed in the layer of approximately 1 mm thickness. As the distance from it increases the microhardness (at a distance of 2.5-3.5 mm) decreases up to 7 GPa with its further increase to maximal values 10 GPa at a distance of 4.5-5.5 mm. Then microhardness decreases to values of the base material. Apparently, sharp decrease in microhardness is connected with the contact zone of single- and repeated-metal deposition.

Analysis of tribological tests showed the following. If the wear performance of material is 95.10*10-6 $mm^3/N{\cdot}m$, then after a single deposition – 0.69*10-6 $mm^3/N{\cdot}m$, and after a double deposition – 0.62*10-6 $mm^3/N{\cdot}m$. Consequently, deposition leads to the increase of wear resistance in 153 times after a double deposition and in 138 times after a single one. In its turn, the coefficient of friction of the base material is 0.259, of the single deposited material is 0.104, and of the repeatedly deposed material is 0.132, i.e. it decreases by almost 2.5 times.

The specified increased strength and tribological properties of the deposited metal are conditioned by the state of the material structure, namely, its phase composition and the state of the defect substructure.

By the methods of X-ray structural analysis it was found that material of a single deposition contains the following phases: α-Fe – 63.2%, Fe_3O_4 – 28.8%, Fe_3C – 6.0%, CrC – 2.0%; double deposition: $M_{23}C_6$ – 52.3%, α-Fe – 46.7%, Nb_6C_5 1.0% (in wt.%). Thus, in the material of single and double deposition a multiphase structure is formed. The basic phase (excluding α-phase) in the single deposition is iron oxide of Fe_3O_4 composition.

The basic phase of the repeatedly deposited coating are carbides based on the special elements (carbides Nb, W and Cr). Formation of the double deposition leads to a substantial increase in the volume fraction of carbide phase and the complete absence of oxide phase.

To establish the factors responsible for the increased properties by the methods of transmission diffraction electron microscopy the analysis of the defect substructure of base material and deposition was carried out.

Studies of the base material revealed the structure, which was formed as a result of quenching and subsequent tempering of steel (tempering of martensite) (Figure 1). Fragmented crystals of lamellar morphology, formed during the martensitic transformation, were found in the α-phase grains. Particles of carbide phase are found in the volume and along the boundaries of crystals.

Structure of subgrain type is found more rarely, the typical image of which is given in Figure 2.

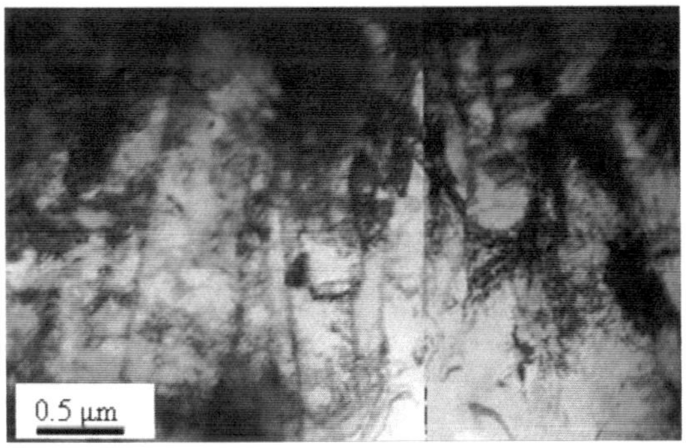

Figure 1. Electron microscope image of the structure of Hardox 450 steel.

Figure 2. Structure of the subgrain type in Hardox 450 steel.

The structure of the metal contact zone of the deposited coating and the basic material shares many similar characteristics with the structure of the initial steel. Martensitic structure is revealed. In its crystals the dislocation substructure in the form of randomly distributed dislocations and dislocation tangles can be seen (Figure 3).

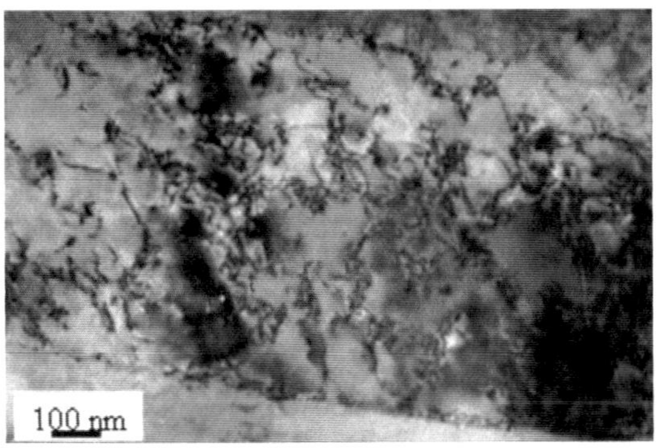

Figure 3. Randomly distributed dislocations and dislocation tangles formed in the structure of α-phase of the contact zone of deposition and base metal.

The typical electron microscope images of the structure of the repeatedly applied coating are presented in Figure 4. It can be seen that the formed material is multiphase. In Figure 4a the inclusions of submicron size, which are difficult to electrolytically polish, are indicated by arrows. The results of the X-ray phase analysis show that these inclusions belong to the carbide phase (carbides of the composition $M_{23}C_6$ ((Cr, Fe, W)$_{23}C_6$)). The other phase is well polished and obviously is a solid solution on the basis of α-Fe (Figure 4b).

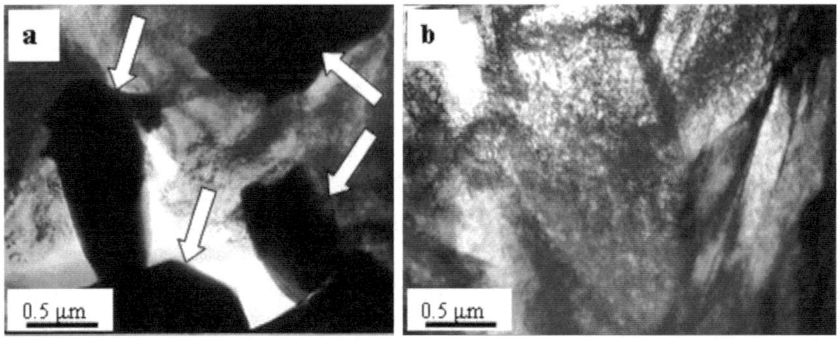

Figure 4. Electron microscope image of the structure of the repeated deposition on the steel.

The characteristic electron microscope image of the structure of α-phase of the repeated deposition is given in Figure 5. The lamellar structure formed as an effect of martensitic transformation is clearly visible.

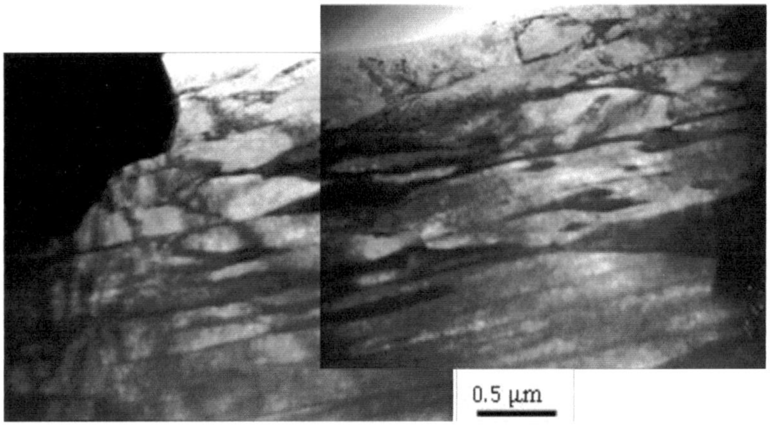

Figure 5. Lammelar structure in the repeated deposition on the steel.

The dislocation substructure in the shape of polydimensional networks is found in the crystals of martensite (Figure 6a). In some cases transformation twins are founded in the volume of martensite crystals (Figure 6b). Niobium carbide NbC nanoparticles of the second phase – are found inside and on the boundaries of martensite crystals.

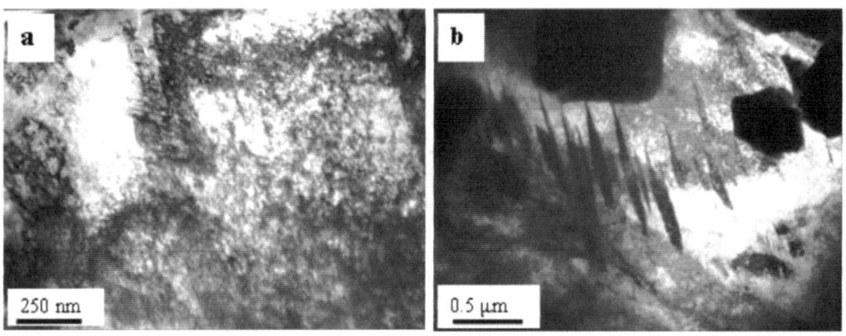

Figure 6. Polydimensional networks and transformation twins in the structure of the repeated deposition on the steel.

4. Conclusions

The research of the phase composition, defect substructure, tribological properties and microhardness of the singular and double coating deposited on the steel by C, V, Cr, Nb, W wire is carried out. It is shown that wear resistance of the deposited metal is about 153 times higher compared to Hardox 450 wear resistance; coefficient of friction of the deposited metal is 2.5 times lower than the steel friction coefficient. It is shown that formation of the second deposited coating does not have a great influence on the tribological characteristics of the material. It is found that the increased mechanical and tribological properties of the deposited layer are conditioned by formation of a multiphase submicron and nano-dimensional structure. The hardening of this structure is associated with the formation of a martensite structure of α-matrix and the presence of high (more than 50%) volume fraction of carbide phase impurities based on Fe, Cr, W and Nb.

5. Acknowledgements

The research was performed under the grant of the Russian Science Foundation (project No. 15-19-00065).

6. References

[1] D'Ans, Pierre and Degrez Marc 2015 *Surface and Coatings Technology* **276** 349–359
[2] Yang Xincheng, Ma Xiaolong, Moering Jordan, Zhou Hao, Wang Wei, Gong Yulan, Tao Jingmei, Zhu Yuntian and Zhu Xinkun 2015 *Materials Science and Engineering: A* **645** 280–285
[3] Rao Lei, Wang Shuang-jun, Zhao Jian-hua, Geng Mao-peng and Ding Gang 201) *J. of Iron and Steel Research, International* **21** 869–877
[4] Słoma J, Szczygieł I and Sachajdak A 2011 *Archives of Civil and Mechanical Engineering* **11** (2) 437–449
[5] Chen Xizhang, Yuan Qibing, Madigan Bruce and Xue Wei 2015 *Corrosion Science* **96** 178–185
[6] Chen Xizhang, Shen Zheng, Chen Xing, Yucheng and Huang Qunying 2011 *Fusion Engineering and Design* **86** 2943–48
[7] Chen Xizhang, Fang Yuanyuan, Li Peng et al 2015 *Materials & Design* **65** 1214–21
[8] Yilbas, Bekir Sami et al 2013 *Laser Surface Processing and Model Studies* (Berlin: Springer)
[9] Senthilkumara B and Kannanb T 2015 *Measurement* **62** 127–136
[10] Kapralov E V, Budovskikh E A, Gromov V E and Ivanov Yu F 2015 *Russian Physics Journal* **58** (4) 471–477
[11] Kapralov E V, Raykov S V, Budovskikh E A, Gromov V E, Vashchuk E S and Ivanov Yu F 2014 *Bulletin of the Russian Academy of Sciences: Physics* **78** (10) 1015–21
[12] Grnezh B 2008 Mining **3** (79) 34–38
[13] Zuev L B, Danilov V I et al 2009 *Physics of the Solid State* **51** (6) 1137–41
[14] Olesyuk O V, Konovalov S V and Romanov D A 2014 *Modern problems of science and education* **2** 85

IOP Conf. Series: Materials Science and Engineering **150** (2016) 012042 doi:10.1088/1757-899X/150/1/012042

ELECTRO-EXPLOSIVE DOPING OF VT6 TITANIUM ALLOY SURFACE BY BORON CARBIDE

T Yu Kobzareva[1], V E Gromov[1], Yu F Ivanov[2], E A Budovskkh[1] and S V Konovalov[1]

[1] Department of Physics n.a. Prof. V.M. Finkel, Siberian State Industrial University, 42 Kirova Street, Novokuznetsk, 654007, Russia

[2] Institute of High Current Electronics, Siberian Brunch of the Russian Academy of Sciences, 2/3 Akademichesky Ave, Tomsk, 634055, Russia

E-mail: gromov@physics.sibsiu.ru

Abstract. The studies carried out in this work target detection of changes in the surface layer of titanium alloy VT6 after electro-explosive alloying (EEA) by boron carbide. EEA of VT6 titanium alloy surface is the plasma alloying formed during the electric explosion of foil with the sample powder of boron carbide. Carbon fibers with weight 140 mg were used as an explosive conductor. Sample powder of boron carbide B4C was placed in the area of explosion on the carbon fibers. It was revealed that EEA of the surface layers of titanium alloy samples VT6 leads to the modification of the layer, thickness of which changes from 10 μm to 50 μm. Heterogeneous distribution of alloying elements was found in the treatment zone by the methods of X-ray microanalysis. A significant difference in their concentration in the identified layers leads to difference in their structural and tribological behaviour. It was revealed that after electro-explosive alloying the microhardness of titanium alloy VT6 significantly increases. Electro-explosive alloying leads to the formation of a structure of submicro- and nano-scale level. It allows strength and tribological properties of the treated surface to be increased.

1. Introduction

Nowadays, titanium and its alloys is one of the advanced structural metals. More than one hundred types of titanium alloys are known. However, only about 30% of titanium alloys have found their practical application in different areas: aviation, rocket engineering, shipbuilding, chemical and medicine industries [1-3].

Titanium and its alloys are characterized by absence of coldbrittleness, high plasticity, durability and corrosion resistance, especially in oxidizing and chlorinated mediums and they also have high antifriction properties [4]. Beyond that point these materials have low wear resistance, high tendency to sticking, high index of friction in pair with most materials [5]. The listed disadvantages of titanium alloys limit their usage in the manufacture of details exposed to friction.

To improve the surface properties of the products from titanium and its alloys and to protect them from the external influences (temperature, pressure, corrosion, erosion, etc.), different coatings are applied [6-10]. Coatings application solves two important problems: change in the original physical and chemical properties of the treated surface, as well as the restoration of properties, sizes, mass of the influence surface under operation conditions.

One of the advanced methods of modification of the metals and alloys surface is the method of electro-explosive alloying (EEA), in which the surface is influenced by the pulsed plasma jets formed during the discharge of capacitive energy storage through the conductive material. Working medium of the plasma accelerator serves for heating and alloying of the material surface layer [11]. The main advantage of EEA over analogues methods of surface treatment is that any conducting material (thin foils of metals and alloys, graphitized carbon fibers and other fibers) can be used as a plasma forming substance.

Powder weights of different substances can be placed into the explosion area. They are carried by the formed jet to the irradiated surface partially transformed into the plasma state [12]. Thus, EEA allows the melting of the surface layer of the component, liquid-phase alloying by the products of conductors explosion with the subsequent self-hardening to be conducted in a single technological cycle. Industrial discharge-pulse installations having constructive simplicity, high reliability and relatively low cost can be used for implementation of EEA technology.

The aim of this research is to analyze the structural conditions of the modified layer, formed by the methods of electro-explosive alloying.

2. Materials and methods

The alloy based on VT6 titanium was used as a material for research. The chemical composition of the alloy corresponded to State Standard 19807-91.

The treatment of the surface layer was carried out by the methods of electro-explosive alloying [14]. Carbon fibers with weight 140 mg, were used as an explosive conductor. Sample powder of boron carbide B4C with weight 496 mg was placed into the area of explosion.

Electro-explosive laboratory-scale plant of EVU 60/10 type with the following characteristics was used for electro-explosive alloying: absorbed power density during the treatment of material surface ~109 W/m2, pressure in the shock-compressed layer of plasma near the irradiated area 106–107 Pa, processing time ~100 μs.

Conditions for a pulse liquid-phase alloying were set by the value of charge voltage of the accelerator energy storage, diameter of nozzle channel and the distance from its exit to the sample. Formation of the plasma flow was carried out at U=2.4 kV [11, 15].

The studies of the modified material structure were carried out by the methods of scanning electron microscopy. The elemental composition of the surface layer was analyzed by X-ray microanalysis. Microhardness of structural elements was carried out by Vickers method on the unit HV-1000A at the load 1 N. [16].

3. Results and discussion

After electro-explosive alloying of the surface of titanium alloy VT6 by boron carbide, a highly developed relief is formed, that includes lappings, drops, craters, and various surface defects – microcracks, micropores. Typical images of the surface structure after treatment are given in Figure 1. The scale of the structure elements of surface alloying is varied in a very wide range – from hundreds of micrometers (Figure 1a) up to tens-hundreds of nanometers (Figure 1b, c).

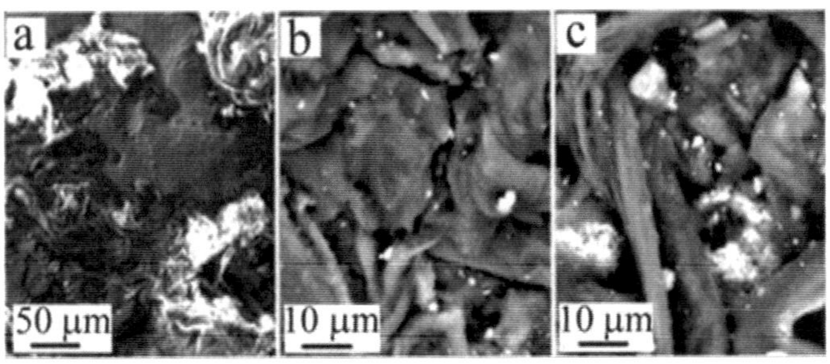

Figure 1. Surface structure of titanium alloy VT6 after electro-explosive alloying by boron carbide.

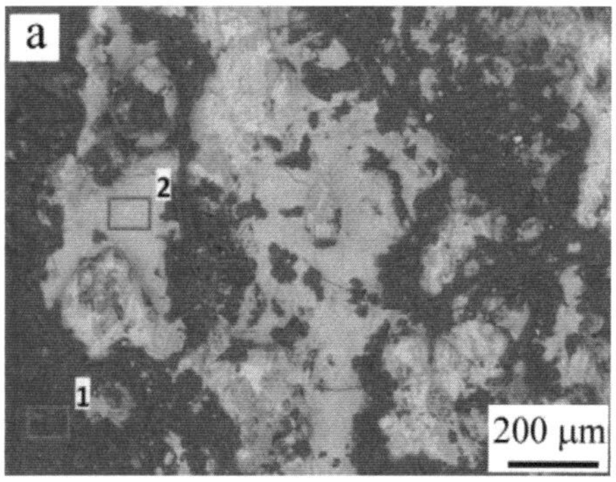

Element	Spectra, at. %	
	1	2
B	7.51	5.03
C	73.87	7.97
O	18.55	22.99
Ti	0.07	59.29
Al	0.0	2.82
V	0.0	1.9

Figure 2. X-ray microanalysis of titanium alloy VT6 surface after EEA by boron carbide. a – surface structure of titanium alloy VT6; b – energy spectrum obtained from area No. 2. The results of X-ray microanalysis of the areas, indicated in (a), are given in the table.

A heterogeneous distribution of alloying elements is observed in the layer of treatment. At a qualitative level it is confirmed by studies of the alloying surface in the back scattered electrons. Black-and-white contrast in Figure 2a proves the formation of areas enriched by the heavy element (titanium) and relatively light elements (carbon and boron). Quantitative analysis of alloying elements distribution on the alloyed surface was carried out by methods of X-ray microanalysis due to differences in elemental composition of different areas.

Analyzing the results from the table located in Figure 2, it may be noted that heterogeneity coefficient of alloying elements distribution in the surface layer (ratio of the total amount of boron, carbon and oxygen in the light and dark areas) reaches 2.8. Thus, electro-explosive alloying of titanium by boron carbide powder leads to the formation of volumes in the surface layer, in which the concentration of alloying elements differ by more than 2.5 times.

After the treatment, the high level of structural heterogeneity of the modified layer over its thickness and the distribution of alloying elements is observed. More clearly it can be seen while studying the structure of cross-sections. Because of the highly developed relief of the surface the total thickness of the alloyed layer varies from 10 µm to 50 µm (Figure 3).

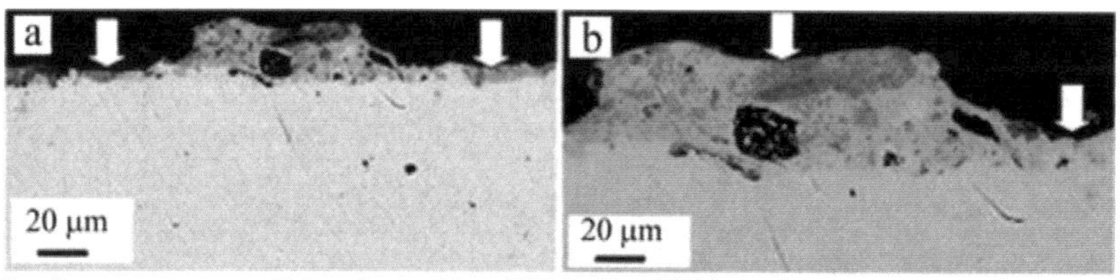

Figure 3. Surface structure of titanium alloy VT6 after EEA. a, b – structure of the cross-section. The EEA surface is indicated by the arrows.

Based on the morphological feature (degree of etchability) it is possible to distinguish at least four layers in the alloyed volume: surface layer (Figure 4, layer 1), intermediate layer (Figure 4, layer 2), transition layer (Figure 4, layer 3), and a layer of thermal influence, smoothly transforming into the bulk of the sample.

It should be noted, that the difference in the degree of etchability is observed not only between the layers but within each layer (Figure 4b). The identified layers are characterized by the substructure, in which the elements sizes vary within 1 µm (Figure 4c, d). Surface and intermediate layers predominantly are of acicular structure. Round shaped areas, differed from the surrounding volume of material in structure and composition are found in the alloyed layer (Figure 4, 5).

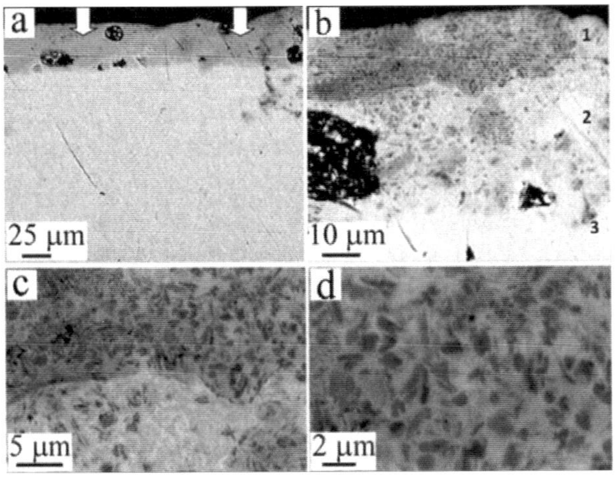

Figure 4. Cross-section structure of titanium alloy VT6 exposed to EEA by boron carbide. a – structure of the cross-section. The EEA surface is indicated by the arrows; b – identified layers. Layers are indicated by figures: 1 – surface layer; 2 – intermediate layer; 3 – transition layer; c – substructure of the surface and intermediate layers; d – substructure of the surface layer.

Different degree of layers etchability shows their difference in phase and elemental composition. The example of the elemental composition analysis of the alloyed layer is given in Figure 5.

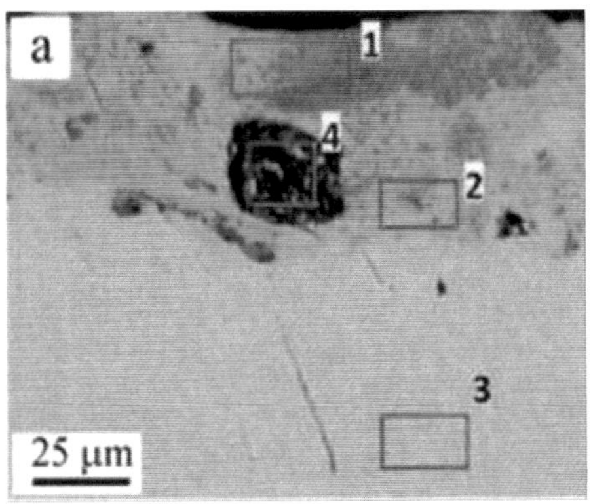

Element	Spectra, at. %			
	1	2	3	4
B	40.75	10.28	0.0	0.0
C	13.22	10.42	4.77	49.93
O	0.0	0.0	0.0	24.29
Si	0.0	0.0	0.0	5.3
Al	3.62	5.1	9.97	1.61
Ti	42.40	57.11	88.28	17.88
V	1.55	1.9	3.02	0.75

Figure 5. The X-ray microanalysis of the cross-section of titanium alloy VT6 after EEA by boron carbide. The table contains the results of X-ray microanalysis of the areas indicated in a (a – surface structure of titanium alloy VT6; b – energy spectrum obtained from area No. 1).

Analyzing the results from the table in Figure 4, a definite pattern in the distribution of alloying elements (carbon and boron) can be revealed. Namely, as the distance from the alloying surface increases (spectra 1-3) the concentration of boron and carbon atoms is reduced. In some cases, round shaped areas (Figure 4, section 4) are found in the alloyed layer, which significantly differs from the surrounding volume of material in structure and elemental composition. The main elements of these areas are carbon, oxygen and silicon (Figure 4, table, spectrum 4).

After EEA with the use of powder sample of TiB2 two layers can be identified (Figure 6) in the microhardness distribution in the depth of the alloying area. The thickness of the surface layer is 20-25 µm and of the transition layer is up to 20 µm. Increase in the weight of B4C powder sample by 8 times leads to the increase of microhardness in the surface layer from 850 to 2400 HV, i.e. by 2.8 times.

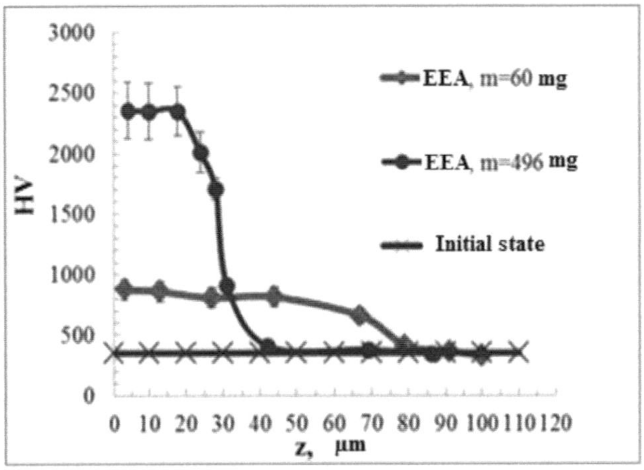

Figure 6. Microhardness distribution in the depth of EEA zone of titanium alloy surface VT6 by boron carbide.

4. Conclusions

Electro-explosive alloying of titanium alloy VT6 by boron carbide powder is carried out. The formation of the modified layer is revealed. Its thickness varies from 10 μm to 50 μm. It is shown that this layer has a gradient structure, characterized by regular change in the elemental composition depending on the distance to the alloyed surface. Microhardness of the surface exposed to EEA by boron carbide increases manifold.

5. Acknowledgements

The research was performed with financial support of the Russian President's grant for the state support of young Russian scientists – doctors of sciences (project MD-2920.2015.8) and state task No. 3.1496.2014/K.

6. References

[1] Chechulin B B, Ushkov S S, Razuvaeva I N and Goldfain V N 1977 *Titanium Alloys in Mechanical Engineering* (M: Mechanical Engineering) p 247
[2] Ilyin A A, Kolachev B A and Polkin I S *Titanium alloys. Composition, Structure, Properties* (M.: VILS – MATI) p 520
[3] Henriques V 2009 *J. Aerosp. Tech. Management* **1** (1) 7–18
[4] Boyer R and Briggs R D 2005 *J. Materials Eng. Perf.* **14** (6) 681–685
[5] Polmear Ya 2008 *Light Alloys: From Traditional Ones to Nanocrystals* (M.: Technosphere) 2008 468
[6] Suslov A G 2000 *Quality of the Surface Layer of Machine Elements* (M.: Engineering) p 320
[7] Legostaeva E V, Sharkeev Yu P, Yakovlev V I, Uvarkin P V and Kryazheva E G 2009 *Proc. 2nd Asian Symp. on Adv. Materials* pp 305–308
[8] Prokhorova A I and Borisova A O 2013 *Metals Technology* **7** 23–25
[9] Swapan K 2006 *Functional Coatings* (Wiley–VCH Verlag GmbH & Co. KGaA) pp 1–18
[10] Grilikhes S Ya and Tikhonov K I 1990 *Electrolytic and chemical coatings* (L.: Chemistry) p 288
[11] Bagautdinov A Ya, Budovskykh E A, Ivanov Yu F and Gromov V E 2007 *Physical Bases for Electro-explosive Alloying of Metals and Alloys* (Novokuznetsk: SibSIU) p 301

[12] Yakushin V L, Jumayev P S, Kalin B A et al 2009 *XIX Int. Meeting on Solid State Radiation Physics* (M.: NII PMT MGIEM (TU)) pp 298–305
[13] Prikhodko V M, Petrova L G and Chudina O V 2003 *Metallophysical Basis for the Development of Hardening Technology* (M.: Mechanical Engineering) p 384
[14] Kobzareva T Yu 2015 *Proc. XIX Int. Conference on Metallurgy: Technology, Innovation, Quality* (Novokuznetsk: SibSIU) pp 288–292
[15] Gromov V E, Kobzareva T Yu, Ivanov Yu F, Budovskikh E A and Bashchenko L P 2016 *AIP Conf. Proc.* **1698** (No. 030006) 5
[16] Karpii S V, Budovskikh E A, Ivanov Yu F et al 2010 *Int. Proc. on Fundamental Aspects of External Fields Influence on Material* (Novokuznetsk: Poligraphic Center) pp 407–41

Plasma metallurgical production of nanocrystalline borides and carbides

G V Galevsky[1], V V Rudneva[1], A N Cherepanov[2], S G Galevsky[3] and K A Efimova[1]

[1] Institute of Metallurgy and Material Science, Siberian State Industrial University, 42 Kirova Street, Novokuznetsk, 654007, Russia
[2] Institute of Theoretical and Applied Mechanics, Siberian Branch of the Russian Academy of Sciences, 4/1 Institutskaya Street, Novosibirsk, 630090, Russia
[3] National Mineral Resources University, 2 21st line V.O, Saint Petersburg, 199106, Russia

E-mail: kafcmet@sibsiu.ru

Abstract. The experience in production and study of properties of nanocrystalline borides and chromium carbides, titanium, silicon was summarized. The design and features of the vertical three-jet once-through reactor with power 150 kW, used in the plasma metallurgical production, was described. The technological, thermotechnical and resource characteristics of the reactor were identified. The parameters of borides and carbides synthesis, their main characteristics in the nanodispersed state and equipment-technological scheme of production were provided. Evaluation of engineering-and-economical performance of the laboratory and industrial levels of borides and carbides production and the state corresponding to the segment of the world market was carried out.

1. Introduction

In 2011 for the first time in the world more than 1.5 billion tonnes of steel – the most demanded construction material were produced. In the structure of constructional materials the percentage of iron-based alloys accounts for 95%, non-ferrous metals and alloys – 4%, the rest – less than 1%. However, the latter group is represented by a wide range of special-purpose materials vital for modern civilization. It includes materials corresponding to such criteria as "infusibility", "superhardness", "heat-resistance" and "refractoriness", which helps to solve many scientific and technological innovation and design problems. Therefore, in this group the materials based on carbides, borides, nitrides, silicides, and their composites occupy an important place. In the development of their national technological base the following steps can be distinguished. In the 50-60's mainly the Institute for Problems in Materials Science and the Institute of Superhard Materials of NAS of Ukraine investigated the properties and developed technologies for their production and use. In the 70-80's a number of territorial research centres on the basis of academic institutions in Moscow, Kiev, Riga, Novosibirsk made efforts to develop the production and implement these materials in the high-dispersed state. The 2000's were marked by the rise of technology, which is still ongoing and covers the development of nanotechnologies and nanomaterials at a new level

At present, Russia is among the leading countries in terms of the volume of conducted nano-technology research, but is significantly inferior in its production of nano-products and their export [1-8]. In accordance with the President's initiative of April 24, 2007 a system of state support for scientific and applied research into these issues was created, providing unification of state resources and private businesses and their concentration, development of the national nanotechnology infrastructure, improving the efficiency of commercialization and transfer of nano-developments, focus of efforts on the most attractive in commercial terms directions. Such directions include functional, structural and compositional nanomaterials.

In the Siberian region of the Russian Federation the foundation of plasma metallurgical nanotechnologies of high superhard materials was laid by the scientific school of Academician M.F. Zhukov. In the early 70's he brought together efforts of several research teams of academic, industrial and educational institutions in Western and Eastern Siberia in the direction of the laboratory production and application of nanopowders for different purposes. The results are reflected in the published afterwards in Russia and abroad multi-volume series "Low-temperature plasma" (Chief-editor Academician of RAS M.F. Zhukov), "Thermal plasma in the technology of new materials" (Science-editor Academician of RAS M.F. Zhukov), "Nanomaterials and nanotechnologies in the production of silicon carbide" (Scientific-editor Prof. G.V. Galevsky), etc. The studies confirmed the operability of plasma technological equipment, the possibility for synthesis of carbides, borides, nitrides in nanostate and their application in new areas. However, socio-economic changes, which took place in Russia at the turn of the 90's, led to the cessation of investments into this area, made it impossible to transit to an industrial level and commercialize the research.

In the XXI century in conditions of nano-technological recovery it seems that the continuation of previous studies for organization of industrial plasma metallurgical production of high-temperature chemical compounds, and identification of priority areas of their application in ceramic, electroplating, metallurgical and other technologies is an important scientific and economic problem relevant to the State program "Development of science and technology for the period till 2020".

The purpose of this paper is to summarize rich experience and the results achieved by Siberian State Industrial University (SibSIU) in creating a profile equipment and development of technologies for production of nano-dispersed borides and chromium carbides, titanium and silicon. The choice of production technology of borides and carbides as a study object is conditioned by a favorable combination of their consumer properties (hardness, refractoriness, wear resistance and heat resistance); availability of raw materials; relative simple furnace synthesis as a base technology; strong demand from consumers, especially as materials coating, coating of functional materials and modifying complexes, increasing the life cycle of products and tools in 3-4 times; a real opportunity to achieve new effects when used in the nanostate.

2. Plasma technological equipment: design, operation, characteristics

For the experimental study of synthesis of borides and carbides and their production SibSIU together with NPF "Polimet" created plasma technological industrial complex (Figure 1), surpassing the experimental variants developed before. The construction of plasmatrons, mixing chamber, the charge dosing device, bag filter are protected by patents of the Russian Federation No. 66877, 107440, 108319, 184916. The main characteristics of the reactor are given in Table 1. More detailed thermal engineering, resource and technological characteristics of the reactor are described in [9]. The plasma technological complex includes a three-jet reactor and systems of electricity, gas, water supply and ventilation, instrumentation and automation equipment, charge dosing, nano-dispersed products capture and neutralization of the exhaust gases.

To generate the plasma flow three arc electric gas heaters (plasmatrons) EDP-104AM with power up to 50 kW each are installed in the mixing chamber at an angle of 30° to the reactor axis. Plasmatrons EDP-104AM operate on the direct current with the following parameters of the electric arc: the arc voltage is up to 250 V, current – up to 200 A. The stabilization of the electric arc – gas-swirling due to the tangential injection of the plasma-forming gas through a special ring. The plasmatron anodes are made of copper, water-cooled with internal diameter 0.008 m with a practically unlimited service life in the presence of cooling and operation in the mixing chamber with an angle of inclination of the plasma jets 30°. Cathodes of plasmatrons consist of water-cooled copper shells and cathode inserts from thoriated tungsten (to reduce the electron work function) with a diameter of 0.003 m with service life of 100-120 hours. Plasmatrons activation is carried out by means of an oscillator.

In contrast to the base model in the plasmatrons EDP-104AM commercially pure nitrogen with oxygen content up to 1.5-2.0% can be used as the plasma-forming gas, that at the moment corresponds to the actual composition of the supplied commercial nitrogen. Plasmatrons power supply is carried by a thyristor converter unit of series AT4-750/600 having a steep volt-ampere characteristic and the following operating parameters: power, kW – 450; rectified voltage, V – 600; rectified current, A – 750; efficiency at a nominal speed, % – 96; supply voltage, kV – 6.

Figure 1. Industrial plasma technological facility.

Table 1.Basic characteristics of the reactor.

Characteristics	Values
Power, kW	150
Reactor type	Three-jet once-through vertical
Type of plasmatron, power, kW	EDP-104A, 50
Plasma-forming gas	nitrogen
Weight of the heated gas, kg/h	32,5
Internal diameter, m	0,054
Reactor volume, m^3	0,001
Lining of the reactor channel	dioxide zirconium
Temperature of the plasma flow, K	5400 (L*=0) –2200 (L=12)
Lining temperature, K	1549 (L=0) –770 (L=12)
Specific electrical power, MW/m^3	2142
Service life:	
- anode	3125

- cathode	112
Contamination of carbides and borides with erosion products, %	
- anode	Cu – 0,0000954
- cathode	W – 0,0000002
L* – relative length of the reactor	

The mixing chamber design provides the effective injection of high-disperse raw materials into the reactor, its mixing with the plasma flow and a practically unlimited service life of the plasmatron anodes. The mixing chamber is connected to a sectional water cooled channel having an internal diameter of 0.064 m. The mixing chamber and the reactor sections are made of stainless steel. Feeding of high-disperse raw materials into the mixing chamber is carried out by water-cooled tuyere. The tuyere is also used for supplying the gaseous hydrocarbon to the reactor. To reduce the radial temperature gradient in the near-wall zone the channel of the reactor is lined inside with high-temperature insulation material.

For dosing of powdered raw material the intermittent dosing device of combined electromechanical and gas-swirl type with removable cylinder – receiver of the powdered raw material intended for low-loose high-disperse materials is used.

The capture system includes a settling chamber where the temperature of the process gases is reduced to 2800-2000 K and up to 10% of nanopowder is captured, and two alternately operating bag filters (trapping up to 85% of nanopowder). Into the settling chamber with the help of a water-cooled probe the reagent, passivating and coagulating nanopowders, are introduced. Filters are made with a water-cooled housing, regeneration of filter sleeve by reverse blowing with a compressed gas (nitrogen). Filter cloth is a gauze from chrome-nickel steel with a twill weave.

Plasma technological complex exceeds the known laboratory and pilot-scale variations in power in 4-5 times, service life in 3-4 times, performance in 2.5-3.5 times.

3. Production and properties of nano-powders of borides and carbides

The study results of plasma technological options of borides and carbides productions and their implementation are described in detail in a number of publications issued at different times including [10-11].

Development and launching of plasma technological production of borides and chromium carbides, titanium and silicon comprises two stages: 1) study of the processes of boride and carbide formation and chemical and physical properties of borides and carbides in the nanostate; 2) commercial development of the investigated technological options.

At the first stage the plasma flows and raw material interaction is simulated, experimental study, discussion of the mechanism of boride and carbide formation, physico-chemical certification of nano-dispersed products take place. Below the equations describing the dependences of borides and carbides content on the basic technology factors (1)-(5) are provided, as well limits for parameters change of borides and carbides formation in the industrial reactor with capacity of 150 kW, the main characteristics of borides and carbides (Table 2) and micrographs (Figure 2).

$$[CrB_2] = -413.53 + 0.09695T_0 + 2.283\,[B] + 0.1736\,\{H_2\} - 0.00058\,T_0\,[B]; \quad (1)$$

$$[Cr_3(C_{0.8}N_{0.2})_2] = -66.12 + 0.03\,T_0 - 0.42\,\{H_2\} - 0.14\,\{N\} - 0.00002\,T_0\,\{N\}; \quad (2)$$

$$[TiB_2] = -412.41 + 0.09489\,T_0 + 2.196\,[B] + 0.1597\,\{H_2\} - 0.00061\,T_0\,[B]; \quad (3)$$

$$[TiC] = 17.3211 + 0.0105\,T_0 - 0.0156\,T_h + 0.1859\,\{CH_4\} - 3.432\,\{H_2\} - 0.4078\{N\}; \quad (4)$$

$$[SiC] = 86.50 + 0.00273\,T_0 - 0.0064\,T_h - 0.144\,\{CH_4\} + 0.00007\,T_h\,\{CH_4\}, \quad (5)$$

where T_0 – initial temperature of the plasma stream, K;
T_h – hardening temperature of borides and carbides formation products;
[B] – content of boron in the mixture (in % from stoichiometrically required);
$\{CH_4\}$ – quantity of hydrocarbon (in % from stoichiometrically required);

{H_2} – hydrogen concentration in the plasma-froming gas, % vol;

{N} – amount of atomic nitrogen in the plasma-froming gas (in % from stoichiometrically required for the formation of hydrogen cyanide).

At the second stage specifications and technical documentation (specifications and manufacturing processes) and equipment-technological schemes of the proposed for implementation in the conditions of NPF "Polimet" technological options for production of borides and carbides are developed; the main technical and economic indicators are identified. Figure 3 shows an example of equipment-technological scheme of production of carbonitride (I) and chromium carbide (II).

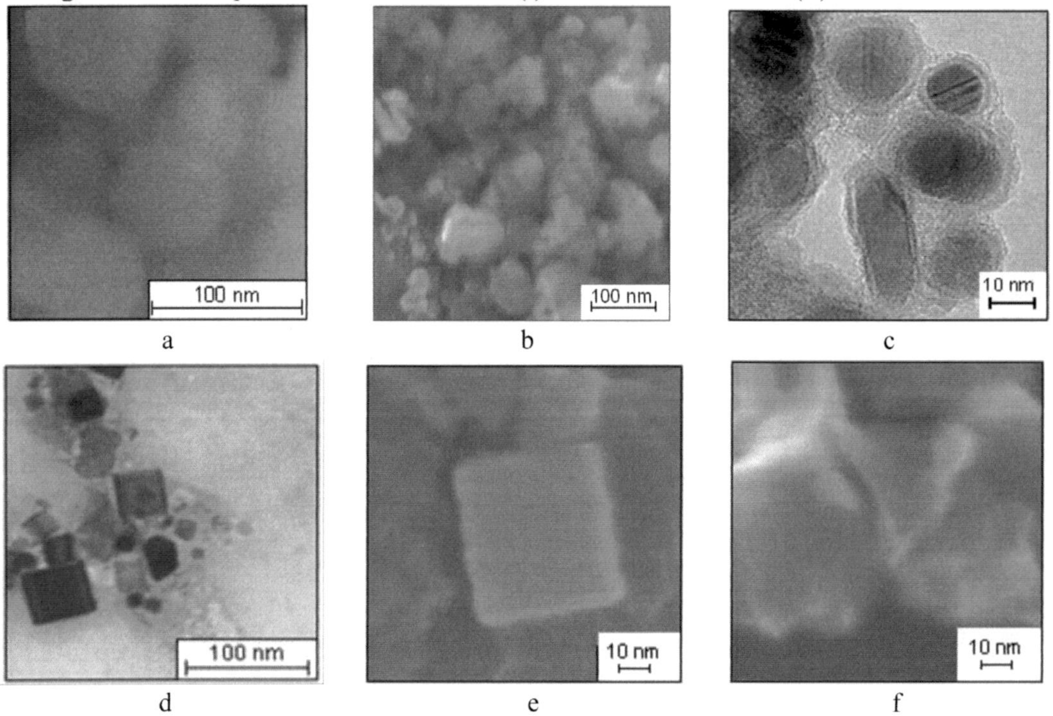

Figure 2. Micrographs of boride and chromium carbonitride nanopowders (a, b), boride and titanium carbide (c, d), silicon carbide (e, f).

Table 2. Permissible limits of change in parameters of borides and carbides synthesis in the industrial reactor with power 150 kW and their main characteristics.

Parameters of synthesis and its characteristics	CrB_2	$Cr_3(C_{0.8}N_{0.2})_2$	TiB_2	TiC	SiC
Composition of the coolant gas,% vol.:					
- nitrogen / hydrogen / methane	74/25/1	99/–/1	74/25/1	99/–/1	99/–/1
Technological variant of synthesis	$Cr+B+H_2$	$Cr+CH_4$	$Ti+B+H_2$	$Ti+CH_4$	$Si+CH_4$
Performanceonfeed, kg/h	3.6	3.1	3.6	3.2	3.0
The amount of boron in the mixture, % from stoichiometric	100-120	–	100-120	–	–
Amount of carbonizer, % from stoichiometric	–	120-140	–	120-140	120-140
Initial temperature of the plasma flow, K	n.l. 5400	n.l. 5400	n.l. 5400	n.l. 5400	n.l. 5400
Hardening temperature, K	2600-2800	2000-2200	2600-2800	2600-2800	2800-3000
Phase composition	CrB_2	$Cr_3(C_{0.8}N_{0.2})_2$	TiB_2	TiC	β-SiC
Content of the main phase, %	92-93	92-93.5	92-93	93-93.5	91-92
Yield of the main phase, %	91-92	90.5-91.5	91.5-92.5	92-92.5	87-90
Performance, kg/h	3.0	3.4	3.4	3.7	4.05
Intensity, kg/h·m^3	1364	2010	1980	2105	2200

Specific surface area, m^2/kg	33000-35000	31000-35000	46000-48000	33000-35000	40000-44000
Size* of the particles, nm	42.0	34.0	36.0	35.0	55.0
Particles shape	Spheroidal	Spheroidal	Spheroidal	Facet, cubic	Facet
Nano-powder oxidation** x10^7, kg O_2/m^2	9.0-9.7	8.0-10.0	5.8-7.6	8.5-9.5	6.5-8.0

* Calculated with value of specific surface;
** Determined after exposure in the air for 24 h.

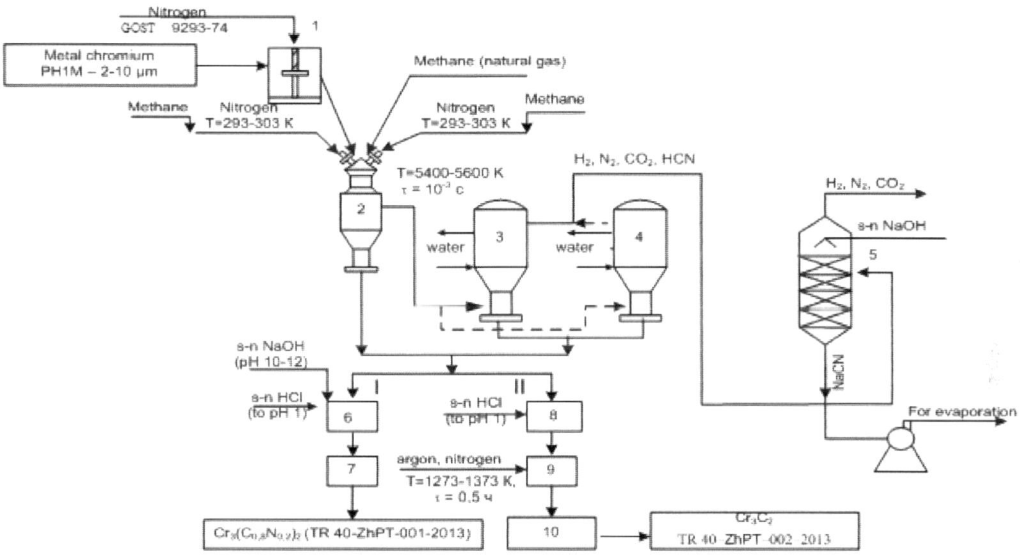

Figure 3. Equipment and technological scheme of carbonitride (I) and chromium carbide (II) production: 1 – charge dosing; 2 – synthesis; 3, 4 – separation of the desired product; 5 – absorptive version of exhaust gases neutralization; 6, 7 – refinement of chromium carbonitride and control of its characteristics; 8, 9 – refinement of chromium carbonitride and its additional carbonization; 10 – control of chromium carbide characteristics.

4. Economic evaluation

The transition to industrial production of borides and carbides nanopowders creates real preconditions for their widespread introduction into the market, at least in the domestic market of nanomaterials. Table 3 presents the comparison of main technical and economic indicators of production of carbide, chromium boride and silicon carbide in the laboratory and industrial conditions.

Table 3. Comparison of technical and economic indicators of nanopowders production in the laboratory and industrial conditions.

Feasibility indicators	CrB_2	Cr_3C_2	SiC
Maximum capacity of the reactor, kW	150/50	150/50	150/50
Content of the main phase,%	95.0*/81.0**	96.0/81.0	99.0/94.0
Oxidation x10^7, oxygen kg/m^2	8-10/18-20	12-14/9-11	0.8/6.7
Productivity, tonne/year per reactor	3.2/0.5	3.6/0.6	3.1/1.8
Intensity, kg/h·m^3	1365/265	1360/245	1210/605
Specific energy consumption, thousand kW·h/tonne	75/145	69/140	74/115
Prime cost, thousand rubles/kg	6.2/15.8	6.6/14.5	6.0/11.0

*/** commercial and laboratory levels

The main developers and manufacturers of nanomaterials based on borides and carbides abroad are scientific-production companies "Nanostructured & Amorphous Materials, Inc." (USA), "Tokyo Tekko Co" (Japan), "Hefei Kaier Nanotechnology & Development Ltd. Co" (China), "NEOMAT Co" (Latvia),
"Plasma Chem GmbH" (Germany) offering them at a price about 1200 USD/kg at a comparable level of quality and dispersion [5-9].

5. Conclusions

The performed analysis confirms the achievement by the plasma technological production of high temperature titanium borides and carbides, chromium, silicon of the industrial level in the usage of a three-jet once-through reactor with power 150 kW as a base model of a technological complex. This transition ensures qualitative and technical and economic parameters comparable with foreign counterparts, and competitiveness of the scientific and technical developments.

6. Acknowledgements

The work was performed at Siberian State Industrial University in the framework of the state task of the Ministry of Education and Science of the Russian Federation No. 11.1531/2014/K.

7. References

[1] Brodov A A, Kazakova T I and Molotilov B V 2006 *Steel in Translation* **36** (5) 67–71
[2] Skachkov O A, Firstov M N et al 2006 *Steel in Translation* **36** (4) 75–80
[3] Werner M, Kohler Th, Mietke S et al 2002 *Book of Abstracts of 6th Int. Conf. on Nanostructured Materials* (NANO 2002) (Piscataway: Rutgers University) p 31
[4] Galevskii G V et al 2007 *Russian J. Non-Fer. Met.* **48** (2) 157–160
[5] *Plasma Chem Nano Powders; Webmaster PIXXL. WEBDESIGN 2008* (Berlin: Plasma Chem Gmbh) www.plasmachem.com
[6] *NEOMAT NANO POWDERS 2008* (Salaspils: Neomat Co) www.neomat.lv
[7] *Nanoceramics Powders 2008* (Hefei Kaier Nanotechnology & Development Ltd. Co.) www.hfkiln.com
[8] *NanoAmor. Nanostructured & Amorphous Materials. Inc. 2008* (Nanostructured & Amorphous Materials Inc.) www.nanoamor.com
[9] Nozdrin I V et al 2011 *Steel in Translation* **41** (8) 644–648
[10] Efimova K A et al 2015 *IOP Conf. Series: Materials Science and Engineering* **91** 1–10
[11] Galevskii G V et al 2015 *IOP Conf. Series: Materials Science and Engineering* **91** 1–5

Protective metal matrix coating with nanocomponents

G V Galevsky[1], V V Rudneva[1], A N Cherepanov[2], S G Galevsky[3] and K A Efimova[1]

[1] Institute of Metallurgy and Material Science, Siberian State Industrial University, 42 Kirova Street, Novokuznetsk, 654007, Russia
[2] Institute of Theoretical and Applied Mechanics, Siberian Branch of the Russian Academy of Sciences, 4/1 Institutskaya Street, Novosibirsk, 630090, Russia
[3] National Mineral Resources University, 2 21st line V.O, Saint Petersburg, 199106, Russia

E-mail: kafcmet@sibsiu.ru

Abstract. Experience of nanocrystalline chromium, titanium, silicon carbides and borides components application as nickel, zinc, chromium based electrodeposited composite coating is generalized. Electrodepositing conditions are determined. Structure and physicochemical properties of coatings, namely micro-hardness, adhesion to steel base, inherent stresses, heat resistance, corrosion currents, en-during quality, and their change during isothermal annealing are studied. As is shown, nanocomponents act as metal matrix modifier. Technological and economic feasibility study to evaluate expediency of replacing high priced nano-diamonds with nanocrystalline borides and carbides is undertaken.

1. Introduction

Increase in materials operation requirements stimulates surface hardening investigations. In this regard, development of environment friendly new material hardening technologies targeted on change in structure-phase state, physical and mechanical properties of surface layers, having advantages in comparison with traditional electroplating technology, thermo-mechanical, thermo-chemical processing, as well as impact methods, based on use of concentrated energy streams is on top of scientific agenda.

In addition, there is a stable interest to high-temperature extra-hard materials (diamonds, borides, carbides, nitrides, etc.) among developers of innovative hardening technologies. Analysis of scientific publications devoted to surface hardening implemented in various technological options, interactions with authors and research executors proves technological and economic feasibility of nanoscale materials (powders with particle size of less than 100 nm) application.

The paper is aimed at summarizing experience gained and the results achieved by the authors in development of technologies of nanodispersed boride, chromium, titanium and silicon carbides application as components of galvanic metal matrix composite coatings.

2. Production of borides and carbides nanopowders

Plated coating is extremely widely used in modern machine engineering technologies. Level of conventional electroplating technologies and equipment design is almost at its peak. However, modern material science challenges investigators with a number of tasks in surface hardening of parts of machines and mechanisms, providing them with higher mechanical, anticorrosion, antifriction and aesthetic properties, which is impossible to do by means of conventional electroplating, that has determined active development of new research in this field.

One of these new research directions is galvanic composite coatings creation (GCC). The group includes coatings deposited as a thin layer on the articles with conductive surfaces made of electrolytes containing internal phase. New composite coatings are designed based on matrix of chromium, nickel, zinc, copper, gold, silver etc. Technology of composite electrodepositing was developed within scientific schools of Professor R.S. Saifullin (KSTU), Professor G.V. Khaldeev (PSU), Professor T.E. Tsupak (Dmitry Mendeleev University of Chemical Technology of Russia), Professor V.Yu. Dolmatov (SPbSTU) [1-5].

In composite electrodepositing micropowders of diamonds, transition metals carbides, borides, nitrides, silicides of 1 - 10 microns are traditionally used. Powders of such grain significantly enhance hardness and wear resistance of composite coatings. However relatively high particle size of powders and relevant uneven distribution of particles in volume of matrix leads to increased porosity of composite and, as a consequence, reduction of thermal and heat and corrosion resistance. These conditions determine feasibility of nano-disperse diamonds, borides and carbides technological application in electrodepositing.

Industrial technologies of synthesis and application of nano-size diamonds, or nanodiamonds, are presented by the authors [6-10]. Production includes detonation synthesis, chemical cleaning and conditioning of the product. Nano-diamonds appeared to be complex objects of triplex structure comprising of diamond nucleus of 4-6 nm size, transition shell around nucleus comprising of carbon X-ray amorphous structures of 0.4 - 1.0 nm thickness, surface layer consisting of carbon, nitrogen, oxygen, hydrogen atoms. Acceptable content of the main impurity – oxidized carbon – is up to 0.5 - 1.0%. Being introduced in materials nano-diamonds always play the role of high intensity structure forming agent, providing dispersion strengthening of composition. Optimal price - quality ratio of the product with nano-diamonds is achieved, according to developers, due to relatively low cost of nano-diamonds as compared to diamonds of static synthesis and positive effect after introducing nano-diamond additives into material, that is usually up to 1.0%. Despite the large number of emerging applications of nano-diamonds, currently leading are three main areas: approximately 70% of nano-diamonds are used in finish polishing, 25% - in electroplating, and about 5% - in oil compositions. Electrochemical composite coatings based on chromium, nickel, tin, zinc, copper, gold, silver, aluminum, iron and various alloys are developed using nano-diamonds. Introduction of nanomaterials in electrodeposited coating significantly increases its micro-hardness, enduring quality, corrosion resistance, improves its exterior appearance, reduces porosity.

Along with advantages and benefits of nano-diamonds application in GCC technology, electroplating experts note a number of problems in their application, namely: extremely high level of dispersion, defined by the way of obtaining, more than demanded field of application; associated delivery of nanodiamonds to the consumer in form of water base suspension, significant tendency to coagulation, rapid loss of sedimentation stability, need of implementation of complex multistage technology of electrolyte preparation and application, limited life of chrome-diamond coatings at high temperatures (over 473 - 573 K) due to destruction reasoned by oxidation of nanodiamond particles, impregnated in the surface layers of chrome matrix, during heating of the operating tool in operation.

Titanium, chromium and silicon borides and carbides nanopowders industrial production technology is described by the authors [11-12]. Manufacturing process includes plasma synthesis and, if necessary, nanopowders refining. Table 1 summarizes parameters of borides and carbides forming in conditions of three-jet vertical once-through reactor of 150 kW capacity and basic characteristics of borides and carbides. It can be seen that, by the main phases content and level of dispersion, introduced materials generally meet requirements on components of metal matrix composite coatings:

- level of dispersion of 30 - 60 nm predetermining creation of micro-regular distribution of particles in the matrix, stability of suspensions of which the coatings are deposited, and the most complete manifestation of particles activity (chemical interaction with matrix, diffusion processes development, material hardening and improvement of its corrosion resistance);

- phase composition homogeneity, determining identity of particles behaviour in electrolyte and coating, local reproducibility of coating properties;
- homogeneity of chemical composition, suggesting minimum amount of impurities, especially free carbon: not always insurmountable technological difficulties are found when using nanocarbides with high content of free carbon;
- controlled state of the particles surface (gas saturation, presence of oxide films, etc.) having significant impact on developing such negative processes in electrolyte -suspension as liquid coalescence;
- high corrosion resistance, ensuring consistent suspensions composition and technological parameters of electrodepositing;
- high thermal oxidation stability, providing such an important operational characteristics of the coating, as scaling resistance.

3. Metal matrix composite coatings with nanocomponents

The main characteristics of coatings with nanocomponents: chrome, nickel - silicon carbide, nickel - carbonitride, chromium boride, nickel - carbide, titanium boride, zinc - chromium boride are given below (Table 2). Terms of electrodepositing, structure and properties of coatings are specified [13-15]. Analysis of the results indicates identity of composite chromium, nickel and zinc plating processes, and reveals the following main factors of raising coatings operational properties:

- borides and carbides nanopowders are represented by particles similar in shape to spherical or oval, having no sharp edges; they have high chemical and ad-sorption activity and, form electrolyte suspensions resistant to sedimentation and coagulation; due to small weight and size effectively are transferred to coated surface;
- during electrodeposition boride and carbide particles suspended in electrolyte interact with growing sediment surface due to hydrodynamic, molecular, electrostatic forces, that leads to formation of composite coating;
- during electrolytic deposition of chromium, nickel and zinc nanoparticles act as crystal seeds from which metal crystallization begins; due to large number of particles, involved in the process, crystallization is of massive multigerm nature and generated coating has small size of structural fragments, distinctive matte colour and are practically non-porous;
- synergy of particles non-inertial mass transfer and massive metals crystallization ensures regular deposition of coatings on equipotential surfaces;
- small size of metal crystallite particles provides precise duplication of surface micro-relief that increases general surface, and composite coating to base adhesive strength;
- improvement of coating quality, i.e. its corrosion resistance and micro hardness is achieved at low content of borides and carbides in coating (0.5 - 1.0%), which makes the process economically feasible.

Consequently, unlike micropowders, borides and carbides nanopowders are not only a filling material or the second phase, they also act as a strong modifier and a structure forming agent in electro-crystallization of metals, which leads to formation of highly dispersed nonporous structure of coating with increased corrosion resistance, micro-hardness and enduring quality.

Table 1. Parameters of borides and carbides forming and basic characteristics of nanopowders.

Parameters of synthesis and its characteristics	CrB_2	$Cr_3(C_{0.8}N_{0.2})_2$	TiB_2	TiC	SiC
Composition of the coolant gas,% vol.:					
- nitrogen / hydrogen / methane	74/25/1	99/–/1	74/25/1	99/–/1	99/–/1
Technological variant of synthesis	$Cr+B+H_2$	$Cr+CH_4$	$Ti+B+H_2$	$Ti+CH_4$	$Si+CH_4$
Performance on feed, kg / h	3.6	3.1	3.6	3.2	3.0

The amount of boron in the mixture, the stoichiometric%	100-120	–	100-120	–	–
Amount of carbonizer % from stoichiometric	–	120-140	–	120-140	120-140
Initial temperature of the plasma flow, K	n.l. 5400	n.l. 5400	n.l. 5400	n.l. 5400	n.l. 5400
Hardening temperature, K	2600-2800	2000-2200	2600-2800	2600-2800	2800-3000
Phase composition	CrB_2	$Cr_3(C_{0.8}N_{0.2})_2$	TiB_2	TiC	β-SiC
Content of the main phase, %	92-93	92-93.5	92-93	93-93.5	91-92
Yield of the main phase, %	91-92	90.5-91.5	91.5-92.5	92-92.5	87-90
Performance, kg/h	3.0	3.4	3.4	3.7	4.05
Intensity, kg/h·m^3	1364	2010	1980	2105	2200
Specific surface area, m^2/kg	33000-35000	31000-35000	46000-48000	33000-35000	40000-44000
Size* of the particles, nm	42.0	34.0	36.0	35.0	55.0
Particles shape	Spheroidal	Spheroidal	Spheroidal	Facet, cubic	Facet
Nanopowder oxidation** x10^7, kg O_2/m^2	9.0-9.7	8.0-10.0	5.8-7.6	8.5-9.5	6.5-8.0

* Calculated with value of specific surface;
** Determined after exposure in the air for 24 h.

Table 2. Basic physical and mechanical properties of composite coatings with nanocomponents.

Coatings	Properties					
	Micro-hardness ±0,3 GPa	Adhesive strength, MPa	Enduring quality**	Oxidation speed** T=1173K	Inherent stresses, MPa	Corrosion currants, μA/cm^2
Chromium	5.4	27.9-28.8	1			
Cr –SiC NP	9.7	29.3-30.4	2.5	*	*	*
Cr –SiC MP	8.9	28.4-29.2	2.2			
Nickel	2.2	29.1-30.3	1	2.55	1.36	0.17
Ni –	4.5	29.8-32.1	1.72	1	0.50	0.02
$Cr_3(C_{0.8}N_{0.2})_2$NP	3.2	28.9-30.2	1.54	1.34	0.78	0.11
Ni – Cr_3C_2MP	4.6	31.2-33.3	1.69	1	0.41	0.02
Ni –CrB_2 NP	3.5	29.8-30.6	1.52	1.17	0.72	0.10
Ni –CrB_2 MP	4.4	30.7-32.8	1.70	1	0.52	0.03
Ni –TiC NP	3.3	29.2-29.6	1.50	1.31	0.78	0.12
Ni –TiC MP	5.3	29.5-32.2	1.76	1	0.54	0.02
Ni –TiB_2NP	3.3	29.1-31.2	1.47	1.24	0.79	0.11
Ni –TiB_2 MP						
Zinc	1.0	*	*	*	*	1.0***
Zn –CrB_2NP	1.2					2.3***

* not evaluated;
** relative values are given;
*** corrosion resistance according to GOST 9.308-85 is provided, relative values
NP - nanopowder
MP - micropowder

To assess possibility of replacing nano-diamonds with nanopowders of carbides and borides in chrome, nickel and zinc plating comparison of characteristics of composite coatings based on chromium, nickel and zinc with diamond, carbide and boride nanocomponents (Tables 3 - 5). It can be seen (Table 3) that chromium carbide coatings have comparable to chromium-diamond coatings enduring quality, micro-hardness, corrosion resistance, longer service life when operating under temperatures above 473 - 573 K.

Table 3. Comparative characteristics of coatings based on chromium – with SiC silicon carbide and diamond nanopowders.

Electrodepositing conditions and achieved results		Technologic options					
		1	2	3	4	5	6
Electrolyte composition, kg/m^3	Chromic anhydride	250	250	250	250	250	260
	Sulphuric acid	3	3	3	3	3	0.3 4
	Trivalent chromium	8	8	8	8	8	-
	Potassium fluorosilicate	-	-	-	-	-	19
	Barium sulfate	-	-	-	-	-	7
	Silicon carbide NP	4	6	7	8	10	-
	Diamond NP	-	-	-	-	-	20
Electrodepositing mode	Temperature, K	328	328	328	328	328	331
	Current density A/dm^2	55	55	55	55	55	55
Results	Enduring quality*	096	1.0	1.05	1.07	1.08	1.0
	Micro-hardness, hPa	8.9	9.0	9.1	9.15	9.15	9.0
	Corrosion resistance *	0.96	1.05	1.07	1.10	1.10	1.0
	Service life when operating at temperatures above 473-573K*	1.1	1.50	1.70	2.00	2.06	1.0
	Cost of $1m^3$ of electrolyte suspension *	0.13	0.16	0.18	0.20	0.23	1.0

* Enduring quality, corrosion resistance, service life when operating at temperatures above 473-573K, cost of $1m^3$ of electrolyte suspension are given relatively to option 6, for which values of these indexes are taken as 1.0.

Table 4. Comparative performance indicators of nickel-based coatings with diamond (ND) and chromium carbide (KBKh) nanopowders.

Electrodepositing conditions and achieved results		Options		
		Ni	Ni+ND	Ni+KBKh
Electrolyte composition, kg/m^3	Nickel sulfate heptahydrate	245	245	245
	Boric acid	30	30	30
	Sodium chloride	20	20	20
	Sodium fluoride	6	6	6
	Diamond NP	-	10	-
	Chromium carbide NP	-	-	10
Electrolyte suspension preparation mode	Duration of nanopowder and electrolyte mixing time, h	-	40–80	0.5
	Duration of ultrasonic treatment at 20 kHz, h	-	1–2	-
	Cathodic current density kA/m^2	-	0.01–0.02	0.01–0.02
	Duration of electrolyte suspension processing, h	-	2	2
Electrolysis conditions	Electrolyte temperature, ° C	50–55	50–55	50–55
	Electrolyte pH	5.0–5.5	5.0–5.5	5.0–5.5
	Cathodic current density kA/m^2	0.5	0.5	1
	Volume of nickel plating baths, m^3	0.6	0.6	0.6
	Electrolyte agitation	no	present	present
Results	Coating thickness, μm	40	40	40
	Nanopowder content in coating,%	-	0.74	0.77
	Micro-hardness (P = 0,49N) ± 0,21, hPA	2.0	5.2	6.0
	Increase of enduring quality, relative units	1.0	3.0	3.5

Increase of corrosion resistance, relative units	1.0	16.7	12.2
Service life of threadlike sensor of stranding machines, months	4	11.5	12
Cost of 1m^3 of electrolyte suspension relative units	-	1	7.5

When replacing nano-diamonds with nanoscale boride and chromium carbide at nickel and zinc plating, fine grained structure coatings, almost nonporous with comparable corrosion resistance are obtained (Tables 4 and 5). Consequently, nanopowders of carbides and borides can be used in chrome, nickel and zinc composite electroplating.

4. Economic evaluation

Table 5. Comparative performance indicators of zinc-based coatings with diamond nanopowders (ND) and chromium carbide (KBKh).

Electrodepositing conditions and achieved results		Options		
		Zn	Zn +ND	Zn +KBKh
Electrolite composition, kg/m^3	Zinc oxide	10	10	10
	Sodium hydroxide	100	100	100
	Organic additive	4	4	4
	Diamond NP	-	10	-
	Chromium carbide NP	-	-	10
Electrolyte suspension preparation mode	Duration of nanopowder and electrolyte mixing time, h	-	40–80	0.5
	Duration of ultrasonic treatment at 20 kHz, h	-	1–2	-
	Electrolyzing cathodic current density kA/m^2	-	0.01–0.02	0.01–0.02
	Duration of electrolyte suspension processing, h	-	2	2
Electrolysis conditions	Electrolyte temperature, ° C	25–28	25–28	25–28
	Cathodic current density kA/m^2	0.2	0.2	0.5
	Volume of nickel plating baths, m^3	0.5	0.5	0.5
	Electrolyte agitation	no	yes	yes
Results	Coating thickness, μm	10	10	10
	Nanopowder content in coating,%	-	0.70	0.64
	Micro-hardness (P = 0.2N) ± 0.10, hPA	1.00	1.30	1.20
	Corrosion resistance, h	40	95	90
	Service life of mold knifes, months	1	4	4
	Cost of 1m^3 of electrolyte suspension relative units	-	650	85
Composition of solution for immunization by phosphate treatment, kg/m^3	Zinc oxide	10	10	10
	Sodium phosphate	50	50	50
	Sodium nitrate	10	10	10
Phosphate treatment conditions	Temperature ° C	70–75	70–75	70–75
	Duration, h	0.5	0.5	0.5
Results	Corrosion resistance, h	52	205	200
	Service life of mold knifes, months	1	10	10

* Corrosion resistance of coating is provided based on GOST 9.308-85 by means of neutral salt spray testing (5% solution of sodium chloride at temperature 35°C).

electrolyte suspension preparation is significantly simplified and accelerated; deposition rate is increased by 2.5 times; cost of 1m3 of electrolyte suspension is reduced by 8-10 times. Replacing of 1 tonne of nano-diamonds by nanopowders of borides and carbides gives economic efficiency of 135 mln rub.

5. Conclusions

Analysis of results of research in nanosize diamonds, borides and carbides application in electroplating metal matrix composite coatings production shows that combination of nanosize state with specifically shaped set of properties inherent to introduced borides and carbides provides sustainable achievement of parametric, concentration, structural and economic effects when they are introduced instead of diamonds, that proves their high competitiveness in domestic and global electroplating technology.

6. Acknowledgements

The work was performed at Siberian State Industrial University in the framework of the state task of the Ministry of Education and Science of the Russian Federation No. 11.1531/2014/K.

7. References

[1] Sayfullin P S 2009 *Proc. of Kaz. Tech. Univ.* **6** 80 – 90
[2] Desyatkova G I, Yagodkina L M et al 2002 *Metals Protection* **38** (5) 525–529
[3] Tsupak T E 2008 *High Productive Processes of Electroplating of Nickel, Alloy Nickel – Phosphorus from Electrolytes Containing Carbon Acids* (Moscow) abstr. PhD thes. p 40
[4] Dolmatov V Yu 2006 *Proc. Int. Conf. on Nanotechnologies – to the Production* (M.: Yanus-K) pp 113–151
[5] Dolmatov V Yu 2003 *Ultra Dispersed Diamonds of Detonation Synthesis* (St. Petersburg: SPbGPU) p 344
[6] Dolmatov V Yu 2005 *Proc. 4th Int. Conf. MP3-IV* (Tsukuba Science City, Ibraki, Japan) (Singapore: Institute of Materials) pp 284–286
[7] Burkat G K, Dolmatov V Y et al 2005 *Composite Chrome-diamond Coating, Coating Layer and Manufacturing Method of thereof* appl. for Pat. Rep. of Korea No. 10-2005-0037999
[8] Loubnin E N, Pimenov S M et al 1999 *New Diamond Frontier Carbon Technology* **9** (4) 273–282
[9] Pyzyr A P, Dolmatov V Yu et al 2004 *Diamond and Related Materials* 13 2020–23
[10] Ilyushin M A, Tselinsky I V et al 2005 *Central European Journal of Energetic, Warsaw, Materials* **1** 21–33
[11] Efimova K A, Galevskii G V et al 2015 *IOP Conf. Series: Materials Science and Engineering* **91** 1–10
[12] Galevskii G V, Rudneva V V et al 2015 *IOP Conf. Series: Materials Science and Engineering* **91** 1–5
[13] Rudneva V V Galevskii G V 2007 *Steel in Translation* **37** (3) 224–227
[14] Rudneva V V, Galevskii G V and Kozyrev N A 2015 *IOP Conf. Series: Materials Science and Engineering* **91** 1–11
[15] Efimova K A, Galevskii G V et al 2015 *IOP Conf. Series: Materials Science and Engineering* **91** 1–8

Institute of Physics
Dirac House, Temple Back
Bristol BS1 6BE UK

ISSN: 1757-8981
ISBN 978-1-5108-3102-5